# Agro-Ecological Intensification of Agricultural Systems in the African Highlands

There is an urgent need to increase agricultural productivity in sub-Saharan Africa in a sustainable and economically-viable manner. Transforming risk-averse smallholders into business-oriented producers who invest in producing surplus food for sale provides a formidable challenge, both from a technological and socio-political perspective.

This book addresses the issue of agricultural intensification in the humid highland areas of Africa – regions with relatively good agricultural potential, but where the scarce land resources are increasingly under pressure from the growing population and from climate change.

In addition to introductory and synthesis chapters, the book focuses on four themes: system components required for agricultural intensification; the integration of components at the system level; drivers for adoption of technologies towards intensification; and the dissemination of complex knowledge. It provides case studies of improved crop and soil management for staple crops such as cassava and bananas, as well as examples of how the livelihoods of rural people can be improved.

The book provides a valuable resource for researchers, development actors, students and policy makers in agricultural systems and economics and in international development. It highlights and addresses key challenges and opportunities that exist for sustainable agricultural intensification in the humid highlands of sub Saharan Africa.

**Bernard Vanlauwe** is the director for Central Africa and Nautral Resource Management at the International Institute of Tropical Agriculture (IITA), based in Nairobi, Kenya.

**Piet van Asten** is a systems agronomist at IITA-Uganda working on sustainable intensification of perennial-based cropping systems (coffee, banana, cocoa) in Africa's humid zones, based in Kampala, Uganda.

**Guy Blomme** is a Bioversity International scientist working on germplasm and integrated disease management for more resilient and productive banana-based cropping systems in east and central Africa, based in Kampala, Uganda.

# Agro-Ecological Intensification of Agricultural Systems in the African Highlands

Edited by
Bernard Vanlauwe,
Piet van Asten and Guy Blomme

Routledge
Taylor & Francis Group
LONDON AND NEW YORK

First published 2013
by Routledge
2 Park Square, Milton Park, Abingdon, Oxon OX14 4RN

Simultaneously published in the USA and Canada
by Routledge
711 Third Avenue, New York, NY 10017

*Routledge is an imprint of the Taylor & Francis Group, an informa business*



*British Library Cataloguing-in-Publication Data*
A catalogue record for this book is available from the British Library

*Library of Congress Cataloging-in-Publication Data*
International Conference on "Challenges and Opportunities for Agricultural Intensification of the Humid Highland Systems of Sub-Saharan Africa" (2011: Kigali, Rwanda)
Agro-ecological intensification of agricultural systems in the African highlands/
edited by Bernard Vanlauwe, Piet van Asten, Guy Blomme.
pages cm
Includes bibliographical references and index.
1. Agricultural intensification – Africa, East – Congresses. 2. Agricultural intensification – Africa, Central – Congresses. 3. Upland agriculture – Africa, East – Congresses. 4. Upland agriculture – Africa, Central – Congresses. 5. Integrated agricultural systems – Africa, East – Congresses. 6. Integrated agricultural systems – Africa, Central – Congresses. 7. Food security – Africa, Sub-Saharan – Congresses. 8. Uplands – Africa, Sub-Saharan–Congresses. 9. Consortium for Improving Agriculture-based Livelihoods in Central Africa – Congresses. I. Vanlauwe, B. (Bernard) II. Asten, Piet van, 1972– III. Blomme, G. IV. Consortium for Improving Agriculture-based Livelihoods
in Central Africa. V. Title.
S472.A354I58 2011
338.1620967–dc23
2013015398

ISBN: 978-0-415-53273-0 (hbk)
ISBN: 978-0-203-11474-2 (ebk)

Typeset in Bembo
by Florence Production Ltd, Stoodleigh, Devon, UK

MIX
Paper from responsible sources
FSC® C013056
www.fsc.org

Printed and bound in Great Britain by
TJ International Ltd, Padstow, Cornwall

# Contents

# Illustrations

## Figures

## Tables

# Foreword

These proceedings provide an important milestone for a collaborative initiative that started in 2006, the Consortium for Improving Agriculture-based Livelihoods in Central Africa (CIALCA). The Consortium operated in an area that was largely devoid of international agricultural research expertise and collaborative research efforts, partly driven by a history of civil strife in Burundi, the Democratic Republic of Congo and Rwanda. What had started as an honest attempt to coordinate research activities led by three independent Consultative Group on International Agricultural Research (CGIAR) centres resulted in visible change of the understanding of farming systems, the identification of appropriate technologies for system intensification and a significant impact on farmers and the research-for-development community in the Great Lakes region. It also resulted in CIALCA being integrated in the new CGIAR Research Program on the Humidtropics, a global programme that aims at sustainable intensification of farming systems in the humid and sub-humid tropics.

This book presents key results from an international conference on agriculture in the Central African region. The Kigali meeting was probably the first international conference on agriculture in the Central African region for many years, possibly decades. The objectives of the meeting were: (1) to exchange, generate and document knowledge on the key challenge of ensuring food security and improving livelihoods of smallholders in one of Africa's most densely populated areas and fastest-growing populations; (2) to expose the science results generated by CIALCA and others to the development community; (3) to bring the huge challenges in the Great Lakes region to the attention of the international research community; (4) to enable the creation and strengthening of research-for-development partnerships in a region that had lost much of its research-for-development capacity; (5) to advance the science of agricultural systems analysis, a field that's of utmost importance in the new CGIAR.

The outputs of the meeting are organized around a number of important themes for system intensification, starting with CIALCA's own experiences and enriched with numerous experiences from other initiatives and countries in the East and Central African highlands. We hope that the lessons learned will be relevant for research and development partners operating in the African highlands towards the betterment of rural populations while conserving the threatened natural resource base.

We would like to thank the Directorate General for Development of Belgium for its past, present and continued belief that collaborative research and knowledge exchange can contribute meaningfully to the improvement of smallholder farmers' livelihoods in the Great Lakes region.

N. Sanginga
Director General
International Institute for Tropical Agriculture (IITA)

R. Echeverria
Director General
International Centre for Tropical Agriculture (CIAT)

E. Frison
Director General
Bioversity International

# Abbreviations

| | |
|---|---|
| AABNF | African Association of Biological Nitrogen-Fixation |
| AE | agronomic efficiency |
| AEI | agro-ecological intensification |
| AHI | African Highlands Initiative |
| AR4D | agricultural research for development |
| ATE | average treatment effect |
| BBTD | Banana bunchy top disease |
| BCR | benefit–cost ratio |
| Bt | *Bacillus thuringiensis* |
| CATALIST | Catalyze Accelerated Agricultural Intensification for Social and Environmental Stability |
| CGIAR | Consultative Group on International Agricultural Research |
| CI | conditional independence |
| CIALCA | Consortium for Improving Agriculture-based Livelihoods in Central Africa |
| CIAT | International Centre for Tropical Agriculture |
| CPU | central processing unit |
| CRP | CGIAR Research Program |
| CV | coefficient of variation |
| D&D | delivery and dissemination |
| DAP | diammonium phosphate |
| DRC | Democratic Republic of Congo |
| DRI | Dietary Reference Intake |
| EAC | East African Community |
| EAHB | East African highland cooking and beer bananas |
| ELISA | enzyme-linked immunosorbent assay |
| FCS | food consumption score |
| FD | functional diversity |
| FFQ | food frequency questionnaire |
| FFS | farmer field schools |
| FGD | focus group discussion |
| FHIA | Fundación Hondureña de Investigación Agrícola |
| FPR | farmer participatory research |

| | |
|---|---|
| FSR | farming systems research |
| FYM | farmyard manure |
| GHG | greenhouse gas |
| GIS | geographic information systems |
| GM | genetically modified |
| IAR4D | Integrated Agricultural Research for Development |
| IARC | international agricultural research centre |
| ICIPE | International Centre for Insect Physiology and Ecology |
| ICM | integrated crop management |
| ICRAF | World Agroforestry Centre |
| IDPM | integrated disease and pest management |
| IFDC | International Fertilizer Development Center |
| IITA | International Institute for Tropical Agriculture |
| INRM | integrated natural resources management |
| IP | innovation platform |
| IPM | integrated pest management |
| ISFM | integrated soil fertility management |
| ITC | International Transit Centre |
| JEA | joint exposure and adoption rate |
| K | potassium |
| KRC | Knowledge Resource Centre |
| LAI | leaf area index |
| LEAD | Livelihoods and Enterprises for Agricultural Development |
| LFD | lateral flow device |
| LGP | length of growing period |
| LKPLS | Lake Kivu Pilot Learning Site |
| M&E | monitoring and evaluation |
| MAC | mid-altitude climber |
| masl | metres above sea level |
| MDGs | Millennium Development Goals |
| MGIS | *Musa* germplasm information system |
| MOP | muriate of potash |
| MTA | material transfer agreement |
| MVP | Millennium Villages Project |
| N | nitrogen |
| NAP | national action plan |
| NARES | public sector agricultural research and extension organization |
| NARI | National Agricultural Research Institute |
| NARO | National Agricultural Research Organisation |
| NGO | non-governmental organization |
| NRM | natural resource management |
| OECD | Organisation for Economic Co-operation and Development |
| OFSP | orange-fleshed sweet potato |
| P | phosphorus |
| PA | Peasant Association |

| | |
|---|---|
| PCR | polymerase chain reaction |
| PEM | protein–energy malnutrition |
| PPP | public–private partnership |
| PRA | Participatory Rural Appraisal |
| PSB | population selection bias |
| PVS | participatory variety selection |
| RAAKS | rapid appraisal of agricultural knowledge systems |
| RAB | Rwanda Agricultural Board |
| RCT | randomized control trial |
| RO | relay organization |
| RP | rock phosphate |
| SLA | sustainable livelihoods approach |
| SPSS | Statistical Package for Social Sciences |
| SSA | sub-Saharan Africa |
| SSA-CP | Sub-Saharan Africa Challenge Programme |
| T&V | training and visit |
| TC | tissue culture |
| TLA | total leaf area |
| TLU | tropical livestock unit |
| TSP | triple super phosphate |
| VCA | value chain analysis |
| VCG | vegetative compatibility group |
| VCR | value–cost ratio |
| VPC | vegetatively propagated crop |
| WTO | World Trade Organization |
| WUE | water use efficiency |
| XW | Xanthomonas wilt |

# 1 Agro-ecological intensification of farming systems in the East and Central African highlands

*B. Vanlauwe,*[1] *G. Blomme*[2] *and P. van Asten*[3]

[1] *International Institute of Tropical Agriculture, Nairobi, Kenya*
[2] *Bioversity International, Kampala, Uganda*
[3] *International Institute of Tropical Agriculture, Kampala, Uganda*

## Introduction

This book contains the keynote presentations and others directly aligned to the main theme of the international conference, organized in Kigali, Rwanda, 24–28 October 2011: *Challenges and opportunities for agricultural intensification of the humid highland systems of sub-Saharan Africa.* Three important keywords feature in the general theme: intensification, systems and African highlands.

With the anticipated growth of the population in sub-Saharan Africa (SSA), the intensification of agricultural production is a prerequisite in the more densely populated areas. Even in places where land is still relatively abundantly available, for example, in the Congo basin, the intensification of existing agricultural land is required to conserve forests and marginal lands to protect biodiversity and reduce greenhouse gas emissions (Burney *et al.* 2010; Defries and Rosenzweig 2010; Gockowski and Sonwa 2010). Following arguments about sustainability, efforts to intensify agriculture are required to embrace the principles of maximizing the efficiencies of agricultural resource use and minimizing the environmental footprints of agriculture (Mueller *et al.* 2012).

Smallholder farmers are managing farming systems consisting of different production units with often substantial interactions among them in terms of material, cash and labour flows. Smallholder systems are characterized by large internal variability in plot-level productivity and soil fertility status. This is partly caused by the different soil types within landscapes, but particularly by a preferential allocation of resources, including organic inputs and labour, to specific production units, especially when access to production factors is limited (Tittonell *et al.* 2005a). Within smallholder farming communities, the level of access to production resources for individual households is variable – often categorized in 'farmer typologies' – with most households experiencing moderate to severe resource constraints (Tittonell *et al.* 2005b). The farmers' production objectives and total resource endowments affect their decisions regarding investments in specific production units. These decisions are further influenced by the economic, social, institutional and political environments in which the farmers operate. Entry points aiming at intensifying agriculture should be aligned to these drivers of agricultural production. Whereas traditional technology development and dissemination was rather top-down and linear, the trend to develop and tailor technology options to the farmers' objectives and constraints using a participatory and farming systems approach has found increasing support (van Asten *et al.* 2009; Neef and Neubert 2011). The development of the new CGIAR Research Programs (CRPs) focusing on system approaches to sustainable intensification is an additional important indicator of the current interest in farming systems as an entry point towards enhancing rural livelihoods.

The East and Central African highlands have relatively high population densities (more than 50 people per square kilometre). The population in East and Central Africa is expected to grow from 387 million in 2010 to 882 million in 2050 (http://esa.un.org/unpd/wpp/unpp/panel_population.htm). The climatic conditions (e.g. length of growing period (LGP) of 180 days) are favourable, but climate change may affect them. By 2050, most of the humid highlands are expected to get wetter, except for the Cameroonian highlands, but the area is also going to experience substantial increases in maximum and minimum temperatures (Figure 1.1), and rainfall distribution will be more erratic. This will exert further pressure on systems that are already vulnerable due to high poverty levels and poor access to markets (Table 1.1), thus putting more strain on already scarce resources.

Within the East and Central African highlands, agro-ecological conditions vary substantially; the wetter areas with an LGP of over 270 days have a higher population density and more land under agriculture (Table 1.2). Inherent soil properties also improve with increases in LGP and the interplay between weather and soil conditions results in major differences in the most prominent farming systems from pastoral and maize mixed in the drylands (LGP <180 days) over root crop and maize mixed in the moist savannas (LGP 180–270 days) to highland perennial and forest-based systems in the humid areas (LGP >270 days). Notwithstanding the marked differences in climate and soil quality, the average yields of maize and cassava are low everywhere, with some higher productivity figures for banana in the wetter areas, though generally figures are far below the attainable yields (Table 1.2).

## Themes of the meeting and the objectives of this book

The conference was organized around four major themes:

1 System components. Farming systems consist of different units, including crop and livestock ventures and the total farm productivity, ecosystem service provision and, ultimately, farmers' well-being depend on the performance of each of these components. Most have specific constraints that prevent them from reaching their potential productivity; addressing these through site- and farmer-specific interventions is crucial to improving rural livelihoods.
2 System integration. Components of farming systems interact with one another and with common property resources, especially in environments where production resources are in short supply. Trade-offs are common between investments in specific system components and particularly for farming households that are less resource-endowed. Models for farming system analysis are important tools for analysing trade-offs and exploring profitable scenarios for the intensification of farming systems.
3 Drivers and determinants for adoption. The adoption of strategies for increased farm-level productivity often requires specific enabling conditions. Such drivers and determinants may operate at different scales and affect specific system components. A clear understanding of those drivers is important to determine adaptive strategies that can contribute to the intensification of important farming systems and prioritize development-oriented investment and policy needs.
4 Knowledge-intensive approaches. System approaches and interventions are often knowledge-intensive and therefore specific dissemination approaches are needed. This is especially relevant for areas with relatively low levels of literacy and formal education. The identification of simple, fast-track interventions that can be disseminated within the lifetime of most projects is needed within the context of more knowledge-intensive approaches. Tensions exist between knowledge-intensive approaches and the need to reach many households.

The objectives of this book are to present the various keynote and other presentations aligned to each of the above themes and to extract the main lessons learned in relation to the agro-ecological intensification of the target systems. The first chapter of each theme summarizes experiences from the Consortium for Improving Agriculture-based Livelihoods in Central Africa (CIALCA – www.cialca.org), which organized the conference. Such lessons are timely, especially in view of the renewed interest in system research and the identification of the East and Central African highlands as a major target area in the context of the Humidtropics CRP, focusing on system intensification in the humid and sub-humid regions of Africa, Asia and Latin America. Most

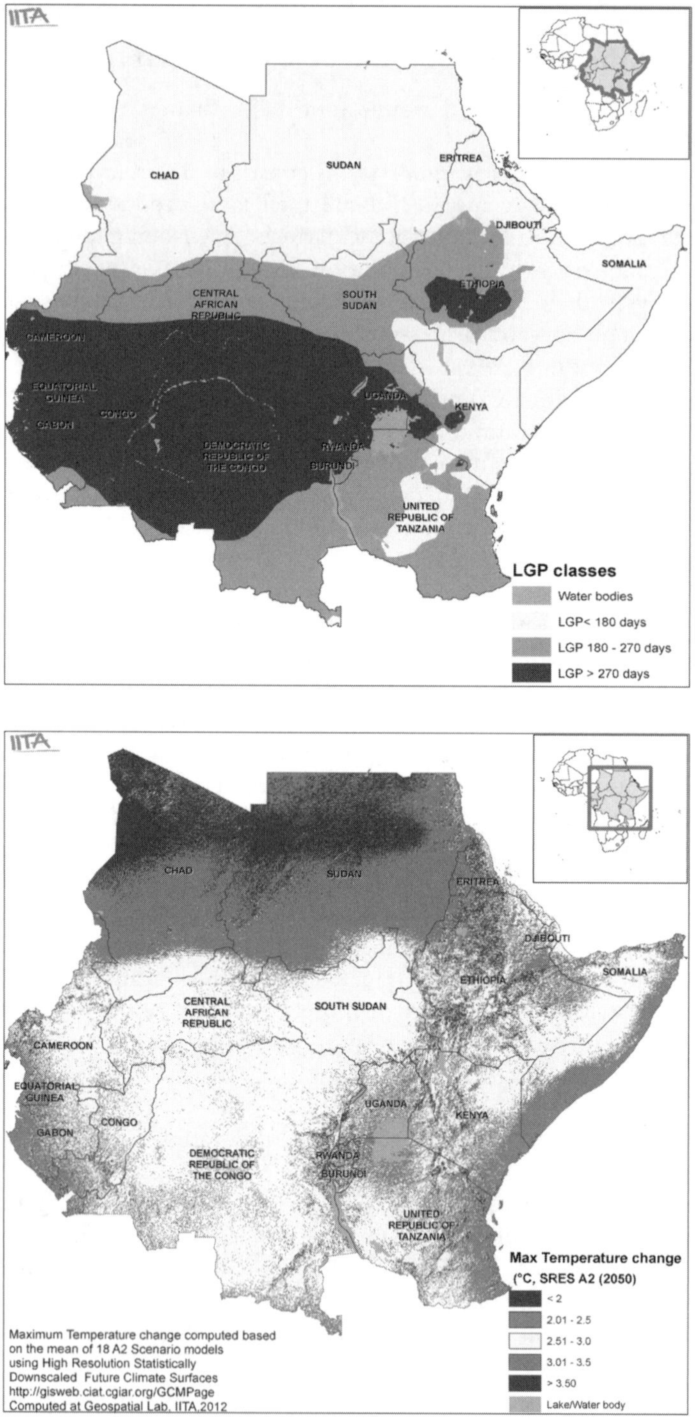


*Figure 1.1* (a) Length of growing period, (b) anticipated changes in rainfall, and (c) maximum and (d) minimum temperature by 2050, following the A2 scenario of the Special Report on Emissions Scenarios (Nakicenovic *et al.* 2000).

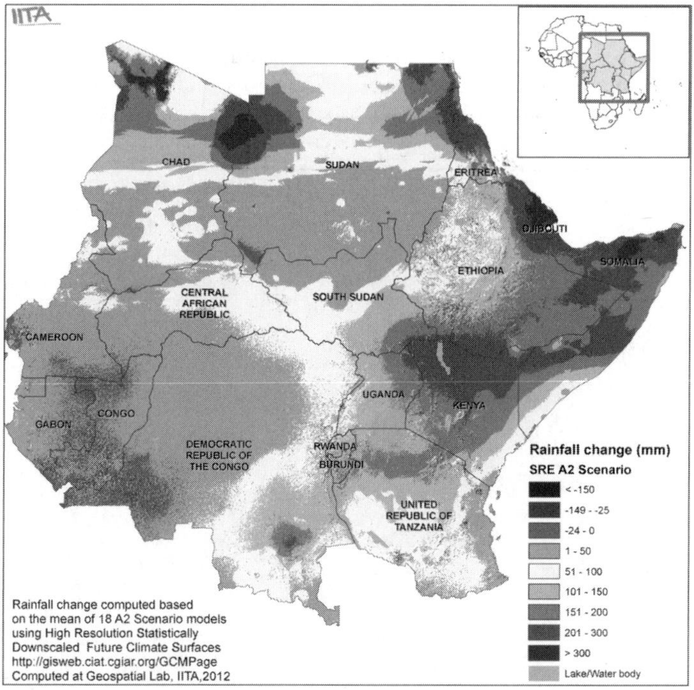
IITA
CHAD
SUDAN
ERITREA
DJIBOUTI
SOMALIA
ETHIOPIA
CENTRAL AFRICAN REPUBLIC
SOUTH SUDAN
CAMEROON
UGANDA
KENYA
GABON
CONGO
DEMOCRATIC REPUBLIC OF THE CONGO
RWANDA
BURUNDI
UNITED REPUBLIC OF TANZANIA
Rainfall change (mm)
SRE A2 Scenario
< -150
-149 - -25
-24 - 0
1 - 50
51 - 100
101 - 150
151 - 200
201 - 300
> 300
Lake/Water body
Rainfall change computed based on the mean of 18 A2 Scenario models using High Resolution Statistically Downscaled Future Climate Surfaces http://gisweb.ciat.cgiar.org/GCMPage Computed at Geospatial Lab, IITA,2012

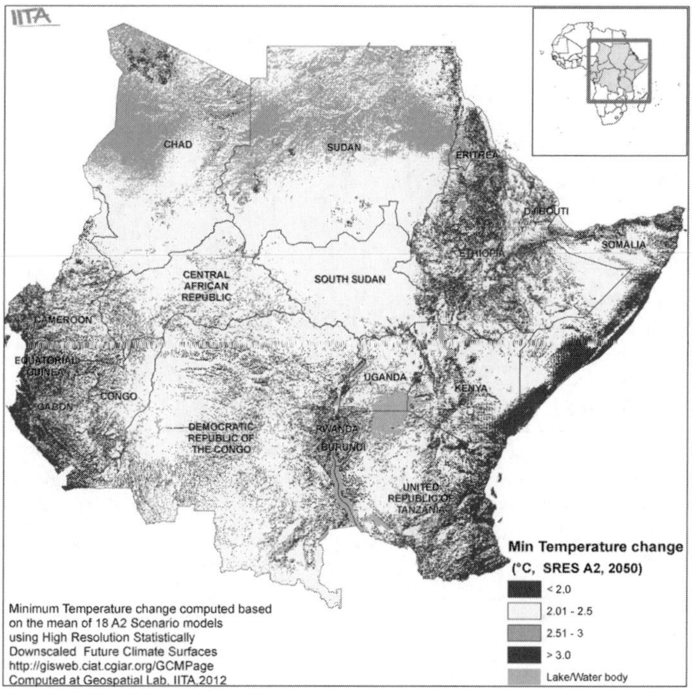
IITA
CHAD
SUDAN
ERITREA
DJIBOUTI
SOMALIA
ETHIOPIA
CENTRAL AFRICAN REPUBLIC
SOUTH SUDAN
CAMEROON
EQUATORIAL GUINEA
UGANDA
KENYA
GABON
CONGO
DEMOCRATIC REPUBLIC OF THE CONGO
RWANDA
BURUNDI
UNITED REPUBLIC OF TANZANIA
Min Temperature change
(°C, SRES A2, 2050)
< 2.0
2.01 - 2.5
2.51 - 3
> 3.0
Lake/Water body
Minimum Temperature change computed based on the mean of 18 A2 Scenario models using High Resolution Statistically Downscaled Future Climate Surfaces http://gisweb.ciat.cgiar.org/GCMPage Computed at Geospatial Lab, IITA,2012

*Table 1.1* Selected characteristics of the East and Central African highlands in relation to the East and Central African areas below 800 metres above sea level (masl) and the rest of sub-Saharan Africa. The data were obtained from the countries depicted in Figure 1.1.

| *Selected characteristics* | *East and Central African highlands (>800 masl)* | *East and Central Africa lowlands (<800 masl)* | *Rest of sub-Saharan Africa* |
|---|---|---|---|
| Total population (million people) (CIESEN *et al.* 2005) | 225.7 | 152.8 | 438.4 |
| Population density (people per $km^2$) (CIESEN *et al.* 2005) | 69.4 | 18.4 | 34.8 |
| Proportion of people living on <1.25 USD per day (%) (Wood *et al.* 2010) | 49.8 | 43.1 | 54.7 |
| Total land area (million ha) (Geospatial Laboratory*) | 325 | 829 | 1,259 |
| Proportion of agricultural land (%) (Fritz *et al.* 2011) | 12.1 | 5.3 | 9.8 |
| Livestock density (ruminants per $km^2$) (FAO 2007) | 21 | 13 | 14 |
| Length of growing period (days) (FAO and IIASA 2007) | 184 | 161 | 116 |
| Percentage of the region with good/poor market access (%) (Geospatial Laboratory**) | 14.3/43.0 | 8.8/50.5 | 15.5/41.8 |

*Notes:*

* Computed from Global Administrative Area (GADM) layer (www.gadm.org)

** Computed from Globecover, Michelin roads for Africa and towns greater than 10,000 people.

*Table 1.2* Selected characteristics of the different agro-ecological zones within the East and Central African highlands. The data were obtained from the countries depicted in Figure 1.1.

| *Selected characteristics* | *Agro-ecological zones* | | |
|---|---|---|---|
| | *LGP <180 days* | *LGP 180–270 days* | *LGP >270 days* |
| Population density (people per km$^2$) (CIESIN *et al.* 2005) | 34 | 62 | 151 |
| Land area (million ha) (Geospatial Laboratory*) | 113.2 | 134.7 | 67.0 |
| Agricultural land (% of total area) | 9.6 | 11.3 | 19.3 |
| Major soils of >8 million ha (million ha) (FAO 1991) | leptosols (45.2); cambisols (18.1); calcisols (8.7) | ferralsols (29.8); cambisols (21.4); leptosols (15.1); acrisols (14.3); nitisols (13.6); vertisols (9.7) | ferralsols (17.3); nitisols (8.8) |
| Livestock density (ruminants per km$^2$) (FAO 2007) | 23 | 19 | 25 |
| Major farming systems with >5 million ha (million ha) (Dixon *et al.* 2001) | pastoral (39.1); maize mixed (33.9); sparse (arid) (21.5); highland temperate mixed (8.8); agro-pastoral millet/sorghum (7.3) | root crop (61.3); maize mixed (49.6); highland temperate mixed (17.6) | highland perennial (23.6); forest-based (18.1); maize mixed (12.9) |
| Average maize productivity (kg ha$^{-1}$) (You *et al.* 2000) | 1,205 | 2,111 | 1,783 |
| Average cassava productivity (kg ha$^{-1}$) (You *et al.* 2000) | 7,555 | 7,756 | 8,900 |

Note:

* Computed from Global Adminstrative Area (GADM) layer (www.gadm.org)

of the chapters directly target the East and Central African highlands, though some provide a broader context for intensifying African smallholder agricultural systems.

## Agro-ecological intensification: conflicting concepts for a generally accepted need

Although the need for intensification, or the production of more outputs/unit of input, whether land, labour or agro-inputs, is no longer questioned, there is no consensus on the ultimate nature and design of such intensified systems and the pathways required to reach there. The Green Revolution strategy certainly prevented widespread starvation in Asia, but many negative environmental side-effects have been demonstrated. These insights have resulted in a more pronounced focus on the resource use efficiency of inputs and on integrating the heterogeneity of smallholder systems within the African Green Revolution – 'a uniquely African Green Revolution', as promoted by Kofi Annan, the Chairman of the Alliance for a Green Revolution in Africa (AGRA – www.agra-alliance.org). Dissenting voices promote ecological intensification as a guiding paradigm, including Mr Olivier de Schutter of the European Union, who states that low external-input techniques, based on the core principles of agro-ecology, including the recycling of nutrients and energy on the farm rather than introducing external inputs, have a proven potential to significantly improve yields (De Schutter 2010).

Keating *et al.* (Chapter 2) demonstrate that agricultural intensification, though not always following principles of sustainability, has actually provided the pathway to global food security to date, except in SSA. Agricultural productivity growth is necessary for future global food security, but this should be accompanied by reductions in demand and in post-harvest losses. Principles of eco-efficiency are proposed as the pillars to future intensification efforts, recognizing the need for external inputs but aiming at maximizing their use efficiencies. For SSA, due to the relatively large availability of unused agricultural land, both intensification and expansion options should be envisaged, although for the humid highlands only the former is a viable option due to the high level of population pressure. Broad-based institutional innovation will be needed to accompany the alleviation of technical constraints to intensification in SSA. Breman (Chapter 3) argues that fertilizer in the framework of integrated soil fertility management (ISFM) is the only viable way forward for the intensification of African agriculture, thus opposing the low external-input techniques. He proposes that the use of external inputs should be 'optimized' rather than 'minimized', and elaborates on subsidies for soil amendments in addition to fertilizer (e.g. lime) as smart alternatives to 'traditional' fertilizer subsidies.

## Technology components for integration in agro-ecological intensification pathways

The core of agro-ecological intensification is a set of improved inputs (e.g. improved varieties, fertilizer), implements (e.g. improved seed planters, weed wipes), and practices (e.g. best agronomic practices, water-harvesting structures) that produce more output per unit of input used relative to traditional practices and whereby the use efficiency of those inputs is maximized. The eco-efficiency concept applies to such farming components where, per unit component deployed, desired outputs are maximized and those undesired are minimized (Table 1.3). Individual components can interact with one another and create more of the desired outputs and less of those undesired. For instance, the application of crop- and site-specific fertilizer to an improved crop variety will enhance crop productivity not only due to the applied nutrients but also due to the higher demand for nutrients by an improved variety compared with a local one (Vanlauwe *et al.* 2012). The ISFM paradigm is explicitly aiming at creating such interactions between individual components towards generating added benefits and maximal input use efficiencies (Vanlauwe *et al.* 2011). Most components require access to external inputs and/or additional knowledge on appropriate input use. Consequently, their deployment at scale will depend on the presence of an enabling environment facilitating this, including, among others, effective extension for knowledge transfer, input value chains accessible to smallholder farmers, and credit schemes. The 'basket' of options providing pathways towards agro-ecological intensification ultimately depends on the specific conditions of biophysical and socio-economic constraints and opportunities in which the farmer operates.

Chapters presented under theme 1 of the conference consider various system components relevant for the East and Central African highlands, including resilient germplasm, improved soil and pest/disease management practices, soil and water conservation practices and seed systems. Pypers *et al.* (Chapter 4) demonstrate that the principles underlying ISFM do apply to crops and systems in the CIALCA region and stress the need for fertilizer and strategies to make this input available to smallholder farmers. Non-responsive soils or soils on which crops do not show substantial increases in productivity after the application of nitrogen–phosphorus–potassium (NPK) fertilizer were also identified, confirming earlier observations of non-responsive soils in other parts of East Africa. Second, several improved banana and legume varieties were identified but, as with fertilizer, access to these through effective seed systems is a precondition to their uptake, especially in the absence of a private seed sector. Pypers *et al.* also highlight the lack of organic resources within Central African farming systems and demonstrate that options to enhance their production on the same piece of land are possible through improved agronomic practices. Specific comments are also made, related to the need for appropriate dissemination tools and approaches for complex knowledge (see Part IV of this

*Table 1.3* Application of eco-efficiency principles to important components of farming systems in the East and Central African highlands and a selection of conditions enabling their deployment

| *Category* | *Example* | *Desired outcomes* | *Undesired outcomes* | *Enabling conditions* |
|---|---|---|---|---|
| Inputs | Crops and soil-specific fertilizer | Higher crop productivity due to crop-specific nutrient availability | Unused fertilizer can be lost to the environment in soluble or gaseous form | Input supply chains; credit to farmers; knowledge folders on appropriate use |
| | Disease-tolerant banana variety | Higher crop productivity from reduced disease pressure | Higher soil nutrient mining in absence of input use; reduced banana intercropping options | Effective seed systems (e.g. macro-propagation) |
| Implements | Improved seed planters | Increased labour use efficiency and reduced drudgery | Decreased germination if not used properly | Local blacksmiths; training on appropriate use; local credit schemes |
| | Weed wipes | Increased effectiveness with lower herbicide application rates and possibility of targeting species within inter-crop systems | Inappropriate storage and application of herbicides | Herbicide supply chains; training on appropriate use |
| Practices | Resource-efficient intercrop systems | Increase in total food/revenue produced per unit area | Increased resource competition affects yields of (some) crop components | Farmers perceive decreasing farm size as an incentive to intercropping |
| | Water-harvesting structures | Improved water and soil retention increases yield and reduces erosion | Structures are labour demanding and may take up space in productive land | Policy and extension support to build and sustain structures |

book). Wairegi and van Asten (Chapter 5) demonstrate that there is scope for fertilizer use in East Africa but the profitability of fertilizer use strongly depends on the choice of crop and the amount and type of fertilizer. Decisions should be based on *ex-ante* input:output prices, crop requirements and residue management. Intensification targeted at one crop can benefit associated crops in intercropping systems. All this should be accompanied by farmer-friendly decision tools on appropriate nutrient management. Zingore and Johnston (Chapter 6) zoom in on aspects of farm-level nutrient management following the principles underlying '4R Nutrient Stewardship' or the need to apply the right sources of nutrients with the right rate at the right time in the right place. The chapter by Swennen *et al.* (Chapter 7) details the biotic constraints affecting banana production in East and Central Africa and ways to mitigate these. Starting from a clear understanding of the pathogen–host-plant–environment disease triangle, specific measures are needed to mitigate these constraints and, as for the above soil management practices, specific interventions equally interact to create greater benefits. In Chapter 8 Smith *et al.* discuss the challenges related to seed systems for vegetatively propagated crops, including banana and cassava, two important food security crops that have become cash crops in East and Central Africa. Specific attention is given to seed health, quality assurance and profitability issues.

## Integration of technical components at the farming system level

From the various conclusions highlighted in the theme 1 chapters, it is clear that single technology components can produce extra outputs and have a higher chance for uptake by smallholder farmers if used in combination with other improved components. Such interactions are embedded in the ISFM paradigm, but also in the sections dealing with banana pest and disease spectra, bio-fertilizer and soil and water conservation practices. Farmers simultaneously manage various production units within their farm. Since each of those units requires specific investments in terms of labour, inputs or capital, the integration of components happens best at the plot or farm level, depending on the nature of the expected interactions. For instance, the application of fertilizer works best in association with improved varieties at the plot level; the integration of fodder legume trees for erosion control can happen at the farm level. Furthermore, aspects of trade-offs, farmers' typologies and within-farm gradients of soil fertility require consideration when decisions are taken on the use of combinations of components, especially since production resources are mostly limited for smallholder farmers.

Chapters presented under theme 2 of the conference cover crop–nutrient, crop–crop and crop–livestock interactions at the farm scale and demonstrate both the occurrence of win–win situations in terms of increased returns to inputs (land, labour) and improved system resilience (e.g. to effects of climate change), particularly in situations where trade-offs in resource allocation are required. Van Asten *et al.* (Chapter 9) summarizes the experiences of CIALCA in the quest for win–win situations when integrating production components in cassava–legume, maize–legume, banana–legume and coffee–banana systems. Although increased efficiencies at the farm level in terms of returns to land, labour or nutrient inputs were often observed, the understanding of the trade-offs in space and time and the socio-cultural and gender constraints to technology uptake need to improve. Tittonell (Chapter 10) stresses that smallholder farming systems are diverse, spatially heterogeneous and highly dynamic: trade-offs emerge when resources are limited and two or more competing objectives are to be simultaneously met. He demonstrates that farm-scale modelling offers ample opportunities for system integration and trade-off analysis to inform systems approaches to agro-ecological intensification and allows the implications of proposed technologies to be studied across spatial and temporal scales. Maass *et al.* (Chapter 11) propose a 'livestock ladder' approach to reconnect farmers in east DR Congo to markets by using small ruminants and accompanying feed systems as entry points, through an innovation platform approach. Final benefits include more cash income and improved nutrition, with specific benefits for women farmers. Giller *et al.* (Chapter 12) demonstrate the flexibility of legumes as entry points towards enhanced livelihoods at the farm scale, following the $(GL \times GR) \times E \times M$ equation, with $GL$ and $GR$ referring to appropriate legume and rhizobial germplasm (both species and varieties or strains), $E$ to environmental conditions (e.g. agro-ecological conditions, market access) and $M$ to management (e.g. planting density, intercropping arrangements). In a last contribution under this theme, Van Rikxoort *et al.* (Chapter 13) focus on the potential of coffee-based systems to adapt to climate change and mitigate the effects, and conclude that climate-smart systems should use intercropping with other food crops and shade plants for a combined result. They also stress the need to include public and private sector partners in the promotion of climate-smart practices.

## Drivers and pathways for achieving impact

Concepts, entry points and pathways related to agro-ecological intensification are dynamic and diverse. Extracting lessons from earlier efforts aiming at facilitating adoption and a positive impact is essential for guiding future efforts. Within this context, the tension between the need to engage with farmers to adapt practices to their specific conditions, on the one hand, and the need to identify 'simple', quick-win solutions to reach scale in (cost-)effective ways, on the other, continues to dominate the debate around scaling agricultural intensification (Linn 2012; Lynam and Twomlow 2012), often with a specific reference to the development of international public goods that include principles for intensification rather than the practices themselves. Furthermore, achieving impact takes time, often far beyond the life of a specific project that initiated the impact pathway. In projects, a good monitoring and evaluation framework that monitors changes in the behaviour of stakeholders that will ultimately result in the desired impact is as crucial as the impact itself. A clearer understanding of the processes leading to impact would allow technologies to be scaled out beyond the initial target areas. Although impact is often measured at the household level, it is equally important to consider the impact of specific interventions on intra-household decision-making and gender-specific benefits from technology adoption. Increases in crop income may pass women by, with consequent high risks for negative impacts on household food and nutrition security.

Ouma *et al.* (Chapter 14) concentrate on the early impact of the CIALCA project and conclude that awareness of CIALCA products is mainly influenced by information access variables, including social networks and participation in technology evaluation; adoption is influenced by binding capital constraints, institutional and location factors and the farmers' perception of the attributes of the technology. Rusike *et al.* (Chapter 15) use value chain approaches to investigate various drivers, operating at different scales that create opportunities for the intensification of grain legumes, important crops for nutrition, income and soil fertility. They also anticipate that strengthened farmers' associations and better targeted research investments will further enhance uptake and livelihood improvements. Two other contributions focus on models for engaging farming communities to achieve more effective impact. The chapter by Remans *et al.* (Chapter 16) argues that crop yields, calories produced per capita, economic output and cost–benefit ratios are insufficient for a holistic evaluation of impact and propose the use of the nutritional functional diversity concept, allowing for the assessment of the degree of redundancy or resilience in the nutrient-provisioning of the system as an important component of system improvement. Degrande *et al.* (Chapter 17) stress the need to work through effective farmers' organizations, in this specific case referred to as relay organizations, but acknowledge that the effectiveness of such organizations is variable. Those with adequate internal human, material and financial capacity and that operate in a more conducive environment seem to perform better. Karltun *et al.* (Chapter 18) demonstrate that the success or failure of an intervention is dependent not only on its content, relevance and implementation strategy but also on local social structures which may vary considerably, even within similar agro-ecological environments.

## Approaches for effective communication on intensification options

The final conference theme dealt with methods and approaches to effectively disseminate the theory and practice underlying agro-ecological intensification, often associated with complex knowledge. An important degree of understanding of this complexity is required by farming communities before adoption can be achieved. Co-learning, farmer-to-farmer dissemination, effective communication and inclusive partnerships are all believed to be crucial principles of any approach aiming at disseminating complex knowledge (van Asten *et al.* 2009; Leeuwis and Aarts 2011; Neef and Neubert 2011). A wide range of communication approaches exist, from rather 'traditional' top-down approaches using radio, television and brochures, through more engaging approaches, such as the training of lead farmers, to end-user-owned approaches, such as farmer field schools. These approaches do not need to be in competition but can be complementary, depending on the existing knowledge of the end-user, the progress already achieved in the adoption pathway and the complexity of the intensification constraints and opportunity targeted.

Van Schagen *et al.* (Chapter 19) conclude that complexity needs to be embraced and coherent approaches for addressing complexity need to be developed, starting from the integration of impact pathways in the project design and implementation phase. Lynam (Chapter 20) reflects on the new drive to focus research on production systems and the need to create tools and approaches to take outputs from such research to scale in cost-effective ways. He argues that different stages of technology dissemination processes and tools exist, from awareness creation to adoption, implying the need for different dissemination methods. Agro-ecological intensification and the current drive for impact accountability drive an integration of research and extension whereby research outputs have a greater chance of being better translated into development outcomes. Buruchara *et al.* (Chapter 21) integrate such farmers' associations into innovation platforms that are the operational units of the Integrated Agricultural Research for Development (IAR4D) approach, promoted by the sub-Saharan Africa Challenge Program. They conclude that the IAR4D approach empowers communities to demand services, participation, ownership and sustainable change and that innovation platforms enhance synergies between institutions whereby an efficient allocation of institutional resources and services avoids unnecessary competition and duplication. Kimaru-Muchai *et al.* (Chapter 22), investigating the effectiveness of various tools for communication and awareness creation, conclude that farmers perceived other farmers and the radio as the most available sources of information, with the socio-economic characteristics of the farming households influencing the preference for specific methods of communication. Mowo *et al.* (Chapter 23) demonstrate innovative ways to integrate farmers' priorities into integrated catchment management approaches, whereby not only are these priorities directly and visibly addressed but other improved natural resource management practices are integrated into the action research and dissemination agenda.

## General conclusions and lessons learned

Based on the various keynote and other oral presentations, the poster presentations and the panel discussions organized around the four major themes of the conference, the following general conclusions and lessons learned were adopted by the participants.

### *Agro-ecological intensification: conflicting concepts for a generally accepted need*

- Growth in agricultural production and productivity is necessary but not sufficient for global food security. Future food security strategies include (1) reducing demand, (2) filling the production shortfall and (3) avoiding losses of productive capacity.
- A medium- and long-term, holistic, multi-functional and systemic view is required in addressing the challenges and aim at treating the causes of low soil productivity, not the symptoms, while ensuring that farmers have short-term benefits as a result of any system change.
- Subsidies and handouts are just one option to facilitate the adoption of new technologies, mainly to raise awareness and to make these technologies affordable to smallholder farmers. Subsidies (1) should be part of a package for better use efficiency, including technical support, business support, market development, institutional development, and facilitation by local organizations and (2) should not be used to push technologies that are not relevant for and/or adapted to local conditions and specific farmers' needs.

### *Technology components for integration in agro-ecological intensification pathways*

- Increased productivity will require investments in nutrients to improve and sustain soil fertility. ISFM offers technologies for managing organic inputs and the efficient use of mineral fertilizers with minimal environmental risks. Successful ISFM interventions must consider trade-offs in the use of labour, and also in financial and nutrient resources.
- Difficulty in getting access to mineral fertilizers is a constraint in many areas of East and Central Africa. The availability of mineral fertilizer needs to be improved for unit costs to be reduced (and made affordable to farmers). Benefits of scale of mineral fertilizer availability are needed if the intensification of farming systems is to be achieved.
- Demonstration to farmers that quality seeds provide better yields and that this translates into profit is essential. Market dynamics can provide a commercial pull for improved seeds. Opportunities exist for investing in the multiplication of improved legume and banana varieties at the community level but, especially for banana, quality assurance is critical.

***Integration of technical components at the farming system level***

- Smallholder farming systems are diverse, spatially heterogeneous and highly dynamic. The integrated analysis of farming systems allows the implications of proposed technologies to be studied across spatial and temporal scales. The integration of legumes into systems and the appropriate allocation of fertilizer need to be based on the identification of 'best-fit' interventions, selected from a 'basket of best-bet options'.
- Different types of innovations need to be identified for different types of farmers. (1) Agricultural labourers: how can labour-intensive agriculture be enhanced? (2) Subsistence farmers: how can risk-reducing agricultural techniques be facilitated? (3) Farmers with surplus potential: how can productive agriculture and market access be enhanced? (4) Farmers with large surpluses: how can we make sure that innovations result in trickle-down effects to the farming community?
- Intercrop systems can increase efficiencies at the farm level, e.g. returns to land, labour and fertilizer. Climate-smart systems can use intercropping to combine adaptation to and mitigation of the effects (e.g. coffee–banana systems).
- The role of livestock as a driver for agro-ecological intensification needs to be exploited. The 'livestock ladder' concept provides a framework to allow an exit from poverty and improve nutrition for poor crop–livestock farmers.

***Drivers and pathways for achieving impact***

- Adoption is influenced by the farmers' perception of the attributes of a technology, capital constraints and institutional support. Social networks and participation in technology evaluation are strong drivers of adoption. A mix of underlying challenges calls for a mix of interventions for different categories of farmers, and the acknowledgement that there is not a 'one size fits all' set of interventions.
- Grain legumes can be important in smallholder farmers' strategies for income, food security, nutrition, natural resource management (NRM) and gender equity, but such interventions are best integrated along effective value chains. It is important to enhance the nutritional diversity of farming systems, based on system diversification, including the diversity of locally important crops.
- Community-based organizations need to be included in agricultural extension efforts. Relay organizations can successfully diffuse innovations to farmers' groups, although continuous training and financial sustainability are crucial. Farmers need to be equipped with simple decision-support tools to aid them in making decisions on various strategies for resource allocation.

- Evaluating the impact of new technologies requires a mix of technical studies, on-farm adaptive research and approaches to learn from the site-specific responses of a specific technology. Socio-economic studies should use state-of-the-art methodology, including randomized control trials, aiming at addressing causality.

### *Approaches for effective communication on intensification options*

- Agro-ecological intensification and impact accountability are driving an integration of research and extension which may lead to a better translation of research outputs into development outcomes.
- Agricultural stakeholders need to consider farmers' socio-economic context in designing extension intervention strategies. The success or failure of an intervention is also dependent on local social structures where traditional institutions may play an important role in interventions and scaling up.
- Innovation platforms are important for relevant, efficient and effective partnerships across various stakeholders' groups. Planning of impact pathways is a necessity from the start. When farmers' priorities are given due consideration (e.g. domestic water availability), their interest in NRM increases.
- Due to the heterogeneity and complexity of smallholder farming systems, local adaptation in scalability needs to be integrated in dissemination approaches, partly building on the genotype × environment × management equation. Specific communication channels should be tailored to the specific technologies being promoted.

## References

Burney, J. A., Davis, S. J. and Lobell, D. B. (2010) 'Greenhouse gas mitigation by agricultural intensification', *PNAS*, vol. 107, no. 26, pp. 12052–12057.

CIESIN, Columbia University, United Nations Food and Agriculture Organization, and Centro Internacional de Agricultura Tropical (2005) *Gridded Population of the World, Version 3 (GPWv3): Population Count Grid, Future Estimates*, Palisades, New York.

Defries, R. and Rosenzweig, C. (2010) 'Toward a whole-landscape approach for sustainable land use in the tropics', *PNAS,* vol. 107, no. 46, pp. 19627–19632.

De Schutter, O. (2010) *Agro-ecology and the Right to Food,* UN Report A/HRC/16/49 to the 16th Session of the Human Rights Council.

Dixon, J. and Gulliver, A. with Gibbon, D. (2001) *Farming Systems and Poverty: Improving Farmers' Livelihoods in a Changing World*, FAO, Rome, and World Bank, Washington, DC.

FAO (1991) *World Soil Resources: An Explanatory Note on the FAO World Soil Resources Map at 1:25,000,000 Scale.* Food and Agriculture Organisation of the United Nations, Rome, Italy. 58pp.

FAO (2007) *Gridded Livestock of the World 2007*, FAO Rome.

FAO and IIASA (2007) *Mapping Biophysical Factors that Influence Agricultural Production and Rural Vulnerability*, FAO, Rome.

Fritz, S., You, L., Bun, A., See, L., McCallum, I., Schill, C., Perger, C., Liu, J., Hansen, M and Obersteiner, M., (2011) 'Cropland for sub-Saharan Africa: a synergistic approach using five land cover data sets', *Geophysical Research Letters*, vol. 38, doi:10.1029/2010GL046213.

Gockowski, J. and Sonwa, D. (2010) 'Cocoa intensification scenarios and their predicted impact on $CO_2$ emissions, biodiversity conservation, and rural livelihoods in the Guinea rain forest of West Africa', *Environmental Management*, vol. 48, pp. 307–321.

Leeuwis, C. and Aarts, N. (2011) 'Rethinking communication in innovation processes: creating space for change in complex systems', *Journal of Agricultural Education and Extension*, vol. 17, no. 1, pp. 21–36.

Linn, F. L. (2012) *Scaling Up in Agriculture, Rural Development, and Nutrition*, International Food Policy Research Institute, Washington, DC.

Lynam, J. K. and Twomlow, S. (2012) *A 21st Century Balancing Act: Smallholder Farm Heterogeneity and Cost Effective Research*, International Fund for Agricultural Development, Rome.

Mueller, N. D., Gerber, J. S., Johnston, M., Ray, D. K., Ramankutty, N. and Foley, J. A. (2012) 'Closing yield gaps through nutrient and water management', *Nature*, vol. 490, pp. 254–257.

Nakicenovic, N., Alcamo, J., Davis, G., de Vries, B., Fenhann, J., Gaffin, S., Gregory, K., Grübler, A., Jung, T. Y., Kram, T., Lebre La Rovere, E., Michaelis, L., Mori, S., Morita, T., Pepper, W., Pitcher, H., Price, L., Riahi, K., Roehrl, A., Rogner, H. H., Sankovski, A., Schlesinger, M., Shukla, P., Smith, S., Swart, R., van Rooijen, S., Victor, N. and Dadi, Z. (2000) *Special Report on Emissions Scenarios: A Special Report of Working Group III of the Intergovernmental Panel on Climate Change,* Cambridge University Press, Cambridge.

Neef, A. and Neubert, D. (2011). 'Stakeholder participation in agricultural research projects: a conceptual framework for reflection and decision-making', *Agriculture and Human Values*, vol. 28, pp. 179–194.

Tittonell, P. B., Vanlauwe, P., Leffelaar, A., Rowe, E. and Giller, K. E. (2005a) 'Exploring diversity in soil fertility management of smallholder farms in western Kenya. I. Heterogeneity at region and farm scale', *Agriculture, Ecosystems and Environment*, vol. 110, pp. 149–165.

Tittonell, P., Vanlauwe, B., Leffelaar, P. A., Shepherd K. D. and Giller, K. E. (2005b) 'Exploring diversity in soil fertility management of smallholder farms in western Kenya. II. Within-farm variability in resource allocation, nutrient flows and soil fertility status', *Agriculture, Ecosystems and Environment*, vol. 110, pp. 166–184.

van Asten, P. J. A., Kaaria, S., Fermont, A. M. and Delve, R. J. (2009) 'Challenges and lessons when using farmer knowledge in agricultural research and development projects in Africa', *Experimental Agriculture*, vol. 45, pp. 1–14.

Vanlauwe, B., Bationo, A., Chianu, J., Giller, K. E., Merckx, R., Mokwunye, U., Ohiokpehai, O., Pypers, P., Tabo, R., Shepherd, K., Smaling, E., Woomer, P. L. and Sanginga, N. (2011) 'Integrated soil fertility management: operational definition and consequences for implementation and dissemination', *Outlook on Agriculture*, vol. 39, pp. 17–24.

Vanlauwe, B., Pypers, P., Birachi, E., Nyagaya, M., van Schagen, B., Huising, J., Ouma, E., Blomme, G. and van Asten, P. (2012) 'Integrated soil fertility management in

Central Africa: experiences of the Consortium for Improving Agriculture-based Livelihoods in Central Africa (CIALCA)', in C. Hershey (ed.), *Issues in Tropical Agriculture Eco-Efficiency: From Vision to Reality,* CIAT, Cali, CO.

Wood, S., Hyman, G., Deichmann, U., Barona, E., Tenorio, R., Guo, Z., Castano, S., Rivera, O., Diaz, E. and Marin, J. (2010) *Sub-national Poverty Maps for the Developing World Using International Poverty Lines:* Available from http://povertymap.info.

You, L., Crespo, S., Guo, Z., Koo, J., Ojo, W., Sebastian, K., Tenorio, M. T., Wood, S. and Wood-Sichra, U. 'Spatial Production Allocation Model (SPAM) 2000 Version 3 Release 2', Available from http://MapSPAM.info (accessed June 2011).

# 2 Agricultural intensification and the food security challenge in sub-Saharan Africa

*B. A. Keating,*[1] *P. S. Carberry*[1] *and J. Dixon*[2]

[1] *Commonwealth Scientific and Industrial Research Organisation, Sustainable Agriculture Flagship*
[2] *Australian Centre for International Agricultural Research, Canberra*

## Introduction

A world population passing seven billion in 2011 and likely to exceed nine billion by 2050, consumption increases, diet shifts and food and land diversion to bio-energy have all combined to place food security high on the political, sociological and scientific agenda (WDR 2008; IAASTD 2009; The Royal Society 2009; Nelson *et al.* 2010; Foresight 2011). Many efforts are now directed at analysing the current food demand and supply scenarios and food price volatility (von Braun and Ahmed 2008; FAO *et al.* 2011) and exploring future scenarios of food demand and supply to 2050 (Agrimonde 2010; Hubert *et al.* 2010; Keating and Carberry 2010; Foresight 2011).

A common conclusion in many of these studies (but significantly not in all) is that global agriculture will have to intensify if estimates of future food demand are to be met. However, a diversity of pathways is inferred in the various uses of the term 'intensification'. Some call for a repeat of the Green Revolution of the period 1960–1980 in which fertilizer, irrigation and agro-chemical inputs combined with responsive germplasm to double or triple yields in many parts of the world; sub-Saharan Africa (SSA) was a notable exception (Evenson and Gollin 2003). Others point to concerns about the environment and sustainability with Green Revolution technologies and practices and make the point that it is 'sustainable intensification' pathways that are needed (Tilman *et al.* 2002). Others focus not so much on the quantum of input use but on the efficiency of a basket of limiting or constrained inputs in the eco-efficiency concept (Keating *et al.* 2010).

Calls for a new paradigm to shape the pathways of agricultural intensification are being made in the form of 'ecological intensification'(Doré *et al.* 2011) and a 'multi-functional' agriculture (Caron *et al.* 2008) that is less dependent on external inputs and more supportive of natural ecological processes and balances. The premise that agricultural production must increase can be questioned if arguments are accepted that food security can be maintained with a growing world population by reducing food waste and restraining the growth of food demands by limiting the livestock protein component of human diets (e.g. the Agrimonde scenario, Chaumet *et al.* 2009).

In this chapter we seek to explore the intensification challenges and opportunities facing SSA. We first examine the global context for food demand and supply to 2050. We then look to agriculture in SSA, its performance during the era of the Green Revolution and its trajectory out to 2050. We consider the drivers for intensification and the likely balance between ecological and sustainable intensification pathways – what mix of ecological, eco-efficient, and agronomic intensification is likely to set African agriculture on a sustainable path?

## Some definitions of intensification

The terminology in the literature around agricultural intensification can be confusing. In Table 2.1 we summarize some key terms and variants used over the last decade with the aim of determining what differences exist in the actions being proposed and the underlying concepts.

The CGIAR Science Forum in 2011 focused on the nexus between agriculture and the environment with sessions explicitly examining ecological intensification (Tianzhi and Bouman 2011). The assertion was made by some that agricultural research is mostly conducted by 'agriculturalists with little expertise on ecology, and most work has typically focused on increasing yield while reducing negative abiotic environmental impacts'. These sessions concluded that ecological intensification is poorly defined but participants were inclined to a view that it encompassed the 'harnessing of nature to increase food security and enhance ecosystem services'. This is not a particularly satisfying definition and there remains controversy over whether the calls for sustainable intensification or ecological intensification are in fact the same or different pathways and if different pathways, what their likely end points will be in terms of food security and environmental health.

In the sections that follow we shall examine selected key historical patterns of agricultural input use and output growth in the context of a growing world population. We shall return to this debate around the theoretical underpinnings of intensification pathways in the light of these data.

## Global context: food demand and supply trajectories

FAO's definition of food security (FAO 1996) reminds us that the physical quantity of food production is but one element of a broader set of drivers, including the global distribution of food, nutritional value of diets, food safety and human health outcomes, inequities in access to food associated with poverty and social disadvantage. This chapter will focus on agricultural production and the physical dimensions of food supply and demand, but it is important

*Table 2.1* Some terminology relevant to the agricultural intensification discourse since 1999

| *Term* | *Definition* | *References* |
|---|---|---|
| Agricultural intensification | Increase in agricultural production per unit of inputs (which may be labour, land, time, fertilizer, seeds, feed or cash). | FAO 2004 |
| Sustainable intensification | Producing more output from the same area of land while reducing the negative environmental impacts and at the same time increasing contributions to natural capital and the flow of environmental services. | Royal Society 2009; Godfray *et al.* 2010, Pretty *et al.* 2011 |
| Ecological intensification 1 | Intensification of cereal production systems [that] satisfies the anticipated increase in food demand while meeting acceptable standards of environmental quality. | Cassman 1999 |
| Ecological intensification 2 | Intensification in the use of the natural functionalities that ecosystems offer.<br><br>'Generally considered to be based essentially on the use of biological regulation to manage agro-ecosystems, at field, farm and landscape scales.' | Chevassus au Louis and Griffon 2008; Doré *et al.* 2011 |
| Eco-efficiency | Eco-efficiency [in an agricultural context] is a multi-dimensional concept relating the efficiency with which a bundle of desired outputs [food, fibre, environmental services] is produced from a bundle of inputs [land, labour, nutrients, energy], with minimal generation of undesired outputs [soil loss, nutrient loss, greenhouse gas emissions]. | Keating *et al.* 2010, 2012 |

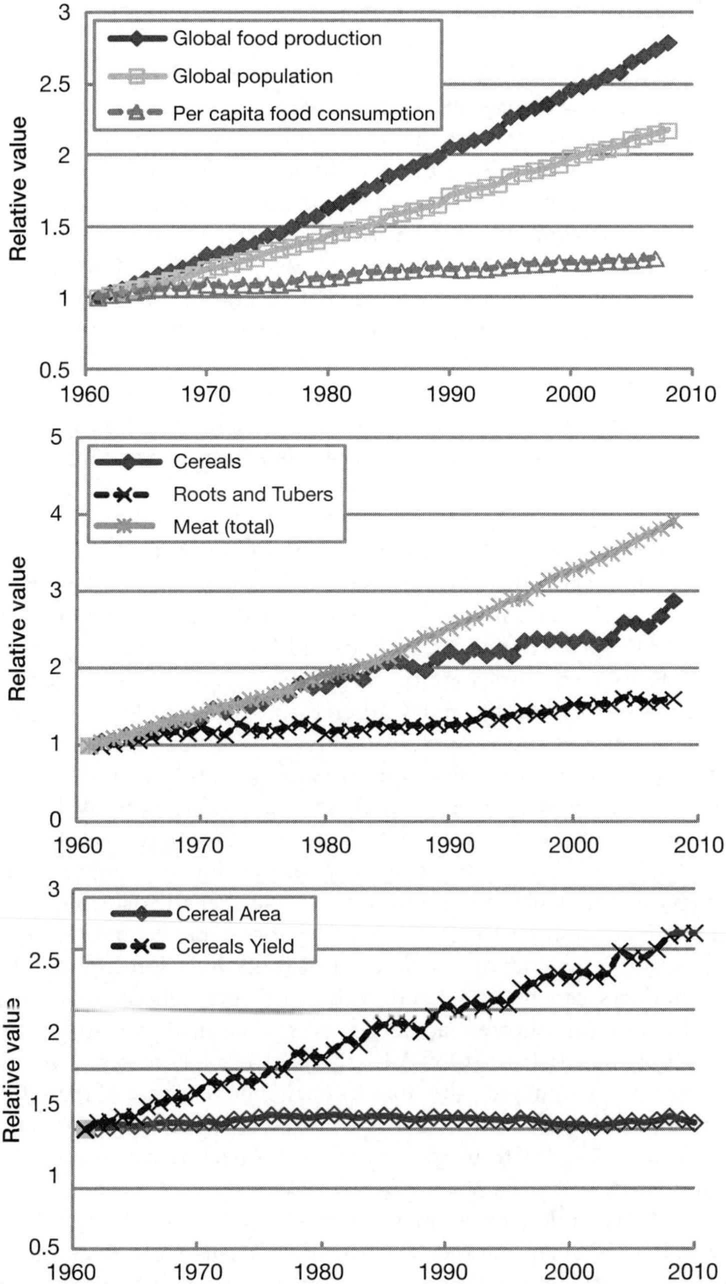


*Figure 2.1* Global agricultural performance metrics (1961–2008): (a) food production, population and per capita food available; (b) relative changes in cereals, root and tuber and meat production; (c) relative changes in cereal yields and areas harvested (source: FAOSTAT 2011).

to note that these are necessary but not sufficient ingredients of a food-secure world. A stark illustration of this is that currently there are over 900 million under-nourished people (FAO 2011) in a world that does, in fact, produce enough food for an adequate diet for everyone.

Looking back from 1961 to 2008, the food calories produced increased by 179 per cent – a rate of increase well above the population increase of 117 per cent. As a consequence, per capita food production rose by 27 per cent (Figure 2.1a). Consumption patterns shifted so that meat, oilseeds, fruit and vegetables grew in significance compared with cereals, while tuber crops declined as a proportion of the global diet (Figure 2.1b). Net land under cultivation grew by only 8 per cent and so these production increases were largely due to 2–3-fold yield increases (Figure 2.1c). Net arable land per capita fell from 0.42 ha to 0.21 ha. These yield increases were achieved with large increases in agricultural inputs; seven-fold increases in N fertilizers, three-fold increases in P fertilizers, and two-fold increases in irrigation water inputs (Tilman *et al.* 2002).

Although most of the world's food supply is produced and consumed within national borders, trade in food provides an important buffer for temporal and spatial production variability. For example, only approximately 13 per cent of the world's cereal consumption is currently traded across national borders (FAOStat 2011). North America, Australia and Europe are sources of net cereal exports while Asia and Africa are the major cereal importing regions.

Looking forward, Keating and Carberry (2010) have developed a set of scenarios of food demand out to 2050 driven by a range of assumptions of population growth, dietary consumption patterns, food diversion to biofuels and food waste along the value chain (Figure 2.2). The more likely scenarios include a world population of 9.2 billion and consumption levels higher than the current values of 3,342 kcal/capita/day for developed countries and 2,779 kcal/capita/day for developing countries. These scenarios suggest that food demand will increase by between 64 and 81 per cent between 2010 and 2050, with the variation dependent on the level of food diversion to biofuels.

These increases can be compared with a 129 per cent increase in global food production (in calories) achieved over a comparable 40-year period between 1970 and 2010 (FAOSTAT 2011). However, in terms of absolute quantities of food demand, the future challenge is greater than the past achievement. Average food demand is projected to increase by 17 petacal/day (2050 compared with 2010) in contrast to the 14 petacal/day average increase estimated over the 40 years from 1970 to 2010.

The almost three-fold increase in global food production achieved since 1961 has not come without environmental costs, and current yields are not necessarily sustainable in all situations. Agriculture accounted for 10–12 per cent of global greenhouse gas emissions in 2005 (Smith *et al.* 2007), and deforestation, much of which has been land clearing for agriculture, accounts for a further 17 per cent (Barker *et al.* 2007). Looking forward to 2050, aggregate greenhouse gas loads need to be reduced by between 50 and 85 per cent if global warming is

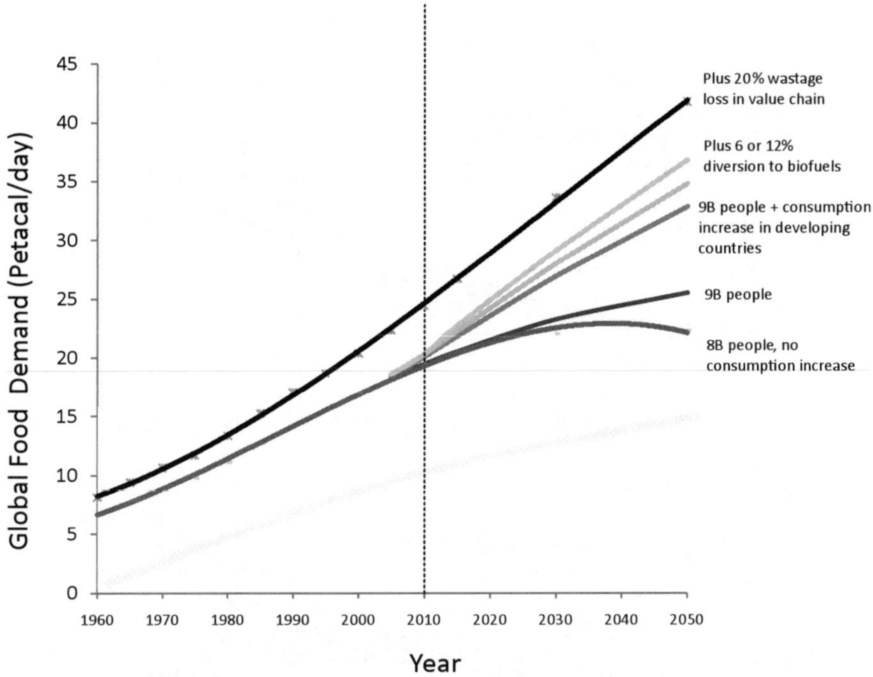


*Figure 2.2* Modelling food demand, 1960–2050 (after Keating and Carberry 2010).

to be constrained to under 2°C. If the global food production system was also to achieve such reductions in emissions at the same time as increasing food output by 50 per cent, the greenhouse gas intensity (tonnes $CO_2$-e/kcal food production) would need to be reduced 3–6-fold from current levels. This is a huge challenge confronting twenty-first-century agriculture; and how it can be met (if at all) is unclear at the present time.

In summary, while the relative growth in food demand looking forward 40 years is less than has been achieved looking back 40 years, the absolute quantities of additional food that will need to be produced from an increasingly constrained natural resource base are greater. When the imperative is considered to be to reduce greenhouse gas emissions to help avoid dangerous climate change and adapt to the unavoidable climate change that is already entrained, we conclude that the challenges to agricultural production of the first half of the twenty-first century are at least as significant as those addressed in the second half of the twentieth century.

## Africa: agricultural performance and prospects

Although agricultural output from Africa more than doubled from 1961 to 2009, the rate of increase was only half of that seen in Asia and South America

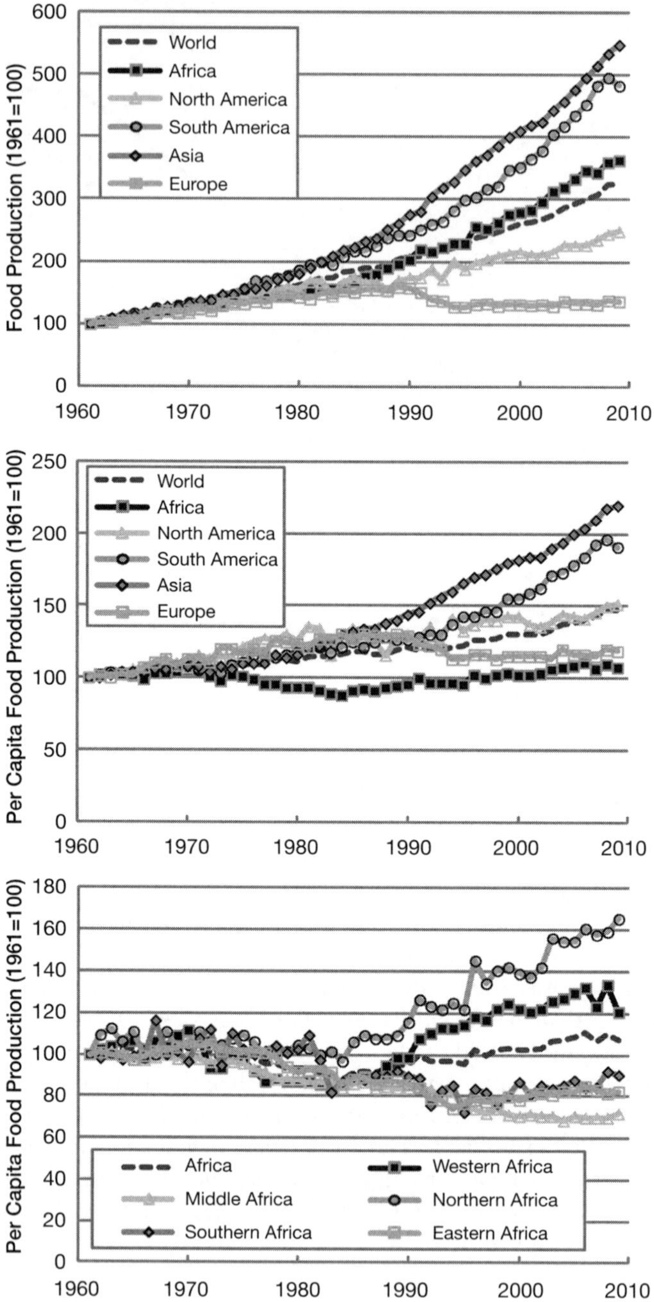


*Figure 2.3* Trends in agricultural performance 1961–2008: (a) index of agricultural output; (b) index of per capita agricultural production; (c) index of per capita food production in African sub-regions (source: FAOSTAT 2011).

(Figure 2.3a). In fact, per capita food production in Africa has changed little over the 49 years from 1961 to 2009. This contrasts with a global increase of almost 50 per cent and a more than doubling of per capita food production in Asia (Figure 2.3b). There are important sub-regional differences within Africa, with real declines in per capita food production in the order of 20–30 per cent in Southern, Central and East Africa (Figure 2.3c).

Unlike most other regions of the world, the increase in agricultural output in SSA was achieved by an expansion of land area, not by yield growth (Figure 2.4).

While the drivers of malnourishment are complex, the lack of growth (and in some cases significant declines) in per capita agricultural output in SSA have coincided with a poor record in reducing hunger. IFPRI (2012) reports a composite global hunger index made up of data on population-wide undernourishment, the under-five-years mortality rate and the prevalence of under-weight children. Values of this index in SSA are among the highest in the world (similar to those in south Asia) and show the least progress in achieving reductions over the last 20 years. Over the last 20–30 years, SSA has become a major food importing region, reflecting the fact that local production has not kept pace with population growth.

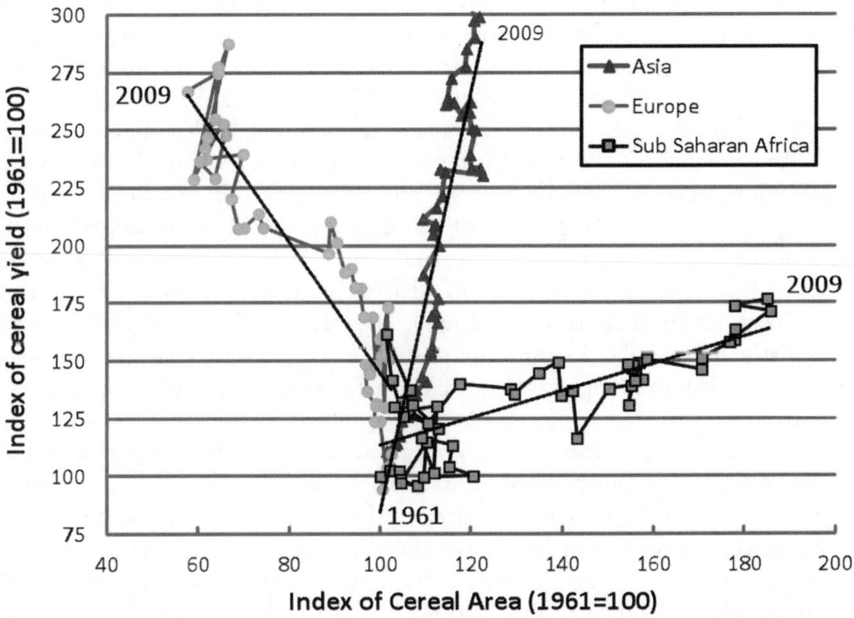


*Figure 2.4* Time trends in the total area of cereal plantings and in cereal yields for selected regions (1961 = 100 for each region). Solid lines join yearly data in sequence from 1961 to 2009. Solid lines without symbols represent linear regressions of yield and area indices.

## Future prospects

Looking forward, the projections for population growth for SSA are the highest in the world. The world's population is projected to rise by 30 per cent to over nine billion in 2050; the population of SSA is forecast to rise by over 100 per cent from an estimated one billion currently to possibly two billion in 2050 (UN 2008).

A host of critical questions arise from consideration of the current status and trends discussed above. These include: How will the growing population of SSA be fed? Will agricultural output keep pace with population growth? Will the existing high levels of under-nourishment continue to grow or will the per capita agricultural output grow, as has been achieved in other parts of the world? What will drive any increased agricultural output – yield growth (through intensification) or land area growth (maintaining low-input and low-output agricultural practices)? Does SSA have the land to support an expansion of arable agriculture? What environmental costs might be associated with bringing new land areas into agricultural production? Could SSA look forward to becoming a 'source' of food exports to other parts of the world (such as Asia) rather than a 'sink' attracting food imports, as is now the case?

An examination of the potential for further expansion of agricultural land in SSA reveals a number of reports that point to large areas of potential arable land suitable for agricultural development. For instance, the FAO TerraStat (2006) database identifies potential arable land as 2–3-fold the current level, with the greatest relative potential of expansion in Central and Southern Africa. Bruinsma (2009) estimates rainfed arable agricultural land in SSA (2005 data) at 320 million hectares, with a potential area of around 1,000 million hectares. The World Bank's (2009) report states that: 'The Guinea Savannah covers some 600 million hectares in Africa, of which about 400 million can be used for agriculture. Less than 10 percent of this area is currently cropped, making it one of the largest underused arable land reserves in the world.' This report suggests that these large areas of potential arable land in the Guinea Savannah zone in countries such as Mozambique, Nigeria and Zambia (among others) have characteristics and potential not dissimilar to regions of Thailand and Brazil that have undergone rapid agricultural expansion and intensification over the last decade.

Others are more cautious in making projections of a large-scale expansion of the footprint of agricultural cultivation in SSA (Young 1999; Kauffman *et al.* 2000), pointing to various sources of error, including the significant soil constraints associated with highly weathered, acid and shallow lateritic soils.

Soil constraints are not the only issue leading some authors to be cautious in assessments of land expansion options for African smallholder farmers. Jayne and Muyanga (2012) report that 25 per cent of the rural population in ten representative countries in SSA reside in areas exceeding 500 persons/km$^2$, estimated by secondary sources as an indicative maximum carrying capacity for areas of rainfed agriculture in the region. The humid highland areas will typically be the areas of greatest population density and in these areas the land expansion pathway is largely excluded by this existing intensity of land use. Even in areas of apparent land abundance, Jayne *et al.* (2012) demonstrate how inequality of access to land and other land tenure-related constraints may pose significant hurdles to the expansion of arable agriculture in practice. Additional constraints related to poor infrastructure and long distances to input and output markets limit agricultural development in some regions with apparently large areas of 'unused land', such as in Sudan and the Democratic Republic of Congo. While some areas of SSA have considerable potential for rainfed agriculture to expand its area footprint, the region is still characterized by generally low agricultural inputs and consequently by low yields. Yield gaps (that is, the difference between what yields are currently achieved on average and what would be technically and economically possible for the agro-ecological conditions if best practice genetics and agronomy were deployed) are generally estimated in the 60–80 per cent range. Neuman *et al.* (2010) compared the actual yields achieved with those estimated via an agro-ecological-economic model and concluded for all SSA countries examined that current yields were in the 15–40 per cent range of attainable yields (that is, yield gaps in the 60–85 per cent range). Using a different approach, the comparison of yields achieved on farmers' fields with those achieved in on-farm demonstration plots (researcher-managed plots with optimal inputs) reveal yield gaps in six countries of SSA in the range of 60–80 per cent (WDR 2008).

## Synopsis: needs, constraints and research implications

The past trends and projections presented above lead to the conclusion that SSA might reasonably aspire to a 2–3-fold increase in agricultural output between 2010 and 2050. A two-fold increase would barely keep pace with the projected population growth. A three-fold increase would begin to redress the declines in per capita food production seen to date and provide some impetus to reduce the current high levels of under-nourishment in the region. It is important to acknowledge again that improved food production is a necessary but not a sufficient driver of hunger alleviation.

So how might a three-fold increase in food production in SSA be achieved over the next 40 years? The yield gap reports considered above would suggest such a target could be met solely from the existing land footprint if a development pathway fully focused on raising inputs and improving practices and technologies (the 'sustainable intensification' pathway). Alternatively, SSA has the land to perhaps double its arable land footprint and raise agricultural output via that pathway (the land expansion pathway); however, uncertainty still exists on the extent to which soil constraints will limit this pathway.

The intensification pathway poses significant technical and institutional challenges and to date progress via this pathway in SSA has been limited in contrast to the rest of the world (see Figure 2.4). The land expansion pathway would come at a significant environmental cost in terms of the greenhouse gas emissions associated with the conversion of forest, savanna and grassland to cultivated agriculture and other negative impacts in terms of loss of biodiversity and degradation of water quality. It is also a more challenging pathway by which to ensure the societal benefits of agricultural development are more equally shared, given the likely mismatch between areas of population and areas of potential agricultural expansion. Bruinsma (2009) suggests that, with appropriate institutional settings and policy support, agriculture that is more commercially oriented may emerge, potentially still smallholder-based and directed towards strengthening the supply of domestic and sub-regional markets for trade in agricultural commodities.

While the intensification pathway would require a significant input of fertilizers and agro-chemicals, the environmental costs of these inputs are likely to be small compared with the impacts of increases in greenhouse gases and loss of biodiversity from the land expansion pathway. In fact, higher-input agronomy, through carefully targeted fertilizer inputs and the adoption of practices such as conservation agriculture would help to reverse the severe degradation of Africa's soils and could help to rebuild soil carbon levels with the benefits of greenhouse gas mitigation that this implies (FAO 2009).

Extreme climate variability and the risk of drought are dominating constraints to dryland agriculture in SSA and efficient use of the often limiting rainfall is critical to successful agriculture in both current and future climates. Unfortunately, the water use efficiency (WUE) of most crops is disappointingly low, due to poor rainfall capture and its utilization. The conundrum for Africa is that lifting soil fertility will readily lift WUE but high rainfall variability drives the farmers' disincentive to invest in fertilizers due to the downside risks. Technologies which improve rainfall capture (conservation agriculture, in-field ridges and pitting, local watershed development, targeted irrigation with limited water at critical times) will increase WUE and have shown promise in many regions of the world. Their adaptation to African smallholder agriculture warrants further exploration. Such examination should include the prospects to expand irrigated agriculture in SSA – for example, over seven million hectares are assessed as potentially available from river-based water resources in the semi-arid tropics, with the largest areas in Nigeria (Niger River) and Ethiopia (Nile River) (FAO 1997).

One can envisage a scenario within which SSA achieves a three-fold increase in agricultural output in 2050 via a mix of both intensification and land expansion activities. It seems essential in any feasible and desirable scenario that agricultural yields are doubled via a mix of technologies and practices, including carefully managed and targeted irrigation, use of agro-inputs such as fertilizers and agro-chemicals, and improved livestock husbandry and health. It is probably also desirable that African agriculture expands in those regions where it is ecologically and economically efficient to do so. Although the challenge of global greenhouse gas abatement would suggest agriculture's net footprint can't expand, it may be feasible and appropriate for less efficient agricultural land in other parts of the world to be transformed into carbon sinks or biodiversity reserves. Perhaps we will see the day when efficient African agriculture can supply global markets. In summary, intensification of agriculture in SSA is an essential ingredient of any agricultural development pathway. There may also be stimuli for agricultural production based on land expansion in particular sub-regions with the necessary agro-ecological resources and institutional support. In both cases, the sustainability of production practices is paramount and eco-efficiency in input usage is critical. Efforts to minimize the greenhouse gas emissions associated with land clearing, cultivation and agricultural production practices are important determinants of long-term sustainability, given the implications of such emissions for climate change. Global analyses by Burney *et al.* (2010) suggest that raising productivity on existing agricultural land through the intensification pathway is likely to be far more efficient in terms of minimizing greenhouse gas emissions per unit of food production than expanding the agricultural land footprint.

While progress over the last 20–30 years has been slow, Binswanger-Mkhize *et al.* (2011) have recently reported on the sense of positive progress taking hold around agricultural development in SSA. They attribute the improved prospects to 'higher international prices, improved macroeconomic and agricultural policies, improved institutional environments for agriculture and rural development, and greater government commitment to agriculture'. Prospects for a greater commercial focus to agriculture are emerging, particularly in response to domestic and regional markets, themselves being stimulated by more rapid economic growth in African economies. When these very experienced authors look to the challenges ahead, they look to the institutional issues rather than the technical issues per se. That is:

> maintaining macroeconomic stability in a challenging global context; further improving agricultural policies and agricultural investment climates; making progress in regional integration, in particular reducing barriers to intra-regional trade, standardizing trade and transport protocols, and standardizing and improving phytosanitary regulations; increasing public expenditures for agriculture and especially for agricultural research and technology dissemination; and improving infrastructure, market access, and access to inputs and credit for farmers.
>
> (Binswanger-Mkhize *et al.* 2011: 21)

If these issues can be addressed in a substantial way, agricultural development in SSA can advance much further in the next 30 years than it has done in the last 30 years.

## Reflections on ecology and agronomy in the intensification discourse

In Table 2.1 we explored the various pathways for intensification being proposed to address the food security challenges of the twenty-first century.

The call for sustainable intensification (and Cassman's (1999) original description of ecological intensification) come more from the agronomic world view – a view in which human intervention in agriculture via the careful use of inputs such as fertilizers and irrigation is essential if the necessary levels of food production are to be achieved without a huge expansion of agriculture's footprint. Phalan *et al.* (2011) have referred to this as a 'land sparing' strategy for the conservation of biodiversity. The ecological intensification pathway has evolved over the last decade to focus more on the natural processes and on agriculture as part of a multi-functional landscape. The call for a multi-functional view of agriculture (when separated from the unproductive capture of this concept in international trade negotiations) is well founded (Caron *et al.* 2008). Ecology is the foundation for agriculture and ecological imbalances will ensure that agriculture is not sustainable in the longer term. There is clear evidence of unsustainable intensification pathways elsewhere in the world and SSA does not want to go down such pathways.

However, the food production challenges of SSA are pressing – yields have to be increased, given the undesired consequences of meeting the 2–3-fold production challenge through the expansion of land areas. There is no question that this intensification pathway has to be sustainable and high levels of eco-efficiency need to be achieved with modest and carefully deployed external inputs. Every opportunity to make better use of existing water or nutrient resources in natural cycles should be taken, and every effort pursued to enhance the agro-diversity of agricultural systems where this adds resilience in the face of climatic, biotic and abiotic stresses while still delivering on productivity goals. However, a sole focus on an ecological intensification pathway (this would align with the Phalan *et al.* (2011) reference to a land-sharing strategy) seems unlikely to deliver the necessary productivity gains for SSA if it does not allow for the careful use of external inputs, in particular fertilizers, to address widespread nutrient imbalances in smallholder farming systems (Vitousek *et al.* 2009).

In 1990, McCown *et al.* (1992) called for a 'fertilizer augmented soil enrichment' strategy and Keating *et al.* (1991) predicted (with simulation models) how modest nutrient inputs and improved residue management could potentially double crop yields in the smallholder maize systems of eastern Kenya. These predictions were confirmed by 20 seasons of trial work by Okwach and Simiyu (1999). Now, 20 years later, we see only limited progress along this sustainable intensification pathway and yet the population in the study area has expanded rapidly. While the agronomic and ecological communities may debate the merits of sustainable versus ecological intensification pathways, there are clearly pressing agricultural development and food security challenges arising in the farming systems of SSA.

Experience elsewhere would suggest that little is likely to change without these technical opportunities being matched by improvements in the enabling institutions (see, for instance, Eichler 1999; Roling 2009; Binswanger-Mkhize *et al.* 2011; Hounkonnou *et al.* 2012), that is, in input and output markets, in knowledge systems and in the sustained commitment of governments to the development of a productive agricultural sector for African smallholder farmers.

## References

Agrimonde (2010) *Agrimonde: Scenarios and Challenges for Feeding the World in 2050*, Paris/Montpellier, INRA/CIRAD.

Barker, T., Bashmakov, I., Bernstein, L., *et al.* (2007) 'Technical summary', in B. Metz, O. R. Davidson, P. Bosch, R. Dave and L. A. Meyer (eds), *Climate Change 2007: Mitigation. Contribution of Working Group III to the Fourth Assessment Report of the Intergovernmental Panel on Climate Change*, Cambridge University Press, Cambridge, and New York, NY.

Binswanger-Mkhize, H. P., Byerlee, D., McCalla, A., Morris, M. and Staatz, J. (2011) 'The growing opportunities for Africa agricultural development', Conference Working Paper 16, ASTI-IFPRI_FARA Conference, Accra Ghana, 5–7 December 2011. www.asti.cgiar.org/pdf/conference/Theme1/Binswanger.pdf (accessed 30 September 2012).

Bruinsma, J. (2009) 'The resource outlook to 2050: by how much do land, water use and crop yields need to increase by 2050?' Technical papers from the Expert Meeting on How to Feed the World in 2050, 24–26 June 2009, FAO, Rome.

Burney, J. A., Davis, S. J. and Lobell, D. B. (2010) 'Greenhouse gas mitigation by agricultural intensification', *Proceedings of the National Academy of Sciences of the United States of America*, vol. 107, no. 26, pp. 12052–12057.

Caron, P., Reig, E., Roep, D., *et al.* (2008) 'Multifunctionality: refocusing a spreading, loose and fashionable concept for looking at sustainability?', *International Journal of Agricultural Resources, Governance and Ecology*, vol. 7, nos 4/5, pp. 301–318.

Cassman, K. G. (1999) 'Ecological intensification of cereal production systems: yield potential, soil quality, and precision agriculture', *Proceedings of the National Academy of Science*, vol. 96, no. 11, pp. 5952–5959.

Chaumet, J. M., Delpeuch, F., Dorin, B., *et al.* (2009) *Agrimonde, World Agriculture and Food in 2050: Scenarios and Challenges for Sustainable Development*, Paris/Montpellier, INRA, CIRAD.

Chevassus au Louis, B. and Griffon, M. (2008) 'La nouvelle modernité: une agriculture productive à haute valeur écologique', *Déméter: Économie et Stratégies Agricoles*, vol. 14, pp. 7–48.

Doré, T., Makowski, D., Malézieux, E., Munier-Jolain, N., Tchamitchian, M. and Tittonell, P. (2011) 'Facing up to the paradigm of ecological intensification in agronomy: revisiting methods, concepts and knowledge', *European Journal of Agronomy*, vol. 34, no. 4, pp. 197–210.

Eichler, C. K. (1999) 'Institutions and the African farmer', Third Distinguished Economist Lecture, CIMMYT, Mexico.

Evenson, R. E. and Gollin, D. (2003) 'Assessing the impact of the Green Revolution, 1960 to 2000', *Science*, vol. 300, pp. 758–762.

FAO (1996) *Declaration on World Food Security*, World Food Summit, FAO, Rome.

FAO (1997) 'Irrigation potential in Africa: a basin approach' *FAO Land and Water Bulletin,* vol. 4. www.fao.org/docrep/w4347e/w4347e00.htm#Contents (accessed 30 September 2012).

FAO (2004) *The Ethics of Sustainable Agricultural Intensification*, vol. 3, FAO, Rome.

FAO (2009) *Food Security and Agricultural Mitigation in Developing Countries: Options for Capturing Synergies*, FAO, Rome.

FAO (2011) *The State of Food Insecurity in the World*, FAO, Rome.

FAO, IFAD, IMF, OECD, UNCTAD, WFP, World Bank, WTO, IFPRI, UNHLTF (2011) *Price Volatility in Food and Agricultural Markets: Policy Responses*, FAO, Rome.

FAOSTAT (2011) http://faostat3.fao.org/home/index.html (accessed 30 September 2012).

FAO TerraStat (2006) www.fao.org/nr/land/information-resources/terrastat/en (accessed 30 September 2012).

Foresight (2011) *Foresight: The Future of Food and Farming: Final Project Report*, Government Office for Science, London.

Godfray, C., Beddington, J. R., Crute, I. R., *et al.* (2010) 'Food security: the challenge of feeding 9 billion people', *Science*, vol. 327, pp. 812–818.

Hounkonnou, D., Kossou, D., Kuyper, T., *et al.* (2012) 'An innovation systems approach to institutional change: smallholder development in west Africa', *Agricultural Systems*, vol. 108, pp. 74–83.

Hubert, B., Rosegrant, M., van Boekel, M.A.J.S. and Ortiz, R. (2010) 'The future of food: scenarios for 2050', *Crop Science*, vol. 50, pp. 33–50.

IAASTD (2009) 'Agriculture at a crossroads', in B. D. McIntyre, H. R. Herren, J. Wakhungu and R. T. Watson (eds), *International Assessment of Agricultural Knowledge, Science and Technology for Development*, Island Press, Washington, DC.

IFPRI (2012) *Global Hunger Index – The Challenge of Hunger: Ensuring Sustainable Food Security Under Land, Water and Energy Stresses*, International Food Policy Research Institute, Washington, DC., www.ifpri.org/sites/default/files/publications/ghi12.pdf

Jayne, T. S. and Muyanga, M. (2012) 'Land constraints in Kenya's densely populated rural areas: implications for food policy and institutional reform', *Food Security*, vol. 4, no. 3, pp. 399–421.

Jayne, T. S., Chamberlin, J. and Muyanga, M. (2012) 'Emerging land issues in African agriculture: implications for food security and poverty reduction strategies', Paper presented as part of Stanford University's *Global Food Policy and Food Security Symposium Series*, 12 January 2012, Stanford, CA.

Kauffman, S., Koning, N. and Heerink, N. (2000) 'Integrated soil management and agricultural development in West Africa: 1 – Potentials and constraints', *The Land*, vol. 4, pp. 73–92.

Keating, B. A. and Carberry, P. S. (2010) 'Sustainable production, food security and supply chain implications', *Aspects of Applied Biology*, vol. 102, pp. 7–20.

Keating, B. A., Godwin, D. C. and Watiki, J. M. (1991) 'Optimization of nitrogen inputs under climatic risk', in R. C. Muchow and J. A. Bellamy (eds), *Climatic Risk in Crop Production: Models and Management for the Semi-arid Tropics and Sub-tropics*, CAB International, Wallingford.

Keating, B. A., Carberry, P. S., Bindraban, P. S., Asseng, S., Meinke, H. and Dixon, J. (2010) 'Eco-efficient agriculture: concepts, challenges and opportunities', *Crop Science*, vol. 50, pp. 109–119.

Keating, B. A., Carberry, P. S., Thomas, S. and Clark, J. (2012) 'Eco-efficient agriculture and climate change: conceptual foundations and frameworks', in *An Eco-efficiency Revolution in Tropical Agriculture*, CIAT. http://ciat.cgiar.org/wp-content/uploads/2012/12/chapter_2_eco_efficiency.pdf (accessed 30 September 2012).

McCown, R. L., Keating, B. A., Probert, M. E. and Jones, R. K. (1992) 'Strategies for sustainable crop production in semi-arid Africa', *Outlook on Agriculture,* vol. 21, pp. 21–31.

Nelson, G. C., Rosegrant, M. W., Palazzo, A., *et al.* (2010) *Food Security, Farming, and Climate Change to 2050: Scenarios, Results, Policy Options*, International Food Policy Research Institute. www.ifpri.org/sites/default/files/publications/rr172.pdf (accessed 30 September 2012).

Neumann, K., Verburg, P. H., Stehfest, E. and Müller, C. (2010) 'The yield gap of global grain production: a spatial analysis', *Agricultural Systems*, vol. 103, pp. 316–326.

Okwach, G. E. and Simiyu, C. S. (1999) 'Effects of land management on runoff, erosion, and crop production in a semi-arid area of Kenya', *East African Agricultural and Forestry Journal*, vol. 65, no. 2, pp. 125–142.

Phalan, B., Onial, M., Balmford, A. and Green, R. E. (2011) 'Reconciling food production and biodiversity conservation: land sharing and land sparing compared', *Science*, vol. 333, no. 2, pp. 1289–1291.

Pretty, J., Toulmin, C. and Williams, S. (2011) 'Sustainable intensification in African agriculture', *International Journal of Agricultural Sustainability*, vol. 9, no. 1, pp. 50–58.

Roling, N. (2009) 'Pathways for impact: scientists' different perspectives on agricultural innovation', *International Journal of Agricultural Sustainability*, vol. 7, no. 2, pp. 83–94.

Royal Society, The (2009) *Reaping the Benefits: Science and the Sustainable Intensification of Global Agriculture*, The Royal Society, London.

Smith, P., Martino, D., Cai, Z., *et al.* (eds) (2007) *Climate Change 2007: Mitigation. Contribution of Working Group III to the Fourth Assessment Report of the Intergovernmental Panel on Climate Change*, Cambridge University Press, Cambridge, and New York, NY.

Tianzhi, R. and Bouman, B. (2011) *CGIAR Science Forum: The Agriculture–Environment Nexus*. www.sciencecouncil.cgiar.org/fileadmin/templates/ispc/documents/Meetings_and_events/Science_Forum/ISPC_science_forum_2011_program_and_abstracts_final_low-res.pdf (accessed 30 September 2012).

Tilman, D., Cassman, K. G., Matson, P. A., Naylor, R. and Polasky, S. (2002) 'Agricultural sustainability and intensive production practices', *Nature*, vol. 418, pp. 671–677.

UN (2008) *World Population Prospects, 2010 Revision*. http://esa.un.org/wpp/index.htm (accessed 30 September 2012).

Vitousek, P. M., Naylor, R., Crews, T., *et al.* (2009) 'Nutrient imbalances in agricultural development', *Science*, vol. 324, pp. 1519–1520.

von Braun, J. and Ahmed, A. (2008) *High Food Prices: The What, Who, and How of Proposed Policy Actions,* International Food Policy Research Institute, Washington, DC.

WDR (2008) *World Development Report 2008: Agriculture for Development*, World Bank, Washington, DC.

World Bank (2009) *Awakening Africa's Sleeping Giant: Prospects for Commercial Agriculture in the Guinea Savannah Zone and Beyond. Directions in Development*, World Bank, Washington, DC, and FAO, Rome.

Young, A. (1999). 'Is there really spare land? A critique of estimates of available cultivable land in developing countries', *Environment, Development and Sustainability*, vol. 1, pp. 3–18.

# 3 The agro-ecological solution!?

## Food security and poverty reduction in sub-Saharan Africa with an emphasis on the East African highlands

*H. Breman*

*International Fertilizer Development Center – Catalyze Accelerated Agricultural Intensification for Social and Environmental Stability*

## Introduction

In his report *Agro-ecology and the Right to Food* (De Schutter 2010), the Special UN Reporter makes the case for ecological farming practices to boost food production, paying special attention to sub-Saharan Africa (SSA): 'Today's scientific evidence demonstrates that agro-ecological methods outperform the use of chemical fertilizers in boosting food production where the hungry live – especially in unfavourable environments.' While sharing his worries about the lack of sustainability of today's agriculture, and in spite of many worthwhile suggestions, I am rejecting the report in its present form. The key reason is that the strongest tool for reaching food security, for reducing poverty and for more sustainable agriculture is neglected: agricultural intensification based on integrated soil fertility management (ISFM). Considering the present and estimated world population, there is no sustainability without external inputs, whereas their optimum use serves better the ecological and economic goals than De Schutter's minimum use (Breman 1990). His report is affected by key weaknesses, as shown in the next section. The potential of the alternative, using fertilizer in an ISFM context, is presented in the following section, using (among others) results of the Dutch-funded International Fertilizer Development Center (IFDC) project in Central Africa's Great Lakes region: Catalyze Accelerated Agricultural Intensification for Social and Environmental Stability (CATALIST).

## Strengths, myths and realities

Who does not share the worries of De Schutter about the sustainability of present agriculture? Indeed, as stressed by the often-cited Pretty, agro-ecological practices and resource-conserving technologies such as integrated pest management, integrated nutrient management, agroforestry, water harvesting in dryland areas and livestock integration into farming systems can be used to improve the stocks and use of natural capital in and around agro-ecosystems, and will make present agriculture more sustainable. Agriculture with a 'human face' requires the participation of smallholder farmers in technology development and diffusion, more attention for agro-ecology in public policies, investment in knowledge, reinforcement of social cohesion, more influence for women farmers and market development (Pretty 2006). The support for the UN report could, however, be much stronger with more nuanced standpoints regarding inorganic fertilizers; sustainable agriculture without them is unrealistic in view of the population density and its increase. A series of related weaknesses makes it difficult to support the report's vision. It exaggerates where cited authors are prudent; it uses weak studies, it does not take ecology and economics seriously, and ISFM, the strongest tool for reaching food security, reducing poverty and generating more sustainable agriculture, is neglected. The last weakness will be treated extensively in the third section; here the others will be illustrated.

### *Exaggeration*

It is not clear why the report goes one step further than most of its references, declaring that organic agriculture can assure food security and alleviate poverty; in terms of adding nutrients to the soil, inorganic fertilizer is not needed; the by-products of animal husbandry, of arable farming, biological nitrogen (N) fixation and fertilizer trees can do the job. Key references do not exclude the use of inorganic fertilizers, e.g. Reij *et al.* (2009), Garrity *et al.* (2010) and Pretty *et al.* (2011). In the cited Ladha and Peoples (1995), biological N-fixation is not presented as an alternative for inorganic fertilizer: e.g. Peoples *et al.* (1995), wondering if biological N-fixation is an efficient source of N for sustainable agricultural production, state supplementary fertilizer-N applications may be required to provide optimal nutrition. Giller and Cadish (1995) state, in relation to the report's most frequently mentioned fertilizer tree *Faidherbia albida*, that there is no direct evidence for a substantial role of N-fixation, and promote P-fertilizer to stimulate N-fixation.

### *Weak references*

#### *Agro-ecological methods outperform the use of inorganic fertilizers*

Among the background documents used for *Africa's Potential for the Ecological Intensification of Agriculture* (Tewolde Berhan Gebre Egziabher and Edwards 2010), two were identified as having a direct link with field research in Africa. One – *Agriculture biologique au Sénegal* – is only a research proposal; the other involves a seven-year test comparing compost and manure on a large number of crops in Tigray, Ethiopia (Edwards and Tewolde Berhan Gebre Egziabher 2007). The study shows that compost indeed in general leads to higher yields.

Calculating average values for all cereals together over the seven-year period, I found that the compost-based average is 2.5 t/ha and the fertilizer-based average is 1.8 t/ha (control 1.20 t/ha). Estimating the amount of N added by 15 t/ha/yr compost,[1] up to twice as much is found as the N from fertilizer, which is 41 kg/ha. Calculating the N-use efficiency from fertilizer-N, a value of only 15 kg/kg is found (kilogram of grain per kilogram of fertilizer-N). Best practices have not been used when applying fertilizer; a value of 25 kg/kg should have been more reasonable. It should have led to a yield of 2.2 t/ha; almost the same as the average compost-based yield obtained with an (almost?) twice as high N gift.

*Cover crops have an extreme potential for fixing nitrogen*

Badgley *et al.* (2007: 86) are used as the reference. They conclude that 'Data from temperate and tropical agro-ecosystems suggests that leguminous cover crops could fix enough nitrogen to replace the amount of synthetic fertilizer currently in use.' The estimated amount of potentially fixed N is obtained by multiplying the global area currently in crop production by the average amount of N available to the subsequent crop from leguminous crops during winter fallow or, in the tropics, between crops. The *average amount of N*, 103 kg/ha/yr (!), is derived from 77 studies worldwide. Questions, such as the availability of P, the length of the period for off-season cover crops and the availability of water and labour, are neglected. Certainly, an article with a different viewpoint has not been exploited, 'Legumes, when and where an option? No panacea for poor tropical West African soils and expensive fertilizers' (Breman and van Reuler 2002). That article shows how difficult it is in West Africa to reach an N-yield of 100 kg/ha by a leguminous crop exploiting the entire growing season; it treats the strong link between the availability of P and the biological N-fixation, and shows – among other things – that the consumer price for 5 kg of beans has to be equal to the price of 1 kg of fertilizer-P to interest farmers in using P.

*Fertilizer trees suppress the dependence on inorganic fertilizer*

The use of fertilizer trees to fix N suppresses the dependence on inorganic fertilizer (Garrity *et al.* 2010). In this context, three simplifications are generally made:

- The total N content of the trees is considered as being obtained through biological N-fixation, neglecting the N-uptake from the soil. As stressed by Giller and Cadish (1995) in relation to *Faidherbia albida*, there is *no direct evidence for a substantial role of N-fixation.*
- Crop yield increase in the direct surroundings of a tree is mentioned while the redistribution of nutrients by the extended root system of the tree is neglected. As shown by Breman and Kessler (1995), the N and P

enrichment of the soil under a tree decreases with increased tree density; less land can be exploited by the roots of individual trees, less redistribution occurs.

- Crop yields obtained in tree crop rotations are mentioned for the first crop after trees have been cut down. Such yields can be two to three times higher than control yields without a tree history. But taking into account the rapid yield decrease in following seasons and the years that the land is occupied only by trees, average yields of one-half of the control without trees are obtained (Breman 2011a).

### *Cheap solutions by simplifying the reality*

Harvesting water is presented as the solution for (semi-)arid land. Thanks to more water, more grass will grow, and thanks to more grass, more livestock can be kept, producing cheap proteins for poor people and manure for crop growth. In this context, Diop (2001) is cited; dangerous forms of belief in *perpetuum mobile*. Where Reij *et al.* (2009) accept that re-greening as such is insufficient to feed the Sahel, Diop and the way De Schutter is using him seem to consider more efficient water use as enough to keep more livestock and to produce more food on dryland. In a series of three books (Penning de Vries and Djitèye 1982; Breman and de Ridder 1991; Breman and Sissoko 1998), the nutrient dependence of crop and livestock production in the Sahel and of the carrying capacity for men and livestock has been shown in detail. Even when harvesting rainwater, 85–90 per cent will be lost by the fact that the growing crop does not find the nutrients required for growth and water absorption.

### *Inaccurate use of a reference*

Defending agroforestry as an alternative to fertilizer subsidies in Malawi's effort for poverty alleviation, a reference is made to a study wondering if agroforestry technologies improve the livelihoods of the resource-poor farmers there. Reading the study, one realizes that the answer is 'no', while the opposite is suggested. Quinion (2008), studying the adoption of agroforestry technologies, concludes that households were found still to be living in extreme poverty and their adoption does not lead to significant livelihood improvements.

### *References with alternative explanations*

Reij *et al.* (2009) describe a farmer-managed, agro-environmental transformation in the West African Sahel; land rehabilitation after decades of drought and overexploitation. Hundreds of thousands if not millions of hectares of land became productive again, and food production is increasing. The work of Reij *et al.* is cited by De Schutter, is cited in papers used as references by De Schutter (e.g. Garrity *et al.* 2010), is used by non-governmental organizations (NGOs), etc., all stressing that agro-ecological miracles are possible. Without denying the importance of the re-greening of the Sahel, it is useful to know that agricultural growth in the region is well correlated with the increasing use of inorganic fertilizer, as shown by Breman (2003) in Mali for a period of 30 years. Fertilizer statistics for Burkina Faso over the last 15 years suggest a similar correlation. Higher productivity of relatively productive land enables the regeneration of less productive land. Instead of using the work of Reij *et al.* to suggest that external inputs are not required, which they do not suggest themselves, it is more constructive to use the opportunity of re-greening for sustainable fertilizer use (see the third section of this chapter).

### *Neglecting ecological and economic laws*

Key references that are used do not exclude the use of inorganic fertilizer; minimizing the use is the tendency, replacing at maximum the nutrients lost otherwise. Such a use of inorganic fertilizer implies that the nutrient content of the soil organic matter stays low, and, consequently, the agronomic fertilizer use efficiency stays low. The reason is competition between crops and soil (micro)organisms for nutrients from applied fertilizer and manure. Therefore, the return on fertilizer investments stays low; fertilizer use is financially less favourable and market-oriented production is less competitive.

Using Garrity *et al.* (2010), Zambian conservation farming efforts are praised and promoted. Minimum tillage and cover crops are key components (Haggblade and Tembo 2003). The author evaluated these conservation farming efforts in 2006. One of the analyses showed an extreme variation in fertilizer use efficiency, using N as an illustration. Only commercial farmers reached the very acceptable level of 35 kg/kg. At 2006 prices, it led to a value–cost ratio (VCR) of 3.7. Average fertilizer use in Zambia on maize had an agronomic N-use efficiency of only 15 kg/kg, leading to a VCR of 1.6. For conservation farming even lower values were observed; VCRs of 0.9–1.3. Using minimum fertilizer rates appeared financially impossible.

## Fertilizer use in an ISFM context

Vanlauwe *et al.* (2010) present a detailed definition of ISFM. Improving crop productivity and maximizing the agronomic use efficiency of the applied nutrients (inorganic and organic) are formulated as the goals. By adding that all inputs need to be managed in accordance with sound agronomic principles, both economic and ecological sustainability are covered as well. Since my first publications about the matter (e.g. Breman 1990), I have stressed that to reach these goals (1) external input use should be added to and not replace benefiting from internal inputs and natural processes, and (2) crop nutrition and soil care should go hand in hand. I agree with De Schutter (2010) that in intensive commercial agriculture these principles are often not the driving forces, but he makes the mistake of thinking that minimizing the use of inorganic fertilizer is the solution. Key arguments mentioned above have been justified in Breman (1990) and Breman and Sissoko (1998): (a) even the African population density is already such that space is lacking to produce enough organic manure; (b) using small amounts of inorganic fertilizer on poor soils is in general accompanied by low agronomic fertilizer use efficiency,[2] and therefore (c) with low production competitiveness. Using ISFM to improve both the soil organic matter concentration and its nutrient content doubles the agronomic fertilizer use efficiency, improving its financial benefits significantly. Whereas it took about 20 years to observe the doubling of fertilizer use efficiency in the Netherlands after the Second World War (when fertilizer use became widespread), it took only five years in the ISFM project of IFDC in West Africa, thanks to an increase in knowledge and experience regarding soil fertility improvement (IFDC 2005). Ecological and economic justifications exist for promoting the optimum instead of minimum use of inorganic fertilizer. The optimum use concerns doses high enough to improve the soil organic matter status leading to high fertilizer use efficiency and accompanying economic benefits. An additional benefit is the relatively low risk of losses of nutrients to the environment. The goal of maximizing the agronomic use efficiency of nutrients requires doses on the straight part of the dose:effect curve, while maximum economic benefits may require higher doses, threatening ecological sustainability. Therefore, doses at the upper end of the straight part of the dose:effect curve are advisable. Maximizing the crop yield at the upper end implies at the end the use of a site-[3] and crop-specific fertilizer formula combined with the right choice of other ISFM components (van Keulen 1982; Breman *et al.* 2012a).

Fertilizer use in an ISFM context means that crops are nourished using inorganic fertilizer in the first place, and that soil amendments are applied to improve and maintain soil qualities. These amendments concern almost always[4] sources of organic matter, and sometimes lime and/or phosphate rock. Too easily, people 'translate' ISFM as combining inorganic fertilizer and manure. Indeed, it is an ISFM option and many examples exist where such a combination is more financially interesting than the use of the two individual components. However, a farmer who can afford to procure inorganic fertilizer can use his organic amendments more intelligently. Lower-quality organic amendments, such as straw[5] can be combined with inorganic fertilizer. The lower mineralization rate will ensure more rapid improvement of the soil's organic matter content, and lower quality and higher availability imply lower costs. At least as important is the fact that the risk of overdosing of inorganic fertilizer, implying lower agronomic efficiency and a less favourable VCR, is more limited. In this way, manure stays available for application on fields for which inorganic fertilizer could not yet be bought. Optimizing the use of inorganic fertilizer in an ISFM context leads to higher yields than the agro-ecological solution without the use of inorganic fertilizer, or with its minimum use. This is illustrated for crops and agroforestry and to the interest of the technology for dry years.

***Arable farming***

It is particularly in relation to arable farming that De Schutter (2010) presents quantitative indications regarding the potential of the agro-ecological solution. Citing Pretty *et al.* (2006), he stresses that more than 12 million farmers in developing countries produce 79 per cent more than others per hectare, thanks to technologies with low external input use that preserve the natural resources. Badgley *et al.* (2007), also using Pretty's data, wonder if organic agriculture is able to feed the world, and answer the question affirmatively. Using almost 300 data sets, they determined the potential of organic agriculture, using yield ratios of organic production:non-organic production for the developed world and for the developing world. The average ratio for the developed world is 0.91; the potential of organic arable farming does not reach entirely that of non-organic or conventional farming. For developing countries, however, organic yields are, on average, 1.71 times those of conventional farming. The potential of organic agriculture seems higher with lower yields.

Yield ratios, not yields, are used in the article. To get an idea of the relationship between yield ratios and yields, I calculated the average yield ratios for cereals.[6] FAO statistics have been used to obtain the average maize yield of all countries for which yield ratios were available. Maize has been chosen, because it is the world's most produced cereal and it has a much wider range of growing conditions than wheat and rice. The obtained relation between yield ratios of organic production:non-organic cereal production and non-organic or conventional maize yields is presented in Figure 3.1. Each dot represents a country; the dimension of the dot increases proportionally with the number of data per country in Badgley *et al.* (2007). Figure 3.1 shows that the yield ratios increase with decreasing yields. With yields higher than 8 t/ha, the potential of organic farming as defined by Badgley *et al.* (2007) is lower than that of conventional farming; with yields below 8 t/ha, organic farming leads to higher yields than conventional farming. The weighted average for cereal production in developed countries is an organic:non-organic ratio of 0.96, which is related to an average maize yield of 8.6 t/ha. For developing countries, the ratio is 1.57, related to an average yield of 1.5 t/ha and organic farming leads, at average, to a cereal yield that is 57 per cent higher than that of conventional farming.

It is on purpose that 'non-organic' and 'conventional' farming are used above, pell-mell. Badgley *et al.* (2007: 88) do the same: 'We assumed that all food currently produced is grown by non-organic methods, as the global area of certified organic agriculture is only 0.3 percent.' In this way, a serious error has crept into the account. The assumption is reasonable for the high yields of developed countries, but entirely unreasonable for yields below 2 t/ha. The high potential of organic farming in developing countries is – at least for an important fraction – obtained by comparing yields of organic farming projects and convinced and well-trained adopters, with yields of subsistence farming and of farmers looking desperately to other sources of income.

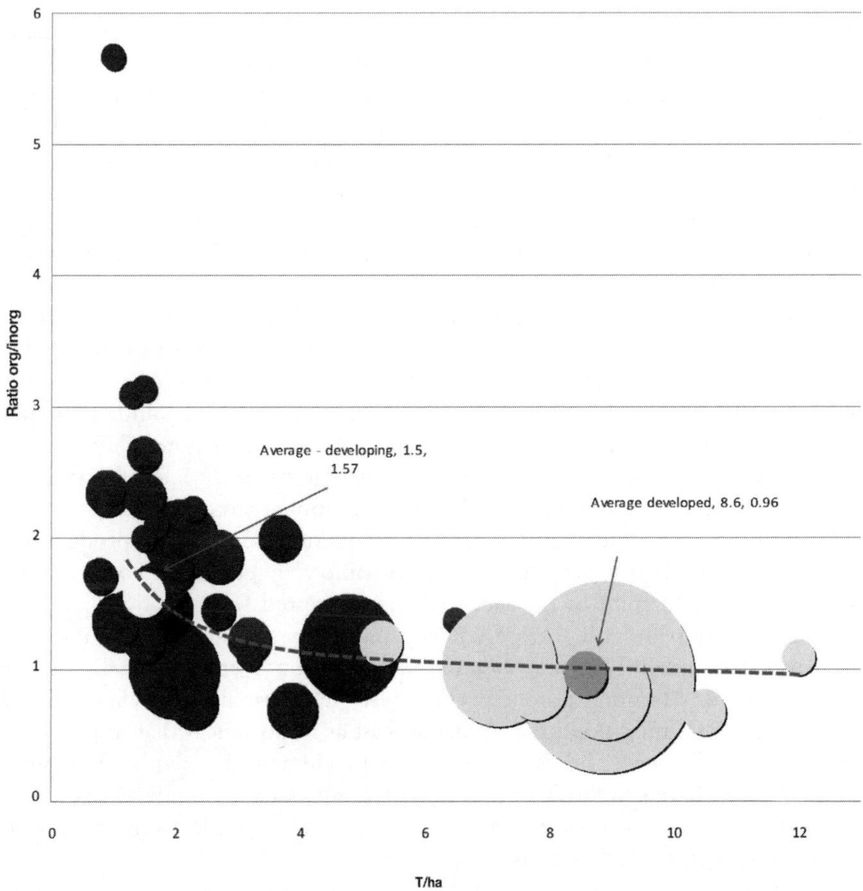


*Figure 3.1* The relation between yield ratios of organic production:non-organic cereal production and non-organic or conventional maize yields (t/ha) for developing and developed countries.

Another weakness in Badgley *et al.* (2007) and in De Schutter (2010) when using these and other data is the absence of the question from where the required manure or compost for organic or agro-ecological farming is to be obtained. Tewolde Berhan Gebre Egziabher and Edwards (2010), who compare the effect of 15 t/ha of compost with the effect of a low dose of inorganic fertilizer do not mention what area is required to produce the compost. They mention, however, the yield of by-products besides products. I calculated the straw production; about 5 t/ha dry matter besides the 3.3 t/ha of cereals. This straw can be used to produce a maximum of 2 t of fresh compost; 13 t has to be found elsewhere, outside the arable land requiring 15 t/ha of compost. Unfertilized land elsewhere will produce less than 2 t/ha of fresh compost or manure![7]

For producing the average 2.3 t/ha of organic cereals in developing countries (1.5 t/ha multiplied by the ratio 1.57), an important acreage of non-arable land is required or of arable land under non-organic agriculture. In this context it is useful to know that Badgley *et al.* (2007) use yields from certified and non-certified organic production *since most food produced is for local consumption where certification is not an issue*. For the East African highlands, the manure of four to five 'cow + calves units' would be required to obtain 2.3 t/ha of cereals, and consequently about 4 ha of rangeland would be required for each hectare of organic arable farming (Breman *et al.* 2012b). Organic farming is an option only in the case of low population pressure.

Optimal fertilizer use, combining inorganic fertilizer with ISFM, is a more realistic alternative for conventional agriculture. Figure 3.2 presents maize yields obtained by thousands of farmers who adopted the ISFM technology, promoted by the CATALIST project[8] in Burundi, Eastern DRC and Rwanda (Breman *et al.* 2012a), in comparison with conventional yields and potential yields of organic farming. The curve presenting the average organic yield potential has been derived from Figure 3.1, multiplying the non-organic (or conventional) yield with the related yield ratio of organic production:non-organic production. As already mentioned above, the potential organic yield equals the conventional yield at 8 t/ha. In spite of the fact that the suggested fertilizer doses are still low, the farmers are still inexperienced, and their soil not yet significantly improved through ISFM, the obtained yields are on average 1.4 t/ha higher than the organic farming potential. The best yields are already twice as high as the organic farming potential. A fact at least as important is that in principle almost no additional land is required for producing the required organic amendments; thanks to fertilizer use, most crops produce enough by-products. Only for potatoes, tomatoes, etc., are by-products not available; using the leafage could cause serious pest and disease problems.

Reaching higher yields while producing (almost) enough organic amendments implies also that the economics of ISFM farming are more favourable than those of organic farming; competition with conventional farming is much easier. One should realize that in poor regions with a very high population density, manure is expensive. In Rwanda, N obtained through manure ($6/kg) is already three times more expensive than N from urea ($2/kg) (Breman *et al.* 2012a). The difference will be even larger for certified organic farming, when the manure or compost also has to be produced through organic farming. Presently in France, biological organic fertilizers are 30–40 per cent more expensive than general organic fertilizers.

### *Agroforestry*

Agroforestry is one of the technologies strongly promoted by De Schutter (2010), using Garrity *et al.* (2010) as a key reference. De Schutter stresses (1) that in the proximity of trees in Zambia as well as in Malawi, maize yields more than 4 t/ha can be obtained, three times higher than yields far from the

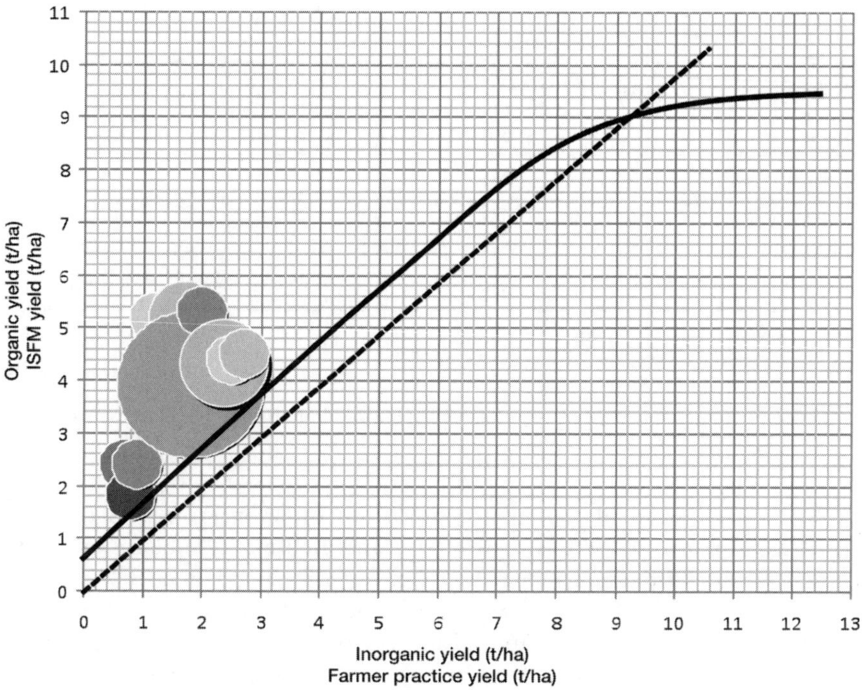


*Figure 3.2* Maize yields of farmers using inorganic fertilizer in an ISFM context* and potential maize yields of organic farming** in relation with conventional maize yields***.

Notes: * Dots (dimension proportional to number of observations per region and year); ** curve; *** the horizontal axis, which is equal to the horizontal axis of Figure 3.1.

trees; and (2) that the N-fixation by those trees suppresses the dependence on fertilizer. Therefore, agroforestry is of particular importance for poor farmers and poor countries, those of SSA, who cannot afford to procure expensive inorganic fertilizer.

Undeniably, under certain circumstances, agroforestry is an interesting source of organic soil amendments, an alternative to manure and compost. But it does not generate miracles, as explained already; it is not a cheap alternative to fertilizer. Planting and maintaining trees cost money and are financially feasible under certain circumstances (for the CATALIST intervention region, see Fleskens 2007; IFDC 2009). Considering the average crop yields of agroforestry fields instead of comparing crop growth in the surroundings of trees and at a distance leads to entirely different conclusions regarding the potential of agroforestry to replace inorganic fertilizer for crop yield increase and for poverty alleviation (Breman and Kessler 1995). Stimulated by De Schutter's report, I made a new review regarding the yield effects of what I call extensive agroforestry, which uses trees to avoid the use of inorganic fertilizer as much as possible (Breman 2011a). In fact, positive yield effects are

*Table 3.1* The effect of fertilizer and trees on maize yield, ten years after installing an agroforestry trial in Togo's coastal zone

| *Treatment* | *Yield (t/ha/year)* | | | | |
|---|---|---|---|---|---|
| | *Grain* | *Straw* | *Leaves* | *Wood* | *Total* |
| Control | 1.3 | 1.9 | 0 | 0 | 3.2 |
| Control + f) | 4.2 | 6.3 | 0 | 0 | 10.5 |
| Parkland | 2.6 | 3.9 | 5.0 | ±6.5★) | ±18★) |
| Parkland + f) | 4.1 | 6.1 | 9.5 | ±13.0★) | ±32★) |

Source: derived from Tamelokpo *et al.* 2007

*Notes*:
f = inorganic fertilizer.
★ ± because the trees were too young to be harvested; wood production has been estimated, not measured as in the case of leaves, straw and grain.

an exception instead of the rule; competition between crops and trees for nutrients, water and light is too strong, and fighting it requires in-depth knowledge, understanding and labour. Yield effects vary between +20 per cent and –50 per cent, the average being approximately –25 per cent. Positive effects are registered for certain parklands (Table 3.1 presents an example) and in the case of crop protection against strong winds. Negative yield effects are usual for alley cropping,[9] small fields surrounded by trees and for tree–crop rotation.

An entirely different picture appears when agroforestry is applied as ISFM technology, aiming at the optimal and financially feasible use of inorganic fertilizer (Breman and Kessler 1995): intensive agroforestry. The main yield increase is obtained because of effective fertilizer use; an extra yield increase of about 25 per cent is due to the tree component of the system. N-fixation plays only a minor role; the main effect of the tree is the improvement of the soil's organic matter status. In fact, leguminous N-fixing trees are suboptimal for this goal; the mineralization rate of their leaves is too high.

Table 3.2 compares the end results of ten years of intensive agroforestry with extensive agroforestry and crop production without trees. Thanks to a strict and well-chosen management scheme, parkland trees alone show a maize yield twice as high as the unfertilized control. In fact, the doubling of maize yield is not thanks to the trees, but rather to yield maintenance. At the start of the trial, ten years earlier, the control yield on the just-cleared old fallow land was more than 2 t/ha. The presence of well-managed trees makes sustainable production possible. Combining trees and fertilizer leads to maize yields similar to those from fertilizer only; more than 4 t/ha, three times more than the control. But in addition, the above-ground tree biomass production doubles, leading to a total biomass production ten times higher than that of the control without fertilizer and trees. Pretty *et al.* (2006) stress the potential carbon sequestration in the case of resource-conserving agriculture, and estimate it at an average of 0.35 t/ha/year of carbon. This amount is insignificant in

*Table 3.2* Yield increase of millet (Niger)* and maize (Rwanda)** during a drought thanks to inorganic fertilizer applied in an ISFM context

| *Country* | *Season* | *Rainfall (mm)* | | *Crop yield (t/ha)* | |
|---|---|---|---|---|---|
| | | *Average* | *Season* | *Control* | *ISFM* |
| Niger | 2001 | 500 | 370 | 0.4 | 2.9 |
| Rwanda | 2011-A | 450 | 230 | 0.0 | 2.1 |

Sources: * Fofana *et al.* 2008; ** Breman 2011b.

*Note*: A = The first of the two annual rainy seasons or crop production seasons.

comparison with the total carbon-sequestration of fertilized maize in the parkland of Table 3.1.

## *ISFM in years of drought*

As long as one is convinced that water is the limiting factor in dryland and during droughts, the agro-ecological solution with its attention for soil-covering vegetation and organic matter management (e.g. Reij *et al.* 2009; Garrity *et al.* 2010), the maximum from natural resource rehabilitation and maintenance is not obtained. It is useful to realize that even in drylands and during droughts, nutrients are often still more limiting than water. Neglecting soil nutrient deficiency in regions and seasons of drought implies neglecting cheap alternatives for irrigation (Breman *et al.* 2003). Table 3.2 presents two illustrations showing that fertilization is the cheapest way of irrigation; thanks to increased nutrient availability, more of the available water can be used by plants. Both cases concern on-farm, farmer-managed trials.

In Niger, in a very dry year, farmers who adopted inorganic fertilizer and ISFM produced crop yields seven times more than the national average. Farmers using conventional agricultural practices during the 2011-A season in northeast Rwanda (Nyagatare) did not harvest maize at all. Trying to guide agronomists and decision-makers in their choice regarding irrigation investments, Stroo (2010) developed the decision-support system 'apparent or real drought?' using illustrations from the drier eastern part of Rwanda. De Schutter (2010) blocks the sight of decision-makers on fertilizer in an ISFM context as an alternative for irrigation, and puts them on the wrong foot. The cause is the neglect of the role of soil nutrients, their contribution to the carrying capacity of the land, and therefore the limited potential of the agro-ecological solution. The East African highlands, the focus of the present conference, are characterized by high rainfall. In many cases, a transition to lower land is inherent in the transition to drier regions. The use of fertilizer in the transition zones from humid to (semi-)arid land is less risky than it is thought to be and is in the interest of drylands dominated by pastoral use and humid land exploitation, dominated by arable land use (Breman, in press).

## Conclusion

The agro-ecological solution is not able to contribute significantly to food security and poverty reduction in poor countries, such as those of SSA. Too many of the arguments in the report of the UN's Special Reporter on the Right to Food (De Schutter 2010) that suggest the opposite are weak or wrong. And while it may be true that agro-ecological agriculture can increase significantly the present low crop yields of developing countries, the use of inorganic fertilizer in an ISFM context is capable of reaching much higher yields at more favourable cost–benefit ratios. Much less space is required for such agriculture and it improves the resilience of agriculture in dry years. Agroforestry that views improving the agronomic fertilizer use efficiency is an example of an ISFM-based production system.

## Acknowledgements

The author thanks Scott Mall for editing the manuscript. He thanks IFDC-CATALIST for the opportunity to represent the project at the CIALCA conference.

## Notes

1 In view of the population density (80 inhabitants/km$^2$), it is difficult to believe that 15 t/ha can be made available for all crop land. Compost use alone will not be sustainable.
2 Hill-placed small P doses on seriously P-deficient sandy soils form an exception, but such micro-doses are in the long term unsustainable.
3 In practice one will often be obliged to accept the formula developed per agro-ecological zone.
4 Soils can be so rich in organic matter that for a number of years the lack of organic amendments will not have any measurable negative consequence.
5 'Lower quality' is used in relation to the fertilizer replacement quality of manure. Organic amendments used in ISFM do not need the high nutrient content, and the low lignin and phenols content of high-quality manure (Palm *et al.* 2001). Straw from fertilized crops is preferable; temporary N binding will be lower.
6 Cereals have been chosen instead of maize to obtain enough data. Using only the maize data for developed countries generates an average yield ratio organic:non-organic equal to that of all cereal data; for developing countries, the ratio for maize is 2.2, somewhat higher than the 1.6 obtained for all cereals.
7 The extreme demand for space in case of agriculture without inorganic fertilizer is for SSA illustrated by the disappearance of nature in most countries, as SSA uses less fertilizer than any other world region. In SSA, elephant density increases exponentially with increasing fertilizer rate on crop land (Smaling *et al.* 2006).
8 'Catalyser l'Intensification Agricole Accelerée pour la Stabilité Sociale et Environnementale', a Dutch-financed project implemented by the IFDC.
9 In Breman (2011a) a publication is treated, describing yield depressions between 20 and 50 per cent for two million hectares of alley cropping in China.

## References


Badgley, C., Moghtader, J., Quintero, E., Zakem, E., Chappell, M. J., Avilés-Vazquez, K., Samulon, A. and Perfecto, Y. (2007) 'Organic agriculture and the global food supply', *Renewable Agriculture and Food Systems*, vol. 22, pp. 86–108.

Breman, H. (1990) 'No sustainability without external inputs', in *Beyond Adjustment. Africa Seminar*, Maastricht, the Netherlands. Ministry of Foreign Affairs, Den Haag, the Netherlands, pp 124–133.

Breman, H. (2003) 'The fertility triangle, motor for rural and economic transformation in the Sahel', *Symposium on Sustainable Agriculture in the Sahel: Lessons and Opportunities for Action*. 1–4 December 2003. Syngenta, Bamako (Mali).

Breman, H. (2011a) *Comment assurer la sécurité alimentaire en RDC? S'appuyer sur de l'agroforesterie et des engrais!* Report IFDC-CATALIST, Kigali (Rwanda).

Breman, H. (2011b) *ISFM in a Season of Drought*, Report IFDC-CATALIST, Kigali (Rwanda).

Breman, H. (in press) *Collaboration agropastorale sahélienne: Intégration Cultures et Élevage comme business*, René Dumont Séminair, November 2012, Paris (France).

Breman, H. and de Ridder, N. (eds) (1991) *Manuel sur les pâturages des pays sahéliens*, ACCT, Paris/CTA, Wageningen/KARTHALA, Paris.

Breman, H. and Kessler, J. J. (1995) *Woody Plants in Agro-ecosystems of Semi-arid Regions (with an Emphasis on the Sahelian Countries)*, Springer-Verlag, Berlin.

Breman, H. and Sissoko, K. (eds) (1998) *L'intensification agricole au Sahel*, KARTHALA, Paris.

Breman, H. and van Reuler, H. (2002) 'Legumes, when and where an option? No panacea for poor tropical West African soils and expensive fertilizers', in B. Vanlauwe, J. Diels, N. Sanginga and R. Merckx (eds), *Integrated Plant Nutrient Management in Sub-Saharan Africa,* CAB International, Wallingford.

Breman, H., Maatman, A. and Wopereis, M. (2003) *African Soils Step-motherly Served by Nature, Governments and Donors*, AEPS, Rosebank and USAID, Washington, DC.

Breman, H., Hatangimana, Th., Kamale Kambale, J.-M., Kavira, S., Lindiro, R., Nzohabonayo, Z., Simbashizubwoba, C. and Ujeneza, N. (2012a) *Proposition pour l'amélioration des recommandations d'intensification agricole sur la base des résultats des tests participatifs et des démonstrations de fertilisation*, IFDC-CATALIST, Kigali (Rwanda).

Breman, H., Köster, H. and Rukundo, R. (2012b) *Alimentation des ruminants dans les systèmes de production mixtes au cœur de la Région des Grands Lacs d'Afrique Centrale*, Vol. II, IFDC-CATALIST, Kigali (Rwanda)

De Schutter, O. (2010) *Agro-ecology and the Right to Food*, UN Report A/HRC/16/49.

Diop, A. M. (2001) 'Management of organic inputs to increase food production in Senegal', in N. Uphoff (ed.), *Agroecological Innovations: Increasing Food Production with Participatory Development*, Earthscan Publications, London.

Edwards, S. and Tewolde Berhan Gebre Egziabher (2007) *Impact of Compost Use on Crop Yields in Tigray, Ethiopia, 2000–2006 Inclusive*, FAO, Rome.

Fleskens, L. (2007) *Prioritizing Rural Public Works Interventions in Support of Agricultural Intensification*, Consultancy report, IFDC-Rwanda.

Fofana, B., Wopereis, M. C. S., Bationo, A., Breman, H. and Mando, A. (2008) 'Millet nutrient use efficiency as affected by natural soil fertility, mineral fertilizer use and rainfall in the West African Sahel', *Nutrient Cycling in Agroecosystems*, vol. 81, pp. 25–36.

Garrity, D. P., Akinnifesi, F. K., Ajayi, O. C., Weldesemayat, S., Mowo, J. G., Kalinganire, A., Larwanou, M. and Bayala, J. (2010) 'Evergreen agriculture: a robust approach to sustainable food security in Africa', *Food Security*, vol. 2, pp. 197–214.

Giller, K. E. and Cadish, G. (1995) 'Future benefits from biological nitrogen fixation: an ecological approach to agriculture', in J. K. Ladha and M. B. Peoples (eds), *Management of Biological Nitrogen Fixation for the Development of More Productive and Sustainable Agricultural Systems*, Kluwer Academic Publishers in cooperation with IRRI, Dordrecht.

Haggblade, S. and Tembo, G. (2003) *Early Evidence on Conservation Farming in Zambia*, EPTRD discussion paper 108, IFPRI, Washington, DC.

IFDC (2005) *Development and Dissemination of Sustainable Integrated Soil Fertility Management Practices for Smallholder Farmers in sub-Saharan Africa*, IFDC, Muscle Shoals.

IFDC (2009) *Sustainable Energy Production Through Woodlots and Agroforestry in the Albertine Rift (SEW)*, IFDC, Bujumbura.

Ladha, J. K. and Peoples, M. B. (eds) (1995) *Management of Biological Nitrogen Fixation for the Development of More Productive and Sustainable Agricultural Systems*, Kluwer Academic Publishers in cooperation with IRRI, Dordrecht.

Palm, C. A, Gachengo, C. N., Delve, R. J., Cadisch, G. and Giller, K. E. (2001) 'Organic inputs for soil fertility management in tropical agroecosystems: application of an organic resource database', *Agriculture, Ecosystems and Environment*, vol. 83, pp. 27–42.

Penning de Vries, F. W. T. and Djitèye M. A. (eds) (1982) *La productivité des pâturages sahéliens: Une étude des sols, des végétations et de l'exploitation de cette ressource naturelle*, PUDOC, Wageningen.

Peoples, M. P., Herridge, D. F. and Ladha, J. K. (1995) 'Biological nitrogen fixation: an efficient source of nitrogen for sustainable agricultural production?', in J. K. Ladha and M. B. Peoples (eds), *Management of Biological Nitrogen Fixation for the Development of More Productive and Sustainable Agricultural Systems*, Kluwer Academic Publishers in cooperation with IRRI, Dordrecht.

Pretty, J. (2006) 'Agroecological approaches to agricultural development', RIMISP 20 Aniversario. Contribution by Rimisp-Latin American Center for Rural Development (www.rimisp.org) to the preparation of the *World Development Report 2008: Agriculture for Development*.

Pretty, J., Noble, A. D., Bossio, D., Dixon, J., Hinne, R. E., Penning de Vries, F. W. T. and Morison, J. I. L. (2006) 'Resource-conserving agriculture increases yields in developing countries', *Environmental Science and Technology*, vol. 40, no. 4, pp. 1114–1119.

Pretty, J., Toulmin, C. and Williams, S. (2011). 'Sustainable intensification: increasing productivity in African food and agricultural systems', *International Journal Agricultural Sustainability*, vol. 9, pp. 5–24.

Quinion, A. F. (2008). 'Contribution of soil fertility replenishment agroforestry technologies to the livelihoods and food security of smallholder farmers in central and southern Malawi', Masters Thesis, Stellenbosch University, South Africa.

Reij, C., Tappan, G. and Smale, M. (2009) 'Agroenvironmental transformation in the Sahel: another "Green revolution"', IFPRI Discussion Paper 00914 in *Millions Fed: Proven Successes in Agricultural Development*. www.ifpri.org/millionsfed

Smaling, E., Toure, M., de Ridder, M. N., Sanginga, N. and Breman, H. (2006) *Fertilizer Use and the Environment in Africa: Friends or Foes?* Background paper presented for the African Fertilizer Summit, 9–13 June 2006, Abuja, Nigeria. NEPAD, Johannesburg/IFDC, Muscle Shoals.

Stroo, J. I. L. (2010) 'Drought, reality or appearance: why and how ISFM works', Bachelor Thesis Earth Sciences – Physical Geography, Free University, Amsterdam.

Tamelokpo, A., Mando, A. and Breman, H. (2007) *Influences des éléments manquants et la gestion du sol sur les rendements de maïs et le recouvrement de N, P et K dans des systèmes d'agroforesterie dans la savane côtière ouest africain*, Rapport IFDC-Afrique, Lomé, Togo.

Tewolde Berhan Gebre Egziabher and Edwards, S. (2010) *Africa's Potential for the Ecological Intensification of Agriculture*, FAO, Rome.

van Keulen, H. (1982) 'Graphical analyses of annual crops response on fertilizer application', *Agricultural Systems*, vol. 9 pp. 111–126.

Vanlauwe, B., Bationo, A., Chianu, J., Giller, K. E., Merckx, R., Mokwunye, U., Ohiokpehai, O., Pypers, P., Tabo, R., Shepherd, K. D., Smaling, E. M. A., Woomer, P. L. and Sanginga, N. (2010) 'Integrated soil fertility management: operational definition and consequences for implementation and dissemination', *Outlook on Agriculture*, vol. 39, pp. 17–24.

# Part I

# System components

# 4 CIALCA interventions for productivity increase of cropping system components in the African Great Lakes zone

*P. Pypers,*[1] *W. Bimponda,*[2] *E. Birachi,*[3] *K. Bishikwabo,*[4] *G. Blomme,*[5] *S. Carpentier,*[6] *A. Gahigi,*[7] *S. Gaidashova,*[7] *J. Jefwa,*[1] *S. Kantengwa,*[4] *J. P. Kanyaruguru,*[8] *P. Lepoint,*[9] *J. P. Lodi-Lama,*[4] *M. Manzekele,*[2] *S. Mapatano,*[10] *R. Merckx,*[6] *T. Ndabamenye,*[7] *T. Ngoga,*[7] *J. J. Nitumfuidi,*[2] *C. Niyuhire,*[11] *J. Ntamwira,*[2] *E. Ouma,*[12] *J. M. Sanginga,*[4] *C. Sivirihauma,*[13] *R. Swennen,*[6] *P. van Asten,*[14] *B. Vanlauwe,*[1] *N. Vigheri,*[15] *and J. M. Walangululu*[16]

[1] *Tropical Soil Biology and Fertility – International Center for Tropical Agriculture (Kenya);* [2] *Institut National pour l'Etude et la Recherche Agronomiques, DR Congo;* [3] *International Center for Tropical Agriculture, DR Congo;* [4] *Tropical Soil Biology and Fertility – International Center for Tropical Agriculture, DR Congo;* [5] *Bioversity, Uganda;* [6] *KULeuven, Belgium;* [7] *Institut des Sciences Agronomiques, Rwanda;* [8] *Consortium for Improving Agriculture-based Livelihoods in Central Africa, Burundi;* [9] *Bioversity, Burundi;* [10] *Diobass, DR Congo;* [11] *Institut des Sciences Agronomiques du Burundi, Burundi;* [12] *International Institute of Tropical Agriculture, Burundi;* [13] *Bioversity, DR Congo;* [14] *International Institute of Tropical Agriculture, Uganda;* [15] *Université Catholique du Graben, DR Congo;* [16] *Université Catholique de Bukavu, DR Congo*

## Introduction

A cropping system can be defined as a community of crops managed within a farm unit to achieve a specific goal of the farmer (FAO 1995). Its productivity can be measured as the output (food, fodder or other raw materials) per unit of land area, labour invested, or input resources applied. In this chapter we consider cropping systems at the level of individual fields, consisting of a number of components: (1) the crop varieties grown, (2) the mineral nutrient resources applied, (3) the organic matter applied (external or generated within the system), and (4) crop management and agronomic measures (e.g. crop arrangement and density). This chapter focuses on the technologies developed or evaluated by CIALCA to improve individual system components of single crops with the aim of increasing productivity, measured as crop yield per unit land area. Interactions with management decisions, dependent on farm-level factors such as household wealth, production objectives and labour availability, are not considered here.

CIALCA conducted baseline and characterization studies in ten mandate areas across DR Congo, Rwanda and Burundi to understand the current conditions and limitations for crop production and develop relevant technologies for improving system components. Agro-ecological conditions are diverse, with altitudes varying between 400 masl in Bas-Congo and over 2,000 masl in the higher parts of Gitega and Sud-Kivu, and rainfall amounts varying between 900 mm/yr in Kigali-Kibungo and 1,700 mm/yr in Sud-Kivu, following a bi-modal rainfall pattern (Farrow *et al.* 2007). Farmers increasingly often report unreliable rainfall as a major constraint, particularly the late arrival of the rains and frequent intra-seasonal droughts. The main soil types are ferralsols and acrisols that have inherently good physical properties but poor chemical conditions and low nutrient stocks. In all mandate areas, except Bas-Congo, average slopes are steep, varying between 11 and 24 per cent and, as a result, soil erosion is a major threat to the long-term sustainability of cropping systems. Major crops are banana, cassava, maize and grain legumes (common beans, soybean and/or groundnut). Few households use improved varieties of banana (<19 per cent), groundnut or soybean (<6 per cent), cassava (<16 per cent) or maize (<24 per cent), but improved varieties of beans are used by 5–91 per cent of households (CIALCA 2010). Mineral inputs are rarely applied. Organic inputs are used by 20–66 per cent of households, but the quantities applied are often low (maximum average application rates of 700 kg/ha to banana and 250 kg/ha to beans) and mostly consist of composted crop residues. Farmyard manure (FYM) is scarce because of the low numbers of livestock. Farmers report poor soil fertility and lack of FYM as the most important limitations for crop production. Characterization studies, based on soil and leaf analyses in legumes and banana and pot experiments in maize, revealed deficiencies in N, P, K and Mg, and soil acidity as major soil constraints (CIALCA 2007, 2008). However, large variability was found in soil conditions, and this is typically a result of the differential allocation of labour and nutrient resources (Tittonell *et al.* 2007). Crop yields therefore vary, even between fields within a single farm. This heterogeneity is an important aspect of smallholder production systems, and must be considered in the development and extension of improved technologies for crop and soil management.

The development and evaluation of technologies is conducted within the context of integrated soil fertility management (ISFM). Vanlauwe *et al.* (2010) defined ISFM as

> a set of soil fertility management practices that necessarily include the use of fertilizer, organic inputs, and improved germplasm combined with the knowledge on how to adapt these practices to local conditions, aiming at maximizing agronomic efficiency (AE) of the applied nutrients and improving crop productivity. All inputs need to be managed following sound agronomic principles.

Full application of ISFM is knowledge-intensive and based on a number of components, including the use of improved germplasm, appropriate fertilizer and crop management, the combined application of organic and mineral inputs and local adaptation. The objective was to apply these ISFM 'principles' to improve the productivity of the individual components of cropping systems. In this chapter, practical examples are given of how ISFM is applied to achieve this in the different mandate areas across Central Africa. Table 4.1 provides an overview of the ISFM-based technology options tested.

## Improved legume germplasm: evaluation, selection and multiplication

Investments in soil fertility may not prove cost-effective if local, degenerated and disease-susceptible varieties are used. In most of the CIALCA intervention areas, farmers have no access to improved varieties (CIALCA 2010). New varieties should have a superior performance in terms of yield and resistance to biotic and abiotic stresses in comparison with local varieties, but should also meet farmers' expectations in terms of duration, taste and tradability. In addition, CIALCA seeks to introduce varieties with traits favourable for improved soil fertility and human nutrition. First, a large set of varieties of beans, soybean, groundnut, cowpea and pigeon pea was collected and evaluated on-station for yield, disease resistance and biomass production. Best-performing varieties were retained for testing in on-farm trials, with the aim of (1) assessing grain and biomass yield, (2) evaluating the response to FYM or fertilizer and (3) obtaining detailed feedback from farmers. Between 8 and 25 varieties of bush beans, climbing beans, soybean, groundnut, cowpea and pigeon pea were evaluated with at least two farmers' associations per action site during at least two seasons in the mandate areas of Kigali-Kibungo, Umutara, Sud-Kivu, Gitega and Bas-Congo. Agronomists made observations on biomass production, disease resistance and grain yield. At mid-podding and after harvest, participatory evaluations were conducted whereby male and female members of farmers' groups selected their most-preferred varieties based on their own criteria.

In all mandate areas, varieties of beans and soybean with grain yields superior to those of local varieties were identified (CIALCA 2007, 2008). Different varieties performed better than the local material in poor soils rather than in fertile soils, or if FYM was added, and performance differed between mandate areas or action sites. Recommendations should therefore take into account not only the altitude and rainfall regime but also local soil conditions. In their

*Table 4.1* An overview of legume- or banana-based ISFM technologies tested in the CIALCA mandate areas

| *Component* | *Technologies* | *Mandate areas* |
|---|---|---|
| Improved germplasm | • *Grain legumes* (beans, soybean and groundnut): new varieties characterized by high yield, resistant to biotic or abiotic stresses, traits desired by farmers and/or benefits for soil fertility or human health | BC, SK, KK, UM, GI |
| | • *Cassava*: new varieties characterized by high yield and resistance to mosaic disease | BC, SK, KK, GI |
| | • *Banana and plantain*: new exotic hybrids and Ugandan highland hybrids characterized by increased pest/disease resistance and improved yield | NK, SK, KK, GI, KR, RZ, GK |
| Mineral fertilizer use | • NPK application in legume–cassava intercrops, targeting responsive soils | BC, SK, KK, GI |
| | • Small doses of DAP to beans, targeting responsive soils | UM, KK |
| | • NPK application in cassava monocrops | BC |
| | • Fertilizer application in climbing beans, maize or soybean–maize rotations or intercrops | SK, GI |
| Organic matter management | • *In situ* biomass production through integration of grain legumes in cropping systems | BC, SK, KK, UM, GI |
| | • Rotational benefits of varieties of high biomass-yielding soybean or climbing beans on subsequent maize crop | SK, GI |
| | • Improved fertilizer AE through combined application of FYM and DAP in beans | UM, KK |
| | • Short *Mucuna* fallows preceding maize cultivation | BC |
| | • Strategic use of high-quality FYM in combination with fertilizer for climbing beans in poor or degraded soils | SK, GI |
| | • Combined application of green manures and mineral fertilizer to cassava monocrops or legume intercrops | BC |
| | • Mulching combined with zero-tillage in banana systems in combination with inter-cropping common beans | NK, SK, KK, GI, KR, RZ |
| Agronomic measures | • Optimal crop arrangements and densities in legume–maize and legume–cassava intercrops | BC, SK, KK, UM, GI |
| | • Water harvesting using tied ridging | UM |
| | • Erosion control using progressive terraces with *Calliandra* hedgerows | SK |
| | • Optimal planting densities and plantation management in banana systems | KK, GK |

*Note*: BC = Bas-Congo, GI = Gitega, KK = Kigali-Kibungo, SK = Sud-Kivu, UM = Umutara, RZ = Rusizi, KR = Kirundo, GK = Gisenyi-Kibuye.

selection, farmers considered a range of traits related to the maturity period, appearance, tradability and taste (Table 4.2), and typically identified different objectives for different varieties. Due both to the agro-ecological diversity as well as the farmers' preferences, a relatively large set of varieties was retained for multiplication, especially for bush beans. Some of the bio-fortified varieties were also retained. Analyses conducted confirmed higher levels of iron and zinc in the grains, but also strong genotype by environment (G × E) interactions (CIALCA 2007, 2008).

Some varieties of soybean and climbing beans with a general level of performance were selected for further testing and to illustrate that a low harvest index holds important benefits for soil fertility. Positive rotational benefits were demonstrated in a series of trials in Sud-Kivu, where maize yields were higher after improved varieties of beans (29–34 per cent after VCB81012) or soybean (20–24 per cent after Maksoy 1n) than after local varieties. Contributions from N-fixation were higher for improved varieties because of the higher biomass yield, rather than because of more effective fixation. Farmers, however, gave little importance to biomass yield (except when leaves are consumed), and had more interest in early maturing varieties characterized by low biomass yields. In conclusion, the selection of varieties for further multiplication involves some trade-offs and discussion, and in general, a set of varieties was identified for further use in the development of ISFM technologies. Preferred varieties were then promoted and made available through investments in community-led seed multiplication schemes, primarily through informal channels by involving the farmers' associations in the action sites, with facilitation by NGO partners and technical backstopping by the legume programmes of national research partners. At the start of the activity in 2007, the associations decided to organize seed production as a group activity but because of problems with land and labour availability, switched to a decentralized system, whereby members multiply seeds in their own fields and return a proportion of the seeds produced to the association at the end of the season. Although this accelerated the diffusion of the varieties, it required intensive supervision and strict adherence to guidelines to guarantee seed quality and varietal purity (CIALCA 2008). By the end of 2008, most associations had succeeded in producing sufficient seeds to satisfy the needs of their members. Several well-organized associations succeeded in producing large amounts of seeds, kept large seed stocks, sold as a group at an opportune moment and shared profits. Seeds were commonly sold to neighbouring farmers or non-governmental organization (NGO) buyers or at local markets. The new varieties were promoted during field days and exchange visits, involving the local press, policy-makers and NGOs active in the region. Significant improvements in access to improved varieties were achieved in the action sites. In Bas-Congo, more than 60 per cent of households now cultivate soybean, which is a new crop for many farmers. In Sud-Kivu, an increase of almost 50 per cent in households using new varieties of bush and climbing beans was recorded, compared with 20 per cent in control sites (CIALCA 2010). However, difficulties were expressed in finding a market for the seeds at the

*Table 4.2* Traits desired in improved legume varieties based on a total of 309 participatory evaluations of beans, soybean, groundnut and cowpea varieties across all mandate areas

| *Trait* | *Criterion* | *Proportion*† | |
|---|---|---|---|
| | | *Women* | *Men* |
| Yield | Large amount of grains at harvest | 62 | 57 |
| | High number of pods at mid-podding | 33 | 31 |
| | High number of grains per pod | 3 | 4 |
| Biomass production | High vigour or height of the plant stand | 26 | 16 |
| | Greenness and leafiness | 10 | 7 |
| | High branching | 1 | 1 |
| Maturity | Early flowering, podding or harvest | 50 | 51 |
| Adaptation to the agro-ecology | Resistance to drought ('to the sun') | 37 | 32 |
| | Resistance to poor soils (need for manure) | 30 | 26 |
| | Resistance to heavy rainfall | 14 | 21 |
| Grain traits (tradability) | Large grain size | 46 | 55 |
| | Preferred grain colour | 35 | 43 |
| | High luminosity of the grains | 21 | 20 |
| | Good homogeneity (size and colour) | 10 | 12 |
| Taste/preparation | Good taste of the grains | 6 | 4 |
| | Density of the grains | 2 | 5 |
| | Short cooking time required | 2 | 1 |
| Resistance to pest and diseases | Resistance to weevil attacks on grains | 24 | 33 |
| | Absence of leaf spots/disease symptoms | 18 | 19 |
| | Absence of insect attacks during growth | 2 | 1 |

*Note*: † Proportion of farmers considering the trait as a criterion important for selection.

due value, which was often set rather high, and in sustaining seed production as a lucrative business.

## Improved banana germplasm: evaluation, selection and multiplication

Twenty different varieties of banana (including cooking, beer and dessert banana and plantain) were tested and promoted in the CIALCA region with the National Agricultural Research Systems (NARS) of Burundi, Rwanda and the eastern Democratic Republic of Congo (DRC) provinces of Nord and Sud Kivu. Varieties included new exotics and their local checks. Yield was determined by measuring bunch weights, number of hands per bunch and plant height at flowering. Sensory acceptability by farmers was rated for pulp, colour, pulp texture, food aroma, taste and general acceptability. The matrix scoring technique was applied to select the variety preferred by the whole group. Every trait was then subjected to a score out of 20, including the one combining all other sensory decisions, and this was the main determinant of the preferred variety.

The performance of varieties differed among the different ecological zones. There were also sensory differences in acceptability by farmers across the region as different people exhibited different preferences. In areas where cooking banana was traditionally eaten, new varieties were generally rated low. However, in areas where cooking banana was less important and cassava and plantain dominated, some hybrids performed better than the local varieties. There were also trade-offs that farmers observed, as they preferred to grow some varieties for the market and others for food. This was mostly observed in Burundi where East African highland bananas were preferred in sensory tests but the Fundación Hondureña de Investigación Agrícola (FHIA) hybrids were selected using the matrix scoring method, primarily because of their larger bunch size.

Within the region, banana plantations are still largely being established with the use of suckers. This method is slow and has a number of challenges including the transmission of pests and diseases. Some private sector tissue culture (TC) labs are already producing plantlets in Burundi and Uganda, and initiatives for more labs are in the pipeline. Nonetheless, due to the small number and limited capacity of the existing labs, the demand for the plantlets still outstrips supply, and the adoption of TC technology is low. Besides, many farmers still cannot afford their prices (often $0.7–1.0 per plantlet), or are located far from supply centres. Despite these problems, however, TC remains the most effective method for the rapid production of clean planting material.

Community-based means of seed multiplication became high on the agenda. CIALCA introduced macro-propagation in the region, using suckers derived from mother gardens established from TC plantlets. Alongside decapitation techniques (false or full), this allowed for a relatively low-cost and rapid local multiplication. Each CIALCA action site operated as a community-based training and multiplying centre. In addition, Trainings of Trainer events were organized for partners, including NGOs, government agencies, private companies and farmers. Different NGOs started multiplication sites, working with farmers' associations. Private enterprises were also established where individuals sold the plantlets they generated. Challenges faced by these local multiplication centres were:

1. Farmers were accustomed to getting planting material free of charge.
2. The control of pests and diseases could be difficult, since farmers sometimes multiply infected suckers, thereby propagating plant health threats.

Generally, macro-propagation was adopted much faster in places where banana plantations had been affected by diseases such as banana Xanthomonas wilt (BXW) and in places where banana plantations were being newly established than in traditional banana-producing areas with intact plantations. In these places, new establishments are mainly conventionally sucker-based.

## Efficient use of mineral fertilizer

With the exception of Gitega, few households (10 per cent, on average) utilize mineral fertilizer in the CIALCA mandate areas, and those who use it apply only small quantities (9–14 kg/ha, on average). The low use is mostly a consequence of the unavailability, the high unit price and the lack of knowledge on the correct utilization of fertilizer. On-station and on-farm experiments evaluating responses to fertilizer in beans, soybean, maize and cassava provide evidence that profitable yield increases can be obtained. In Sud-Kivu, results from a set of farmers' demonstration trials showed that the use of NPK (17:17:17) in cassava–bean intercropping systems is profitable in the action sites north of Bukavu, with an average benefit–cost ratio (BCR) of \$3.5 per dollar (Pypers *et al.* 2011). In the action sites south of Bukavu, where soils are generally more acidic and less fertile, responses were variable. In farmer-managed trials in these sites, only 56 out of the 180 participating farmers obtained yield increases exceeding 200 kg/ha, the minimal increase needed to obtain a BCR > \$2.0 per dollar. Demonstration trials on soybean–maize rotations show that the use of NPK is profitable, particularly with an improved variety (BCR = \$3.6 per dollar) rather than the local variety (BCR = \$2.4 per dollar). In an on-station trial, BCR values for fertilizer application in a climbing bean–maize rotation varied between \$2.1 and \$5.0 per dollar over a six-season period. In Bas-Congo, application of NPK increased cassava root yields by 42 and 212 per cent in two representative soils, with BCR values of \$2.2–5.7 per dollar (Pypers *et al.* 2012). Profitability was lower in the less fertile site, where also *Imperata* infestation was more severe and demanded more intensive weeding. These findings indicate that fertilizer use can be profitable despite high unit prices, but the fertilizer should be correctly applied to a well-managed crop, and local soil conditions should be taken into account.

This was confirmed in farmer-managed fertilizer response trials in the eastern province of Rwanda (Figure 4.1). A higher fertilizer AE was obtained with less than 20 per cent loss in crop stand during the season, if the bean crop was planted at the optimal plant density within two weeks after the onset of the rains. Maximal AE values were observed when control yields were moderate (700–1,500 kg/ha). Fields with low control yields often have limitations other than a deficiency in N and P; fields with high control yields are typically fertile homestead fields with high availability of N and P. As the subsidized price of diammonium phosphate (DAP) fertilizer is comparable to the price of beans grains on local markets, a minimum AE of 2 kg grain per kilogram of DAP

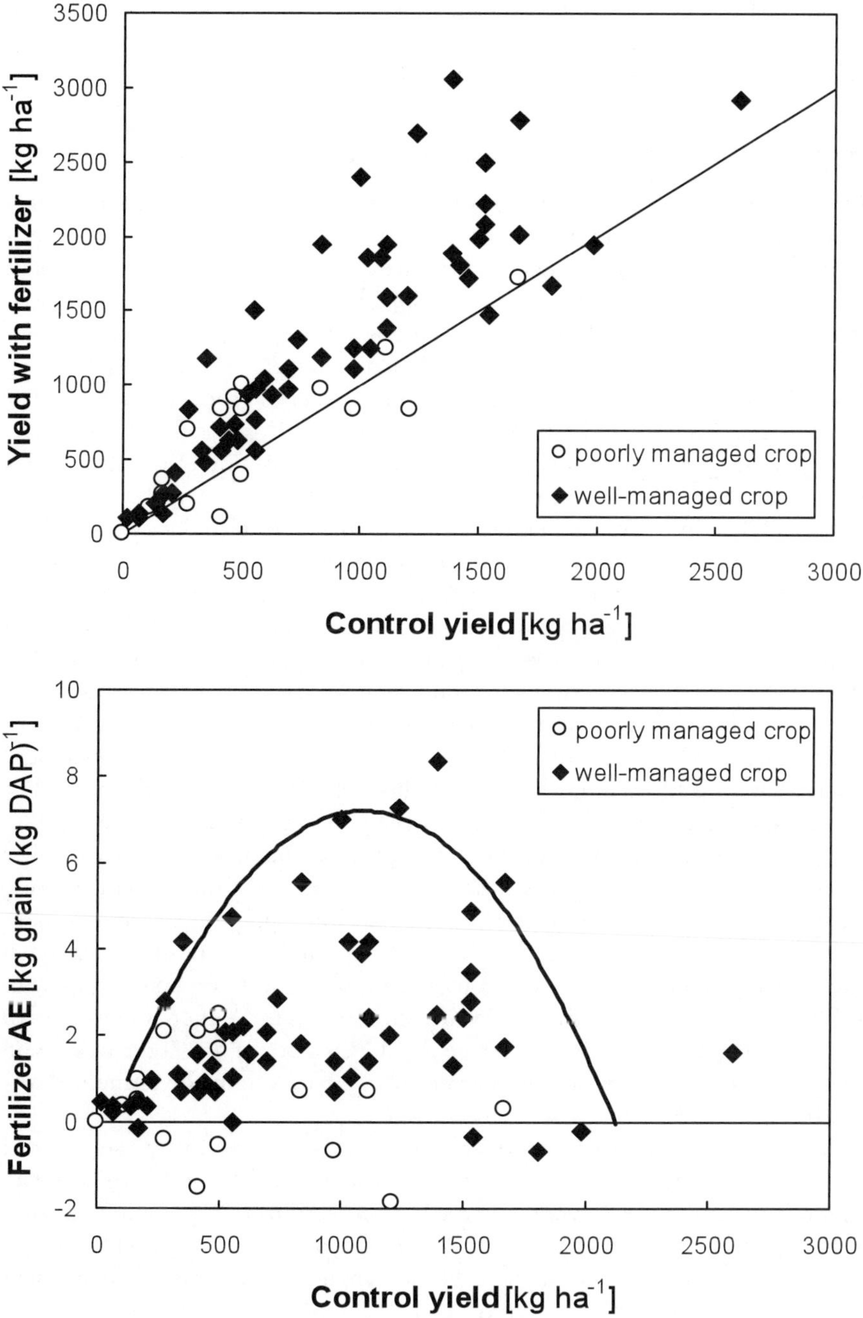


*Figure 4.1* Yield and agronomic efficiency (AE) of DAP fertilizer applied at 200 kg/ha in beans, in function of control yields, obtained from farmer-managed adaptation trials in the eastern province of Rwanda.

is required to obtain a BCR larger than $2 per dollar. Hence, the profitable use of fertilizer requires that it is applied to responsive fields with moderate control yields, and the crop is well managed.

## Improved organic matter management

Organic inputs contain nutrients for crop growth and contribute to the build-up of soil organic matter which in turn improves nutrient availability and retention. When applied at affordable rates, these inputs seldom release sufficient nutrients to obtain optimal crop yields. Smallholders in the tropics are advised to use the combined application of organic and mineral inputs because neither of the resources is usually available in sufficient quantities, and because both are needed in the long term to sustain soil fertility and crop production. A number of trials have been carried out in the CIALCA mandate areas to evaluate technologies using organic inputs, and results illustrate how the efficient management of organic matter contributes to increased crop production.

In field demonstration and farmer-managed trials in eastern Rwanda, fertilizer AE was improved by 157 per cent in maize, and by 24 per cent in beans, when fertilizer was applied in combination with low-quality FYM. An improvement in fertilizer AE is a result of positive interactions between both nutrient resources which may arise from better availability and synchrony in the supply of nutrients to the crop, reduced nutrient losses, improved soil water storage or alleviated other crop constraints (Vanlauwe *et al.* 2001). In Bas-Congo, combining NPK 17:17:17 fertilizer with green manures (*Tithonia* and *Chromolaena*) resulted in an increase in the yield of cassava storage roots but yield increases from mineral and organic inputs were additive (Pypers *et al.* 2012). Improved synchrony of nutrient supply and demand is likely to be less relevant for a root crop such as cassava, as K is most often the growth-limiting nutrient. Like N, K is released rapidly from decomposing organic residues, but the released N is immobilized by soil microbes and K is mostly retained on the exchange complex. Still, the combined application of organic inputs with NPK fertilizer was more profitable than the sole use of either resource, with net benefit increases of 30–50 per cent, BCR values of $5–7 per dollar and marginal rates of return of $4–5 per dollar compared with the farmers' slash-and-burn practice.

In Sud-Kivu, demonstration trials were conducted to evaluate the effect of FYM quality, rate and method of application on yield and fertilizer AE. Highest yields and AE were observed when fertilizer was combined with FYM applied in the planting holes, rather than broadcast (Figure 4.2). Yields were highest when FYM of high quality was applied at a high rate, independent of fertilizer application and method of FYM application. Soils in this region are very acidic and P-deficient. Presumably, when only limited amounts of mineral and organic inputs are available, crops draw most benefits when these resources are concentrated in a small soil volume, where P availability and pH

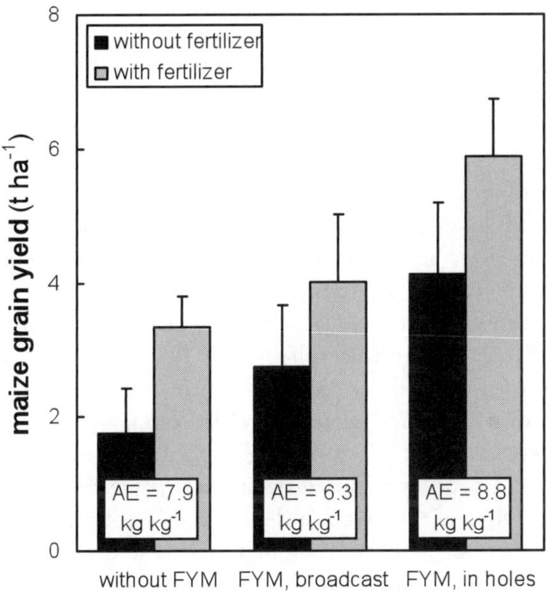


*Figure 4.2* Effect of FYM application method on yield and fertilizer AE, as observed in farmer-demonstration trials in Sud-Kivu, DR Congo.

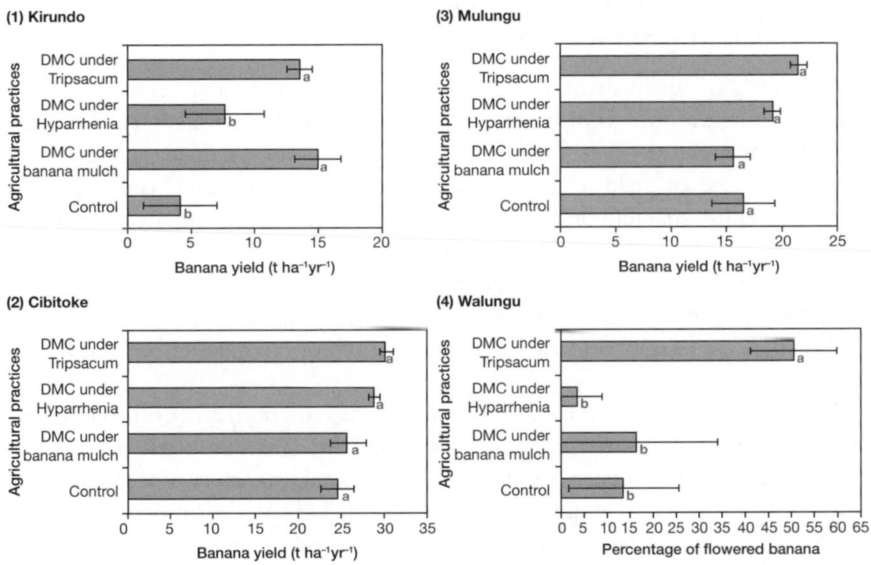


*Figure 4.3* Banana yields (t/ha/yr) under various types of mulch, with control trials representing the farmers' practice of mulch removal and tillage. Observations in Walungu (DRC) showed that *Tripsacum* increased not only yield quantity but also timing (percentage of flowered bananas approximately two years after trial installation).

are then favourable. This example illustrates how both fertilizer and organic inputs are needed in poor soils to optimize crop yields, and how management of organic inputs can contribute to the improved use efficiency of the nutrients applied.

Mulching combined with zero-tillage improves nutrient recycling and reduces losses of precious top-soil in banana systems. This practice prevents damage to the superficial rooting system while decreasing direct evaporation losses from the soil surface and suppressing weeds. Mulch with zero-tillage trials were installed in three locations in Rwanda (Butare, Kibungo, Ruhengeri), two in South Kivu (Walungu, Mulungu) and three in Burundi (Kirundo, Gitega, Cibitoke). Treatments included (1) 'farmers' practice' control plot (tillage and mulch removal), (2) zero-tillage self-mulch, (3) zero-tillage and 25 t/ha/yr *Hyparrhenia*, and (4) zero-tillage and 25 t/ha/yr *Tripsacum laxum*. Zero-tillage and the application of mulch increased yields often by 5–10 t/ha/yr (Figure 4.3). This effect was most pronounced in poor soils. The highest recorded root densities (3–4 roots per square decimetre) were observed on good soils (e.g. Mulungu), whereas much lower densities (1–3 roots per square decimetre) were observed in poor soils (Walungu, Rubona and Gitega). Tillage affects the structural stability of soil and increases susceptibility to erosion. Both tillage and mulching have a positive impact on water infiltration.

## Accompanying agronomic measures for increased productivity

Productivity can be further enhanced through agronomic measures such as water harvesting techniques in drought-prone areas or modifications in crop density and crop arrangements. In eastern Rwanda, intra-seasonal drought spells are a common cause of yield loss, and trials were conducted to evaluate the effect of tied ridging on yield and response to fertilizer. Improvements in yield up to 30 per cent were obtained on-station, but on-farm yield improvements were minimal, and farmers were reluctant to invest the additional labour. Other measures related to crop arrangements in intercropping systems (Pypers *et al.* 2011) or plant density and the management of banana plantations were very successful.

Farmers practise banana planting density management in an effort to enhance total farm production. Farmers often decrease the densities of the banana mat (i.e. a single mother plant with interconnected suckers) in an effort to increase bunch size (i.e. their indicator for yield). However, the effectiveness and profitability of this planting density practice and its relationship with climatic and edaphic factors have never been studied in detail. On-station trials were performed in three contrasting agro-ecological zones in Rwanda as well as on-farm surveys in seven contrasting agro-ecological sites of Rwanda (Ruhengeri, Rusizi, Bugesera, Ruhango, Butare, Karongi and Kibungo). Plant density was positively correlated with rainfall ($r^2 = 0.50$), whereby high plant densities

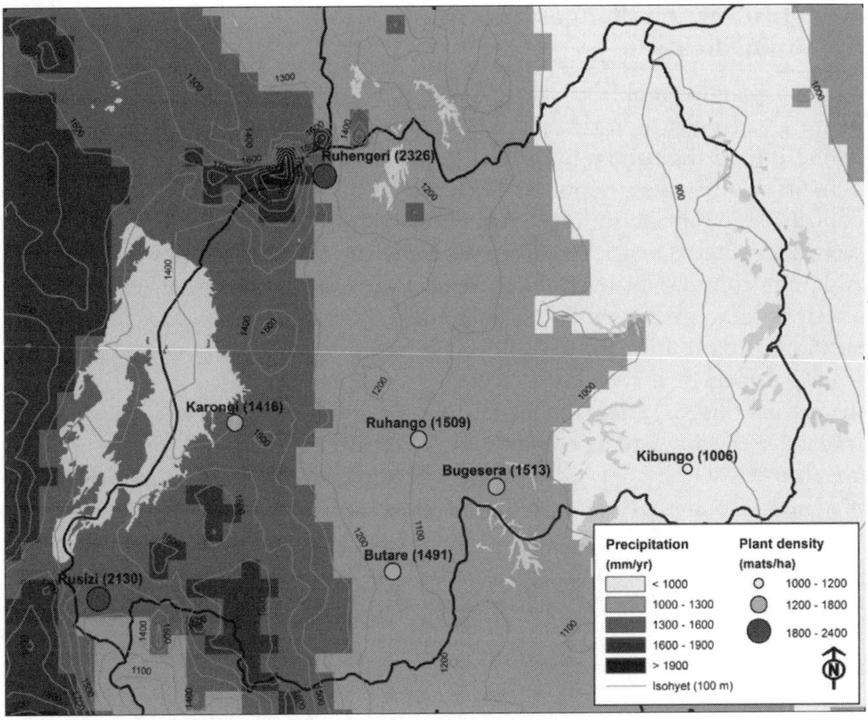


*Figure 4.4* Rwanda, showing that actual on-farm banana plant densities are low, particularly in low-rainfall sites. Actual on-farm plant densities were less than half of the densities that provided the best yields in research-managed plant density trials across Rwanda (Telesphore, pers. comm.), showing that farmers can still make huge yield gains (>30 per cent) by increasing densities in banana monocrop fields.

(>1,500 mats/ha) were found in high rainfall areas (>1,300 mm/yr); the lowest plant densities (1,000–1,500 mats/ha) were found in lower rainfall areas (900–1,200 mm/yr). Yield (t/ha/cycle) increased with an increase in plant density ($r^2 = 0.62$). Lower soil and banana leaf nutrient contents (especially of N, K, Ca and Mg) were observed on weathered soils (acrisols) and were accompanied by low yields compared with those from fertile soils (andosols, nitisols) (e.g. Ruhengeri). Farmers tended to reduce mat densities if they wanted to intercrop and to increase bunch mass to adapt to market preferences. Farmers seldom practise the plant densities generally recommended by extension bodies (3 m × 3 m, i.e. 1,111 mats/ha or 2 m × 3 m; i.e., 1,666 mats/ha) nor do these seem to be very appropriate, as higher densities seem more productive, particularly in areas with high rainfall and relatively good soil fertility.

## Conclusions, challenges and implications for intensification

This chapter exemplifies how the ISFM framework with improved germplasm as an entry point is an effective means to increase crop productivity in the humid highlands. Improved germplasm is essential to obtain maximal benefits from investments in fertilizer and other technologies, and can contribute to long-term improvement in soil fertility. A number of challenges were encountered, in the first place related to the diffusion of improved varieties. Although community-led multiplication and macro-propagation are potentially effective, there are drawbacks related to the quality and tradability of seeds within farming communities. Second, although fertilizer use is effective and profitable if well-managed, it remains a challenge to ensure farmers can gain access to the commodity, as well as the information on appropriate fertilizer management. Initiatives to stimulate development of a fertilizer market and involvement by government will be needed to achieve intensification. Finally, improvements in system productivity often require knowledge-intensive interventions based on technologies with different components. These technologies were first demonstrated in field trials with farmers' groups, and then adapted in farmer-managed trials with individual households. Additional efforts are needed to stimulate farmers to continue experimenting, modifying and fine-tuning technologies, and ultimately to achieve uptake and widespread diffusion.

## References

CIALCA (2007) *CIALCA Annual Report 2007*. Available at: www.cialca.org

CIALCA (2008) *CIALCA Annual Report 2008*. Available at: www.cialca.org

CIALCA (2010) *CIALCA Baseline Survey Report*, CIALCA, led by IITA, Kampala, TSBF-CIAT, Nairobi, Bioversity International, Kampala.

FAO (1995) *Sustainable Dryland Cropping in Relation to Soil Productivity*, FAO, Rome.

Farrow, A., Busingye, L. and Bagenze, P. (2007) *Characterisation of Mandate Areas for the Consortium for Improved Agricultural Livelihoods in Central Africa* (CIALCA). Available at: www.cialca.org

Pypers, P., Sanginga, J. M., Bishikwabo, K. Walangululu, M. and Vanlauwe, B. (2011) 'Increased productivity through integrated soil fertility management in cassava–legume intercropping systems in the highlands of Sud-Kivu, DR Congo', *Field Crops Research*, vol. 120, pp. 76–85.

Pypers, P., Bimponda, W., Lodi-Lama, J.-P., Lele, B., Mulumba, R., Kachaka, C., Boeckx, P., Merckx, R. and Vanlauwe, B. (2012) 'Combining mineral fertilizer and green manure for increased, profitable cassava production', *Agronomy Journal*, vol. 104, pp. 178–187.

Tittonell, P., Vanlauwe, B., de Ridder, N. and Giller, K. E. (2007) 'Heterogeneity of crop productivity and resource use efficiency within smallholder Kenyan farms: soil fertility gradients or management intensity gradients?', *Agricultural Systems*, vol. 94, pp. 376–390.

Vanlauwe, B., Wendt, J. and Diels, J. (2001) 'Combined application of organic matter and fertilizer', in G. Tian *et al.* (eds), *Sustaining Soil Fertility in West Africa*, Soil Science Society of America, Madison, WI, pp. 247–280.

Vanlauwe, B., Bationo, A., Chianu, J., Giller, K. E., Merckx, R., Mokwunye, U., Ohiokpehai, O., Pypers, P., Tabo, R., Shepherd, K., Smaling, E., Woomer, P. L. and Sanginga, N. (2010) 'Integrated soil fertility management: operational definition and consequences for implementation and dissemination', *Outlook on Agriculture*, vol. 39, pp. 17–24.

# 5 Exploring the scope of fertilizer use in the East African region

*L. W. I. Wairegi*[1] *and P. J. A. van Asten*[2]

[1] *Centre for Agricultural Bioscience International*
[2] *International Institute of Tropical Agriculture*

## Introduction

Most small-scale farmers in sub-Saharan Africa (SSA) continue to have poor production (Voortman *et al.* 2003), and food production is below demand (Dorward *et al.* 2004). In the past, growth in food production was achieved through an increase in the area of cropped land, but expansion is no longer feasible in many countries as land has become a constraint. Hence, efforts to increase agricultural production should be achieved mainly through intensification.

Poor soil fertility is a constraint to productivity in the region, as most of the soils are highly weathered (Jaetzold and Schmidt 1982) and highly depleted (Stoorvogel *et al.* 1993). In addition, there is heterogeneity in resources among and within farms, and complexity in cropping systems. The use of inputs in smallholder farming systems is influenced by competition among enterprises for limited resources (Bekunda and Woomer 1996) and the quantities of mineral fertilizers used are not always adequate.

This study attempts to estimate the amount of nutrients removed in crop products, and compares the value of the edible biomass with the cost of fertilizer equivalent to the nutrients taken up by plants in the edible and above-ground biomass.

## Methodology

The study is on the production of maize, beans, banana, cassava, rice and coffee in Burundi, DR Congo (DRC), Kenya, Rwanda, Tanzania and Kenya. These crops occupy approximately 54–80 per cent of the arable land in countries in East and Central Africa (FAO 2011).

Data on nutrient mass fractions of nitrogen (N), phosphorus (P) and potassium (K) in plant parts, and the prices of crops and fertilizers in major markets in 2010/2011 were obtained from the literature, websites and personal communication. Where data were based on fresh edible biomass, they were standardized to dry matter based on the assumption that the moisture content for fresh edible biomass was 14 per cent for maize, 12 per cent for beans, rice and coffee, 86 per cent for banana and 70 per cent for cassava.

Calculations were based on the nutrients required for the production of one tonne of edible dry matter yield. For all crops except banana and coffee the nutrients in the above-ground biomass were calculated based on the total above-ground biomass. For coffee, nutrients in the above-ground biomass included nutrients in the crop and prunings, and nutrients used for the growth, development and maintenance of trees (IFA 1992a). For banana, the nutrients in the above-ground biomass were based on the nutrients taken up by a crop yielding 35 t of fruit (IFA 1992b). The dry matter content of the fingers was estimated to be 18 per cent. Since smallholder farmers normally remove bunches and peduncle from the fields at harvest, the edible yield was assumed to consist of the hands and peduncle. For rice, the edible yield was assumed to be made up of grain and bran.

Based on the assumption that nutrient losses were 50 per cent (N), 85 per cent (P) and 40 per cent (K), the estimated amounts of N, P and K in plant parts were equated to amounts of fertilizers: urea for N, triple super phosphate (TSP) for P and muriate of potash (MOP) for K. Value–cost ratios (VCRs) of the value of edible biomass to the cost of fertilizer equivalent to the nutrients in plant parts were calculated based on prices in 2010/2011. Sensitivity analysis was performed for ratios based on the whole-plant bio-vmass to evaluate the effect on 'profitability' of a 50–100 change in crop and fertilizer prices. Whole-plant biomass included roots for cassava but not for other crops.

## Results

Nutrients in whole plant biomass range from 27.1 kg (rice) to 87.4 kg (coffee) for N; from 3.5 kg (cassava) to 8.7 kg (beans) for P; and from 18.8 kg (cassava) to 118.9 kg (banana) for K (Table 5.1). In the edible biomass, nutrients range from 9.0 kg (cassava) to 36.3 kg (beans) for N; from 0.8 kg (cassava) to 4.2 kg (maize) for P; and from 2.1 kg (rice) to 35.6 kg (banana) for K. Amounts of fertilizer equivalent to the nutrients in plant parts and whole plants range from 87 kg (cassava roots) to 848 kg (whole coffee plants) (Table 5.1).

Prices ($/t) of urea, TSP and MOP were estimated at $543, $747 and $746, respectively, in Burundi; $543, $806 and $705 in Kenya; $540, $675 and $567 in Rwanda; $563, $773 and $773 in Tanzania; and $712, $853 and $853 in Uganda. In DRC the price was $1,250 per ton for all fertilizers.

Commodity prices ($) for 1 t of edible dry matter were estimated to be $398 (maize), $973 (beans), $1,176 (banana), $492 (cassava), $978 (rice) and $2,706 (Arabica coffee) in Burundi; $498 (maize), $831 (beans), $1,176 (banana), $571 (cassava), $1,030 (rice), $2,094 (Arabica coffee) and $1,335 (Robusta coffee) in DRC; $300 (maize), $765 (beans), $1,100 (banana), $443 (cassava), $957 (rice) and $2,873 (Arabica coffee) in Kenya; $372 (maize), $682 (beans), $1,714 (banana), $542 (cassava), $1,372 (rice) and $2,580 (Arabica coffee) in Rwanda;

*Table 5.1* Amount of N, P and K removed in plant parts for the production of 1 t of edible dry matter yield, amount of fertilizer equivalent to nutrients in plant parts and ratio of value of 1 t edible dry matter yield to the value of fertilizer equivalent to nutrients in plant parts in six East African countries

| *Crop* | | *Nutrients (kg)* | | | *Fertilizer equivalent (kg)* | *Value of 1 t edible dry matter yield (cost of fertilizer equivalent)* | | | | | |
|---|---|---|---|---|---|---|---|---|---|---|---|
| | | *N* | *P* | *K* | | *Burundi* | *DR Congo* | *Kenya* | *Rwanda* | *Tanzania* | *Uganda* |
| Maize | Whole plant | 32.6 | 5.7 | 47.7 | 483 | 0.9 | 0.8 | 0.9 | 1.3 | 0.7 | 0.6 |
| | Grain and cobs | 18.4 | 4.2 | 9.1 | 247 | 1.7 | 1.6 | 1.7 | 2.4 | 1.4 | 1.1 |
| | Grain | 17.0 | 4.1 | 6.9 | 231 | 1.8 | 1.7 | 1.8 | 2.6 | 1.5 | 1.2 |
| Beans | Whole plant | 69.6 | 8.7 | 45.7 | 738 | 1.4 (1.8) | 0.9 (1.1) | 1.5 (1.8) | 1.5 (1.9) | 1.5 (1.8) | 1.2 (1.4) |
| | Grain | 36.3 | 4.2 | 22.3 | 368 | 2.9 (3.6) | 1.8 (2.3) | 3.1 (3.7) | 3.1 (3.9) | 3.0 (3.6) | 2.4 (2.9) |
| Banana | Whole plant* | 51.0 | 6.7 | 118.9 | 827 | 2.3 | 1.7 | 1.9 | 3.5 | 1.9 | 1.5 |
| | Whole bunches | 12.1 | 1.4 | 35.6 | 214 | 8.8 | 6.7 | 7.5 | 13.8 | 7.1 | 5.7 |
| Cassava | Whole plant | 34.8 | 3.5 | 18.8 | 327 | 1.6 | 1.4 | 2.0 | 2.8 | 2.0 | 1.2 |
| | Stems, roots (minus leaves) | 25.8 | 2.7 | 12.1 | 190 | 2.8 | 2.4 | 3.4 | 4.8 | 3.4 | 2.0 |
| | Roots | 9.0 | 0.8 | 6.7 | 87 | 6.2 | 5.2 | 7.7 | 10.6 | 7.5 | 4.4 |
| Rice | Whole plant | 27.1 | 4.6 | 36.0 | 388 | 2.7 | 2.1 | 3.6 | 5.9 | 2.9 | 2.6 |
| | Brown grain and husks | 15.7 | 3.3 | 3.6 | 189 | 5.6 | 4.4 | 7.2 | 11.7 | 6.0 | 5.3 |
| | Brown grain | 14.3 | 3.1 | 2.1 | 172 | 6.1 | 4.8 | 7.8 | 12.8 | 6.6 | 5.8 |
| Coffee | Whole plant* | 87.4 | 6.2 | 81.7 | 848 | 3.5 | 2.0 (1.3) | 5.2 | 5.2 | 4.3 | 3.1 (2.0) |
| | Unpulped cherries | 38.6 | 2.9 | 46.0 | 412 | 7.1 | 4.1 (2.6) | 10.5 | 10.8 | 8.8 | 6.4 (4.1) |
| | Green bean | 24.4 | 1.9 | 19.8 | 233 | 12.7 | 7.2 (4.6) | 18.7 | 19.0 | 15.7 | 11.4 (7.3) |

*Notes:* * Nutrients in whole plant for banana and coffee refer to uptake. Fertilizer equivalents are based on the amounts of urea (46%N), TSP (20%P) and MOP (51%K) required to replace N, P and K removed in plant parts, at recovery rates of 50%N, 15%P and 60%K. For beans, figures without parentheses assume that all N is supplied by urea, and figures in parentheses are based on the assumption that 50% of N required by plants is fixed through biological N fixation. For coffee, figures without parentheses are for Arabica coffee and figures in parentheses are for Robusta coffee.

$250 (maize), $756 (beans), $1,100 (banana), $443 (cassava), $790 (rice) and $2,486 (Arabica coffee) in Tanzania; and $216 (maize), $691 (beans), $1,000 (banana), $300 (cassava), $808 (rice), $2,094 (Arabica coffee) and $1,335 (Robusta coffee) in Uganda.

The VCR calculated for fertilizers equivalent to the nutrients removed in whole plants ranged from 0.6 (maize in Uganda) to 5.2 (Arabica coffee in Kenya) (Table 5.1). The ratio calculated for nutrients removed in edible biomass was least for maize (≤2.6) and most for Arabica coffee (≥7.2) in all countries. For beans, this ratio ranged from 1.8 to 3.1 when all N was assumed to be supplied by urea and from 2.3 to 3.9 when 50 per cent of N was assumed to be acquired through biological N-fixation.

An increase in crop prices by 50 per cent was estimated to give ratios for fertilizers equivalent to the nutrients removed in whole plants and was >2 for all crops except maize in all countries, Robusta coffee in DRC, and beans and cassava in Uganda (Figure 5.1). A 50 per cent decrease in crop prices gave ratios for fertilizers equivalent to nutrients in whole plants that were <1 for maize and beans in all countries, bananas in DRC, Kenya, Tanzania and Uganda, cassava in DRC and Uganda, Arabica and Robusta coffee in DRC.

Based on a 50 per cent decrease in fertilizer prices, ratios for fertilizers equivalent to the nutrients removed in whole plants for maize were <2 in all countries except Rwanda (Figure 5.2). For other crops, the ratios were >2 in all countries. A 50 per cent increase in fertilizer prices gave ratios for fertilizers equivalent to the nutrients removed in whole plants that were <1 for maize in all countries, beans in DRC and Uganda, cassava in DRC and Uganda, and Robusta coffee in DRC.

## Discussion

The edible yield contains approximately 24–53 per cent N, 21–68 per cent P and 6–49 per cent K of the nutrients taken up by plants (Table 5.1). Based on the estimated ratios of the value of the yield and the cost of the fertilizer required (Table 5.1), the sale of maize in all countries except Rwanda, unlike the sale of Arabica coffee and rice in all countries, would most likely not generate enough income for the purchase of fertilizer to replenish the nutrients removed by the above-ground biomass. However, if maize stover was left in the field, the sale of maize grain would be most likely to meet the costs of fertilizer required to replenish the nutrients removed in grain and cobs in all countries. Investment in fertilizer to replenish nutrients does not seem attractive to farmers even with an increase in maize prices (Figure 5.1) or a decrease in fertilizer prices (Figure 5.2)

This study suggests that farmers would be better off intensifying production in crops other than maize, but maize is a major food crop grown in SSA. Hence, there is a need to explore ways of improving production. This could be through practices that improve fertilizer recovery and use efficiency, combined with the adoption of improved germplasm.

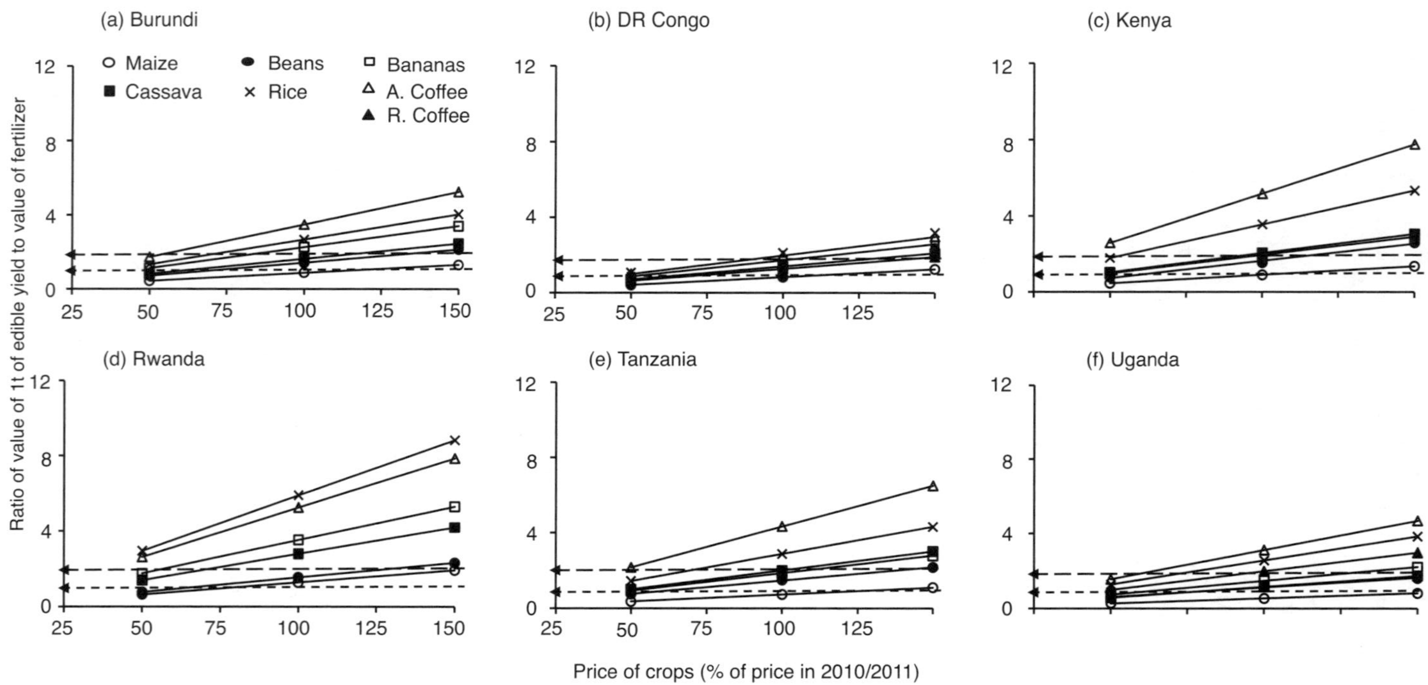


*Figure 5.1* Relationship between a change in the price of crops and the ratio of value of 1 t of edible yield to the cost of fertilizer equivalent to the nutrients removed in whole plants.

*Note*: A. coffee denotes Arabica coffee; R. coffee denotes Robusta coffee. Fertilizer prices at 100 per cent represent average prices for 2010/2011. The dashed lines with arrows represent ratios equal to 1 (short dashes) and 2 (long dashes).

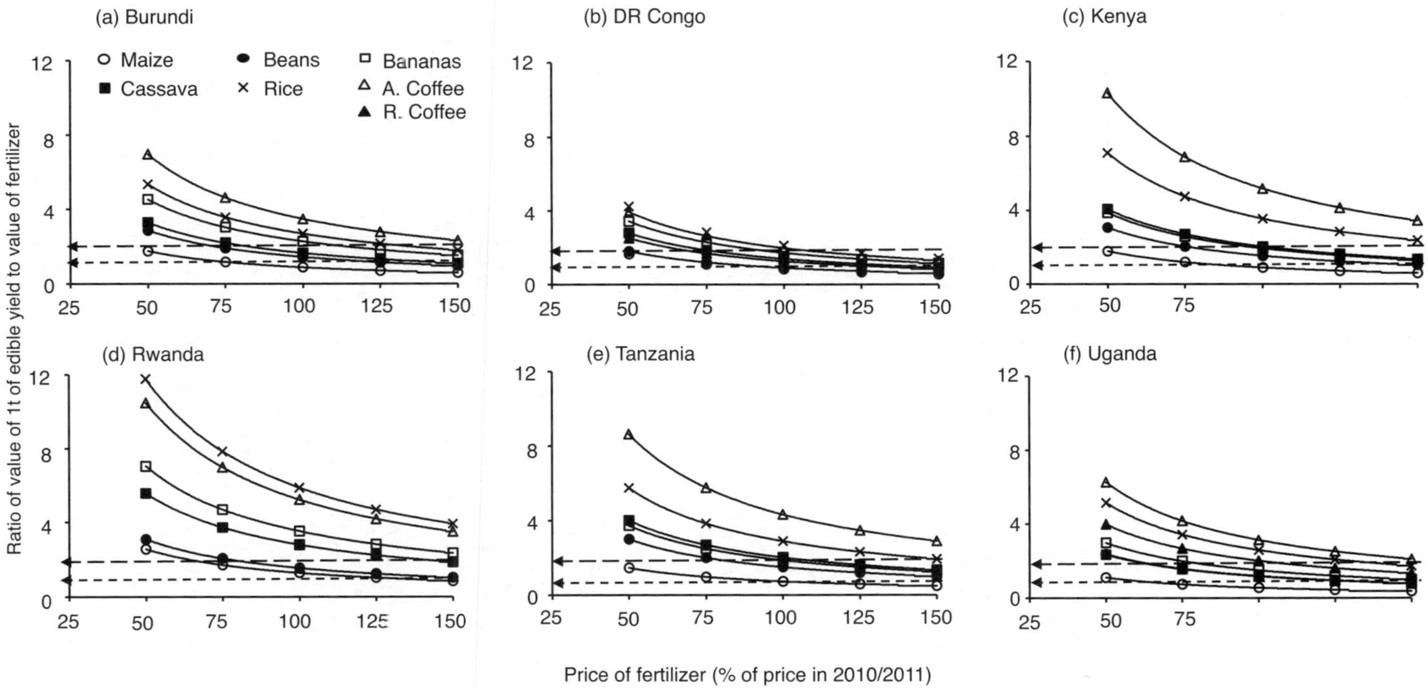


*Figure 5.2* Relationship between a change in the price of fertilizers and the ratio of value of 1 t of edible yield to the cost of fertilizer equivalent to the nutrients removed in whole plants.

*Note*: A. coffee denotes Arabica coffee; R. coffee denotes Robusta coffee. Fertilizer prices at 100 per cent represent average prices for 2010/2011. The dashed lines with arrows represent ratios equal to 1 (short dashes) and 2 (long dashes).

Although this study explores the amounts of nutrients removed by crops and relates this to the amounts of fertilizers that would be needed to replenish these nutrients, it is an over-simplification of fertilizer requirements. The study does not take into consideration nutrient stocks in the soil, crop responses to fertilizer use and crop mixtures. The study assumes that farmers would be willing to replenish nutrients but that they would be unwise to invest in fertilizers if there are no benefits.

## Conclusion

The diversity and complexity of smallholder farming systems in SSA complicates any development of recommendations for use by farmers. Even with the shift from 'blanket' recommendations to site-specific recommendations, farmers growing more than one crop must decide which crop or crops to intensify. It is clear from the study that there is a need and scope for fertilizer use in the East African region, but the choice of crop for intensification and decisions on the amount and type of fertilizer should depend on input/output prices, crop residue management and crop response. It is also clear that there is a need to equip farmers with simple decision-support tools that can aid them in making decisions on which crop to intensify, the types of inputs to use and appropriate management practices. The decision-support tools to guide farmers on how best to invest their scarce resources within the farming system will lead to the improved use of soil inputs, better soil health and improved agricultural production.

## References

Bekunda, M. A. and Woomer, P. L. (1996) 'Organic resource management in banana-based cropping systems of the Lake Victoria Basin, Uganda', *Agriculture, Ecosystems and Environment*, vol. 59, pp. 171–180.

Dorward, A., Kydd, J., Morrison, J. and Urey, I. (2004) 'A policy agenda for pro-poor agricultural growth', *World Development*, vol. 32, pp.73–89.

FAO (2011) 'FAOSTAT'. Available from: http://faostat.fao.org (accessed 28 February 2011).

IFA (1992a) 'Coffee [*Coffea arabica* L. (Arabica coffee); *Coffea canephora* Pierre ex Froehner (Robusta coffee); *Coffea liberica* Bull ex Hiern. (Liberica coffee); *Coffea excelsa* Chev. (Excelsa coffee)]', in W. Wichmann (ed.), *IFA World Fertilizer Use Manual*, International Fertilizer Association, Paris.

IFA (1992b) 'Banana', in W. Wichmann (ed.), *IFA World Fertilizer Use Manual*, International Fertilizer Association, Paris.

Jaetzold, R. and Schmidt, H. (1982) *Farm Management Handbook of Kenya*, Vol. II, Government Printer, Ministry of Agriculture, Nairobi.

Stoorvogel, J. J., Smaling, E. M. A. and Janssen, B. H. (1993) 'Calculating soil nutrient balances in Africa at different scales. I. Supra-national scale', *Fertilizer Research*, vol. 35, pp. 227–235.

Voortman, R. L., Sonneveld, B. G. J. S. and Keyzer, M. A. (2003) 'African land ecology: opportunities and constraints for agricultural development', *Ambio*, vol. 32, pp. 367–373.

# 6 The 4R Nutrient Stewardship in the context of smallholder agriculture in Africa

*S. Zingore*[1] *and A. Johnston*[2]

[1] *International Plant Nutrition Institute, Africa Program*
[2] *International Plant Nutrition Institute, Canada*

## Introduction

There is a growing awareness of the need to improve nutrient use efficiency to achieve sustainable increases in crop productivity in sub-Saharan Africa (SSA). Given the low levels of fertilizer use and poor soils, fertilizer use must increase if the region is to reverse the current trends of low crop productivity and land degradation. There are renewed efforts to re-launch the Green Revolution by raising fertilizer use through improved marketing, a supportive policy and changes in the socio-economic environment so as to increase availability at prices affordable by smallholder farmers. However, these efforts will have limited impact, unless the fundamental issues of providing the crops with adequate and balanced nutrients under variable soil fertility and climatic conditions are effectively addressed. Nutrient management research over the past decades in SSA has shown that an integrated soil fertility management (ISFM) approach offers the best prospects to increase crop productivity from the limited amounts of fertilizers and organic resources available. ISFM embodies

> soil fertility management practices that include the use of fertilizer, organic inputs and improved germplasm combined with the knowledge on how to adapt these practices to local conditions, aiming at maximizing agronomic use efficiency of the applied nutrients and improving crop productivity.
>
> (Vanlauwe *et al.* 2010)

The ISFM concept recognizes the need for fertilizer as an entry point for the ecological intensification of smallholder farming systems in SSA, and places the major emphasis on the use of nutrient inputs following sound agronomic principles. The 4R Nutrient Stewardship developed by the fertilizer industry focuses on applying the right fertilizer source at the right rate, at the right time in the growing season, and in the right place (IPNI 2012). It provides an essential basis for defining strategies for optimizing the use of nutrients within the ISFM framework. It also provides a framework to evaluate the implications of intensification for sustainability, considering the multi-functionality of crop production systems.

## The 4R Nutrient Stewardship

Applying the right source of plant nutrients at the right rate, at the right time, and in the right place is the core concept of 4R Nutrient Stewardship (IPNI 2012). These four 'rights' are all necessary for the proper management of plant nutrition to sustainably increase the productivity of crops. Sustainability consists of economic, social and environmental dimensions. All three need to be included in the assessment of any nutrient management practice to determine whether or not it is 'right'. The fertilizer rights – source, rate, time and place – are connected to the goals of sustainable development (Figure 6.1). Given the problems of malnutrition, high poverty levels and the prevalence of soil degradation in SSA, the performance indicators related to productivity, profitability and environmental sustainability should take precedence in the region.

Specific scientific principles guide the development of practices determining right source, rate, time and place. Some examples of the key principles and

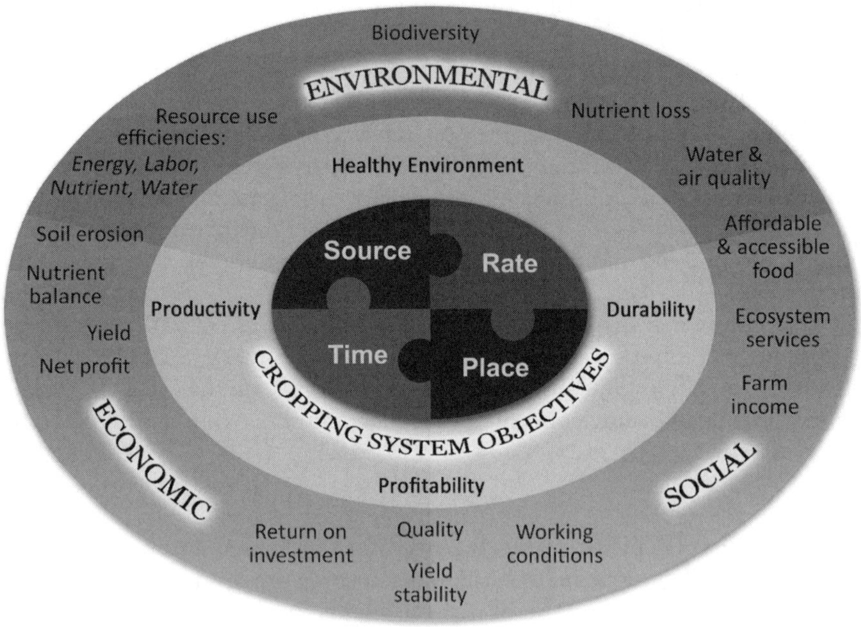


*Figure 6.1* The 4R Nutrient Stewardship and outcome indicators reflecting the social, economic and environmental aspects of the performance of the plant–soil–climate system; priority depends on context-specific conditions and stakeholder objectives.

*Table 6.1* Examples of key scientific principles and associated practices that form the basis of the 4R Nutrient Stewardship

| | *'Right'* | | | |
|---|---|---|---|---|
| | *Source* | *Rate* | *Time* | *Place* |
| *Examples of key scientific principles* | • Ensure balanced supply of nutrients<br>• Suit soil properties | • Assess nutrient supply from all sources<br>• Assess plant demand | • Assess dynamics of crop uptake and soil supply<br>• Determine timing of loss risk | • Recognize crop rooting patterns<br>• Manage spatial variability |
| *Examples of practical choices* | • Mineral fertilizer<br>• Livestock manure<br>• Compost<br>• Crop residue | • Test soils for nutrients<br>• Calculate economics<br>• Balance crop removal | • Pre-plant<br>• At planting<br>• At flowering<br>• At fruiting | • Broadcast<br>• Band |

practices are shown in Table 6.1. The principles are the same worldwide, but how they are put into practice locally varies, depending on specific soil, crop, climate, weather, economic and socio-economic conditions. Practices are selected and applied locally in accordance with these principles.

## The 4R Nutrient Stewardship in the context of smallholder farming systems in SSA

### *The context*

Much of the knowledge underlying the principles of the 4R Nutrient Stewardship has come from research in agricultural systems in the developed world or from developing countries in regions where the rates of fertilizer use are high. Applying the 4R Nutrient Stewardship in SSA will require an in-depth understanding of the relatively more variable and complex cropping systems. Unfavourable socio-economic conditions imply that fertilizers are generally unaffordable and poorly accessible to most farmers, leading to a very low fertilizer use rate that is more than five times lower than the global average (World Bank 2008). SSA contains some of the oldest and most inherently infertile soils (Brouwer and Powell 1998), with the exception of small areas of rich volcanic ash in the highland areas. Many of these soils have been used for agricultural production for many decades with little or no addition of nutrient resources, thus leading to soil degradation. The problem of soil fertility depletion is particularly acute in the East and Central African highlands, characterized by very high population densities and intense land use. In addition, the predominance of perennial crops, such as banana, cassava and coffee, contributes to a very high drain of soil nutrients.

Despite the general depletion in soil fertility in smallholder farming systems in SSA, nutrient use and changes in soil fertility are not uniform between regions, farms and fields within farms. Smallholder farmers typically have limited amounts of nutrient resources and prefer to use them on the particular fields closest to homesteads rather than on outlying fields. As a consequence, steep gradients of decreasing soil fertility have developed within farms as the distance from homesteads increases (Giller *et al.* 2006). The preferred application of mulch to plots closest to homesteads has been found to be a major factor driving variability in soil fertility in banana-based cropping systems (Bekunda and Woomer 1996). Nutrient management is further complicated by the mixed cropping that is commonly practised in the highland areas, and crop–livestock interaction within very small farms. Although large areas in the East African highlands receive high rainfall, drought stress has been identified as a key production constraint in some parts of the region and is a factor that must be addressed in adapting the 4R Nutrient Stewardship to the East African highlands.

*The right source.* Determining the right source under the complex cropping systems in SSA requires identification of nutrient resources that supply plant-available forms and suit site-specific soil properties, and recognizing synergisms among nutrient resources. Despite the demonstrated benefits of fertilizer, particularly in banana- and cassava-based cropping systems, fertilizer use has remained low due to poor availability and high costs. Alternative sources of nutrients are limited, as organic resources, including animal manures, are scarce and of poor quality. Their use alone has been found to have little impact on crop productivity, unless the very large amounts that are not practicable for most farmers are used. In the short term, strategies that encourage the judicious use of limited organic and mineral nutrient sources are required to raise productivity. Choice of the right source of nutrients for the highly variable soil fertility conditions is a major factor determining fertilizer use efficiency. Substantial improvements in crop productivity and agronomic fertilizer use efficiency in the short term can be expected in soils that are nutrient deficient, but not degraded. However, the application of fertilizer has little impact on crop productivity in those degraded soils termed 'non-responsive degraded' (Tittonell *et al.* 2008). Variability in soil fertility is a major implication for fertilizer use efficiency in various cropping systems in the East African highlands and elsewhere (Zingore *et al.* 2007; Pypers *et al.* 2011). Improving productivity in degraded soils is a major challenge, as large investments may be required over multiple seasons before meaningful yield responses can be achieved. Strategies that can be explored for rehabilitating degraded soils in various conditions include: (1) managing organic resources to increase soil organic matter; (2) combining balanced nutrient application with adapted crop varieties to increase biomass production and retain crop residues; (3) using soil moisture conservation techniques to increase plant/water availability. Despite the wide variability in soil fertility in SSA, fertilizer recommendations are mostly blanket rates for a crop. For example, in Uganda, the official mineral fertilizer recommendation for highland bananas is a blanket rate of 100 kg N, 30 kg P, 100 kg K and 25 kg Mg per hectare per year (Ssali *et al.* 2003), although studies have shown bananas in some sites have a relatively higher demand for K than for N (Nyombi *et al.* 2010). Inherently infertile soils cultivated continuously with little addition of organic matter exhibit problems of soil acidity and multiple nutrient deficiencies, which reduce the crop response to the fertilizers supplying limited nutrients. Strategic selection of the right source of fertilizer according to the soil's fertility status is therefore crucial to increasing crop productivity and fertilizer use efficiency in smallholder cropping systems in SSA. Another challenge for determining the right sources of nutrient for various highland cropping systems is the complex mixed cropping practised, with banana/coffee, banana/legume, cassava/legume intercropping common. Recommendations for fertilizer sources of nutrients will therefore require taking into account the nutrient requirements not only for sole crops, but also for the various crop combinations.

*The right rate.* The key steps to developing the right rate include making the appropriate assessments of soil nutrient supply, all available indigenous nutrient sources and crop nutrient demand, and in estimating fertilizer use efficiency. Fertilizer rates must target the appropriate input level for the specific nutrients limiting crop productivity, considering both agronomic and economic implications. Many studies conducted in the East African highlands have shown very strong nutrient responses, particularly to K for cassava and banana, with substantial yield increases observed at rates as high as 300 kg of K per hectare. The major determinant for the right rate is economics, with fertilizer use profitable at high rates in areas with good access to markets (Wairegi and van Asten 2010).

*The right time.* Nutrients should be applied to match the crop's seasonal nutrient demand, which depends on planting date, plant growth characteristics and sensitivity to deficiencies at particular growth stages, etc. Timing of nutrient application should also consider the dynamics of nutrient supply through the mineralization of organic resources and the dynamics of soil nutrient loss. For example, leaching losses tend to be higher on sandy than on clay soils. Nutrients are used most efficiently when their availability is synchronized with crop demand. Basal fertilizer application is done at or just after planting to supply NPK and other nutrients required for early crop growth. The major challenge in much of SSA is that fertilizers are often not available to farmers at the right time, resulting in a late application of basal fertilizers and reduced yields. Nutrients that are not easily leached (K and P) from the soil are also applied as basal fertilizers as they remain in the soil during the crop-growing period. N is highly mobile and can be easily lost and additional N is applied at various stages during the crop-growing season. The application of N and K in perennial crops should be split into a number of dressings so as to provide a continuous supply from planting right through to harvest, with smaller and more frequent dressings where the risk of loss by leaching is higher, e.g. on light-textured soils in the humid tropics.

*The right place.* The right place covers the strategic positioning of nutrient supplies so that a plant has access to them. Plant type and variety, placement technologies, tillage practices, plant spacing, crop rotation or intercropping, weather variability and a host of other factors can all dictate the most appropriate method of application. Incorporation is usually the best option to keep nutrients in place and increase their efficiency. There are three main fertilizer placement methods: broadcasting, banding and spot application. In the cropping systems in the East African highlands where farmers mostly use low rates of fertilizer, spot application by placing fertilizers in small amounts in the hills together with the seeds at planting or close to each plant station during the crop-growing season is most effective placement for legume crops. Hole fertilizer application also works more effectively for banana, coffee and cassava.

Nutrient management practices are always nested in cropping systems, and other management and site factors such as tillage, drainage and cultivar selection, which can greatly influence the effectiveness of a specific practice. Factors such as genetic yield potential, weeds and pests and diseases influence the effectiveness of nutrient management practices. Effective performance of the 4Rs will require the support of the associated agronomic practices. Getting the agronomic practice balance correct is often a challenge for most farmers, as a labour shortage often implies suboptimal weeding and a diminished crop response to fertilizer.

## Sustainability implications and future issues

The 4R Nutrient Stewardship approach is an essential tool in the development of sustainable agricultural systems because its application forms the basis of the economic, environmental and social sustainability of cropping systems. Increased crop and labour productivity, improved profitability, improved soil health and decreased environmental pollution are evidence of the direct impact of 4R Nutrient Stewardship at the farm level. When viewed in a wide and integrated way, 4R Nutrient Stewardship has far-reaching effects on the sustainability of agricultural systems. Substantial improvement in agricultural productivity is required to support growing rural and urban populations in the East African highlands. Because of the high population density and strong pressure on land resources, effective intensification on existing crop lands is the only way to increase agricultural production. The 4R Nutrient Stewardship also enhances competitiveness in cropping systems that can allow farming communities to viably transit from subsistence to commercially oriented farming.

The main issues for future research and development related to the 4R Nutrient Stewardship include the following:

- Consolidation of fragmented results from many past research activities on nutrient management in SSA for a better understanding of the balance required among the 4Rs for specific sites and crops.
- Development of approaches for the rapid diagnosis of degraded soils that respond poorly to fertilizer application and practical recommendations for improving their productivity.
- Synthesizing and packaging the 4R information to develop products for the innovative and effective transfer of knowledge on best fertilizer management practices to input suppliers, public and private agricultural service providers, extension services and farmers.
- The process to develop 4R recommendations has to take into account not only the complex factors that affect soil–water–crop relations, but also the socio-economic factors. Decision-support tools are required to effectively integrate the various factors affecting crop productivity in determining site-specific best management practices.

## References


Bekunda, M. A. and Woomer, P. L. (1996) 'Organic resource management in banana-based cropping systems of the Lake Victoria Basin, Uganda', *Agriculture Ecosystems and Environment,* vol. 9, pp. 171–180.

Brouwer, J. and Powell, J. M. (1998) 'Micro-topography and leaching: possibilities for making more efficient use of nutrients in African agriculture', in *Nutrient Balances as Indicators of Productivity and Sustainability in sub-Saharan African Agriculture*, special edition of *Agriculture Ecosystems and Environment*, vol. 71, pp. 229–239.

Giller, K. E., Rowe, E., de Ridder, N. and van Keulen, H. (2006) 'Resource use dynamics and interactions in the tropics: scaling up in space and time', *Agricultural Systems,* vol. 88, pp. 8–27.

IPNI (2012) *4R Plant Nutrition: A Manual for Improving the Management of Plant Nutrition,* International Plant Nutrition Institute, Norcross, GA.

Nyombi, K., van Asten, P. J. A., Corbeels, M., Taulya, G., Leffelaar, P. A. and Giller, K. E. (2010) 'Mineral fertilizer response and nutrient use efficiencies of East African highland banana (*Musa* spp., AAA-EAHB, cv. Kisansa)', *Field Crops Research,* vol. 117, pp. 38–50.

Pypers, P., Sanginga, J. M., Kasereka, B., Walangululu, M. and Vanlauwe, B. (2011) 'Increased productivity through integrated soil fertility management in cassava–legume intercropping systems in the highlands of Sud-Kivu, DR Congo', *Field Crops Research*, vol. 120, pp. 76–85.

Ssali, H., McIntyre, B. D., Gold, C. S., Kashaija, I. N. and Kizito, F. (2003) 'Effects of mulch and mineral fertilizer on crop, weevil and soil quality parameters in highland banana', *Nutrient Cycling in Agroecosystems,* vol. 65, pp. 141–150.

Tittonell, P., Vanlauwe, B., Corbeels, M. and Giller, K. E. (2008) 'Yield gaps, nutrient use efficiencies and response to fertilizers by maize across heterogeneous smallholder farms of western Kenya', *Plant and Soil*, vol. 313, pp. 19–37.

Vanlauwe, B., Bationo, A., Chianu, J., Giller, K. E., Merckx, R., Mokwunye, U., Ohiokpehai, O., Pypers, P., Tabo, R., Shepherd, K., Smaling, E., Woomer, P. L. and Sanginga, N. (2010) 'Integrated soil fertility management: operational definition and consequences for implementation and dissemination', *Outlook on Agriculture*, vol. 39, pp. 17–24.

Wairegi, L. W. I. and van Asten, P. J. A. (2010) 'The agronomic and economic benefits of fertilizer and mulch use in highland banana systems in Uganda', *Agricultural Systems*, vol. 108, pp. 543–550.

World Bank (2008) *World Development Report 2008: Agriculture for Development*, World Bank, Washington, DC.

Zingore, S., Murwira, H. K., Delve, R. J. and Giller, K. E. (2007) 'Influence of nutrient management strategies on variability of soil fertility, crop yields and nutrient balances on smallholder farms in Zimbabwe', *Agriculture Ecosystems and Environment*, vol. 119, pp. 112–126.

# 7 Mitigating the impact of biotic constraints to build resilient banana systems in Central and Eastern Africa

*Rony Swennen,[1] Guy Blomme,[2] Piet van Asten,[3] Pascale Lepoint,[4] Eldad Karamura,[2] Emmanuel Njukwe,[5] William Tinzaara,[2] Altus Viljoen,[6] Patrick Karangwa,[7] Danny Coyne[8] and Jim Lorenzen[9]*

[1] *KU Leuven and Bioversity International, Belgium and International Institute of Tropical Agriculture (IITA), Tanzania;* [2] *Bioversity International, Uganda;* [3] *IITA, Uganda;* [4] *Bioversity International, Burundi;* [5] *IITA, Burundi;* [6] *Department of Plant Pathology, Stellenbosch University, South Africa;* [7] *Rwanda Agriculture Board, Rwanda;* [8] *IITA, Tanzania;* [9] *Bill & Melinda Gates Foundation*

## Introduction

Most edible banana (*Musa* spp.) are triploids which arose from natural crosses between the *Musa acuminata* subspecies providing the A genome, or from interspecific crosses between *M. acuminata* and *M. balbisiana*, the latter providing the B genome. Banana is a major staple food and a source of income for millions of people across East and Central Africa, such as in Uganda with 250 kg/person/year (Simmonds 1966). The crop mainly consists of the East African highland cooking and beer bananas (EAHB) (AAA-EA genome), with a limited amount of cooking (ABB genome) and dessert bananas (AAA and AAB genome) and plantain (AAB genome).

The crop is dominantly produced by smallholder farmers (<2 ha) and yields in East and Central Africa are low (5–30 t/ha/yr; Wairegi *et al.* 2010; Okumu *et al.* 2011) compared with yields of >70 t/ha/yr observed at some experimental stations (van Asten *et al.* 2004) and in some farmers' fields in high rainfall areas (>1,500 mm/yr). The low yield per hectare can primarily be attributed to poor and declining soil fertility, drought stress, suboptimal crop management practices, and a high and sometimes increasing pressure of pathogen and pest damage (Karamura *et al.* 1999; Stover 2000; Gaidashova *et al.* 2009; Wairegi *et al.* 2010; van Asten *et al.* 2011). These problems are augmented as producers establish their plantations with suckers infected with pathogens and pests. Planting material is also moved between villages and districts and even across country borders without any restrictions.

This chapter provides information on the important diseases and pests of banana in East and Central Africa. We describe how the multiplication of healthy planting material can contribute to a vigorous and sustainable sector. Furthermore, the need for improved linkages among researchers, extension officers, the private sector and policy-makers is highlighted to improve the productivity and resilience of banana systems.

## Diseases and pests

The most important diseases include banana bunchy top, black leaf streak, Fusarium and Xanthomonas wilt and pests such as nematodes and weevils. These diseases and pests now occur throughout the region (Table 7.1).

### *Banana bunchy top disease*

Banana bunchy top disease (BBTD) is considered the most serious viral disease (Dale 1987) affecting all varieties and is caused by the BBTV virus (genus *Babuvirus*, family *Nanoviridae*) (Table 7.1). It entered Africa via Egypt in 1901 (Magee 1953), and now occurs in at least 13 countries, including western Rwanda, the Rusizi valley in Burundi, the Congo basin and eastern highlands in the Democratic Republic of Congo (Niyongere *et al.* 2012). BBTV is disseminated over short distances by the aphid vector *Pentalonia nigronervosa* Coquerel (Hemiptera: Aphididae) and requires a latent period of 20 hours between vectors (Anhalt and Almeida 2008). Ninety-nine per cent of new infections occur within 86 m of the primary source of inoculum (Allen and Barnier 1977; Allen 1978), but favourable wind currents can increase this distance for winged aphids. Over longer distances, BBTV is disseminated in infected planting material. The virus is not transmitted mechanically.

BBTV infects plants at any growth stage, with infection depending on temperature (maximum efficiency at 27°C), the life stage of the vector and the plant access period (Anhalt and Almeida 2008). A negative correlation has, for instance, been found between aphid presence and altitude (cooler temperatures) in the Great Lakes region of Africa (Niyongere *et al.* 2012). Temperature also has an impact on the rate of symptom development; these are characteristic (Table 7.1), especially the upright and bunched-up leaves produced at the plant's apex. Affected plants may not produce any fruit (Dale 1987; Su *et al.* 2003).

Management of BBTV is based on two approaches: exclusion and eradication (Table 7.1). However, control measures are not always possible in East and Central African countries where banana is produced by resource-limited smallholders, and where control measures are left to their goodwill (Lepoint *et al.* 2011).

*Table 7.1* Important diseases and pests of banana in East and Central Africa

| *Disease name* | *Pathogen/pest* | *Symptoms* | *Occurrence* | *Dissemination* | *Management* |
|---|---|---|---|---|---|
| Banana bunchy top | BBTV | Dark green leaf streaks, progressive dwarfing, marginal leaf chlorosis, upright and bunched-up leaves at plant apex | DRC, Rwanda, Burundi | Banana aphid, diseased planting material | Removal of affected plants, insecticides to control aphids, herbicides, clean planting material, tissue culture, plant quarantine |
| Black leaf streak | *Mycosphaerella fijiensis* | Yellow spots on leaves gradually turn into brown and black streaks | All lowlands | Infected planting material, wind, rain | Resistant varieties, clean planting material, varietal mixtures, use of healthy and fertile soils, leaf pruning |
| Fusarium wilt | *Fusarium oxysporum* f. sp. *cubense* | Chlorosis, necrosis and wilting progressing from the oldest to youngest leaves, reddish-brown discolouration of vascular tissue in the rhizome and pseudostem | Tanzania, Uganda, Kenya, DRC, Rwanda, Burundi | Infected planting material, soil attached to shoes and implements, water | Resistant varieties, clean planting material, proper quarantine |
| Xanthomonas wilt | *Xanthomonas campestris* pv. *musacearum* | Shrivelling of male bud, premature fruit ripening and colouration of pulp, yellowing and wilting of leaves, oozing of bacteria from infected tissue | Ethiopia, Uganda, Kenya, Rwanda, Burundi, DRC, Tanzania | Infected planting material, insects and birds, contaminated field tools, browsing domestic animals | Escape varieties which lack the male inflorescence or have persistent male bracts and flowers, clean planting material, disinfected tools, early removal of male buds, field hygiene |
| Nematodes | *Helicotylenchus multicinctus*, *Meloidogyne* spp., *Pratylenchus goodeyi*, *Radopholus similis* | Stunting, poor plant growth, narrow and weak stems, foliar chlorosis, root rotting and galling, plant toppling | All countries, *R. similis* in the lowland and *P. goodeyi* in the highlands | Infected planting material | Resistant varieties, clean planting material, paring and hot water treatment of suckers |
| Banana weevil | *Cosmopolitus sordidus* | Stunting, delayed maturation, reduced bunch sizes, snapping, premature death | Uganda, Tanzania, Burundi, Rwanda | Infected planting material | Resistant varieties, clean planting material, chemical control, cultural control (mulching away from the plants, removal and destruction of residual pseudostems), endophyte enhancement |

### *Black leaf streak*

*Mycosphaerella fijiensis*, which causes black leaf streak (black sigatoka) disease, is a hemiobiotrophic fungus that results in significant leaf necrosis, with yield losses between 33 and 76 per cent (Mobambo *et al.* 1993, 1996). The fungus was discovered in West/Central Africa in 1978 and in East Africa in 1987 (Jones 2000). Black leaf streak has now replaced yellow sigatoka, caused by the milder pathogen *Mycosphaerella musicola*, throughout tropical Africa, except at altitudes above 1,800 masl. In the past two decades, the black leaf streak fungus has adapted to cooler conditions, however, and now replaces yellow sigatoka at altitudes higher than 1,500 masl (Jones 2000). Black leaf streak is disseminated by ascospores and conidia through wind, water, leaves and suckers. Spore release is accelerated by the alternation of humid and dry weather conditions. After spore germination, the fungus penetrates the stomata. Infection starts at the lower leaf side at the moment the cigar appears. The different black-streaked symptoms gave this disease its name. Black leaf streak is more severe than yellow sigatoka because symptoms appear much earlier on younger leaves and it develops faster. Therefore, leaf areas are more seriously affected. Premature finger yellowing is also noticed. Yellow Sigatoka affected mostly dessert banana; black leaf streak affects dessert banana, plantain, some ABB cooking banana and highland banana. In export banana fields, black leaf streak is controlled by 15–50 aerial fungicide applications based on a weather forecasting system and symptom evolution (de Lapeyre de Bellaire *et al.* 2010). Highland banana and plantain are rarely treated because smallholders cannot afford these expensive chemicals. In addition, these varieties are often cultivated at a short distance from the homestead and chemical treatment would therefore impose a serious health hazard. Therefore African smallholders rely especially on leaf pruning (Table 7.1). In the medium to longer term, farmers will benefit from the deployment of resistant varieties.

### *Fusarium wilt*

Banana Fusarium wilt (Panama disease), caused by the fungus *Fusarium oxysporum* f. sp. *cubense* (Foc), is the most destructive disease of banana worldwide and a major threat to the dessert and beer banana production in East and Central Africa. The disease was detected in West Africa in 1924 (Stover 1962) and in 1951 in mainland Tanzania, following the introduction of exotic dessert cultivars (Rutherford 2001). Since then, races 1 and 2 have spread throughout the region (Ploetz *et al.* 1990), most likely through infected suckers (Stover 1962). Once introduced into a plantation, the surviving spores (called chlamydospores) can be spread with irrigation water and soil attached to shoes and farm implements (Jones 2000).

In Foc-infested plantations, chlamydospores are stimulated to germinate by root exudates (Stover 1962). Roots of both susceptible and resistant cultivars are infected. In resistant cultivars, early recognition halts infections by the formation of occluding gels and tyloses and cell wall strengthening (Van den

Berg *et al.* 2007) shortly after the pathogen reaches the cortex. In susceptible cultivars, however, the delayed plant response allows the pathogen to progress into the rhizome and pseudostem, where it blocks the flow of water. Fusarium wilt becomes visible when leaves turn yellow from the older to the younger leaves before they wilt and die. When the rhizome and pseudostem are split open, infected vascular tissue becomes visible as reddish-brown to maroon streaks. Upon plant death, chlamydospores are released into the soil, where they can remain dormant for several decades (Stover 1962).

Foc consists of three races and 24 vegetative compatibility groups (VCGs) that can be identified by pathogenicity testing and pairing with a VCG tester set, respectively. All the strains occurring in ECA belong to Foc races 1 and 2, and are included into a VCG complex consisting of VCGs 0124/5/8/12/20. Management of Fusarium wilt is difficult and constrained by a lack of reliable and/or efficient control options (Ploetz *et al.* 1990; Jones 2000), with exclusion and the use of host resistance considered to be the most effective (Table 7.1).

### *Xanthomonas wilt*

Xanthomonas wilt (XW) is caused by the bacterium *Xanthomonas campestris* pv. *musacearum* (Xcm). The disease originated in Ethiopia, where it affects production of both enset (*Ensete ventricosum*) and banana (Yirgou and Bradbury 1974). The first outbreaks in East and Central Africa were reported in central Uganda and north Kivu, eastern DRC, in 2001 (Smith *et al.* 2008; Tripathi *et al.* 2009). Since then, the disease has spread to Rwanda (Reeder *et al.* 2007), Tanzania and Burundi (Carter *et al.* 2010) and Kenya (Mbaka *et al.* 2007).

XW is transmitted by insects, bats and birds that visit infected flowers/bunches from which they contaminate healthy plants and mats (Eden-Green 2004). Similarly, mechanical transmission can be achieved by using contaminated field tools (machetes, hoes) (Eden-Green 2004). When infected suckers are used to establish a new crop, the disease is spread to new fields. The first symptoms after insect transmission are shrivelling of the male bud bracts and decaying rachis followed by fruit discolouration and rotting (Eden-Green 2004) (Table 7.1). After inflorescence colonization, the bacteria move down the real stem into the corm, leaf sheaths and lateral shoots. Foliar symptoms include yellowing and wilting of leaves (therefore the naming of the disease), followed by the plant dying off (Tushemereirwe *et al.* 2004).

The management of Xcm, based on its exclusion from unaffected areas and the eradication of affected plants (Table 7.1), relies on the accurate and rapid diagnosis of the XW pathogen. The yellow bacterial ooze which can be found in the leaf sheaths, stem and leaf petioles, and the discolouration of fruit pulp are typical and easily recognizable symptoms. In addition, PCR technologies can distinguish Xcm from *Ralstonia solanacearum*, the bacterium responsible for *Moko* in other parts of the world (Thwaites *et al.* 2000; Lewis Ivey *et al.* 2010; Adriko *et al.* 2011). A sensitive and robust serological assay based on ELISA and LFD assays (www.dsmz.de) has also been developed against a DRC Xcm.

### *Nematodes*

A complex of nematode pests affect banana in East and Central Africa (Table 7.1); *Radopholus similis* is arguably the most important and most damaging (Karamura 1993; Gowen *et al.* 2005; Jones 2009). *R. similis* is not indigenous to the region and appears to have been introduced into Uganda in the 1960s, very likely on multiple occasions, from where it spread, although introductions could have occurred elsewhere in the region (Price 2006). It is associated with the banana decline encountered from the 1970s with severe losses, especially from toppled plants. *Pratylenchus goodeyi* is indigenous, occurs only in cooler conditions, effectively above 1,400 masl (Kashaija *et al.* 1994; Price and Bridge 1995). It is prevalent on banana above this altitude, causing severe root necrosis where densities are high, but has yet to be consistently linked with high losses (Gaidoshova *et al.* 2009). Other commonly occurring species include *Meloidogyne* spp. and *Helicotylenchus multicinctus*, both of which are widely spread within East and Central Africa (Gowen *et al.* 2005). Nematodes are introduced into new banana fields with infected planting material, which is perhaps the most common form of dissemination.

Nematode feeding results in the destruction of root and rhizome tissue. Damaged tissue becomes necrotic before the roots decay and die, thereby reducing nutrient and water uptake and bunch weights, and increasing the time to harvest. The death of roots additionally affects the plant's anchorage, and particularly in strong winds and with a developing bunch, plant toppling can occur (Sarah 2000; Jones 2009). Reduced plant turgidity, due to suppressed water uptake, can result in the snapping of plant stems during windy conditions in periods of low water availability. Fruit on fallen plants often have no value, resulting in high yield losses (Gowen *et al.* 2005). Common symptoms of severe nematode infestation include stunting, poor plant growth, narrow and weak stems, foliar chlorosis, root rotting and galling and plant toppling (Table 7.1).

*R. similis* poses the primary threat to production, except in the cool conditions above 1,400 masl, where it is replaced by *P. goodeyi*. Otherwise, these two species are regularly found in combination with *Meloidogyne* spp. and *H. multicinctus*. The pathogenic potential of *R. similis* varies, with some populations being particularly aggressive and able to overcome resistance in varieties which are otherwise effective (Plowright *et al.* 2012). However, *P. goodeyi* populations morphologically identified as *P. goodeyi* but differing at the molecular level have been observed causing severe damage in tropical lowland banana in coastal Kenya (Seshu Reddy *et al.* 2007). Nematodes can best be controlled by planting healthy, nematode-free plants (Table 7.1) in a nematode-free environment.

### *The banana weevil*

The banana weevil, *Cosmopolites sordidus* (Coleoptera: Curculionidae), originates from Malaysia and Indonesia but is now a major constraint in East and

Central Africa, especially in smallholder farming systems (Gold *et al.* 1999, 2001). In Uganda it arrived around the turn of the twentieth century (Price 2006). It occurs mostly in combination with nematodes. This fact creates difficulty in apportioning individual importance, but together the two pests contributed to the disappearance of East African highland cooking banana from its traditional growing areas in central Uganda (Gold *et al.* 1999; Price 2006) and western Tanzania (Mbwana and Rukazambuga 1999). This weevil can cause up to 30–50 per cent yield loss but is normally not a problem above 1,500 masl (Rukazambuga *et al.* 1998). Weevils tend not to fly and are mostly introduced into new fields with infected planting material.

Eggs are laid into the rhizomes and pseudostem, where the hatched larvae then burrow, creating a labyrinth of tunnels resulting in stunting, leaf yellowing, reduced bunch sizes and the snapping off of affected plant stems (Gold *et al.* 2001) (Table 7.1). The life cycle takes some 30–40 days, with the destructive larval stage lasting 15–20 days and adults living for up to four years. The available management options include chemical control, biological control, trapping, cultural control and host plant resistance (Gold *et al.* 2001). These methods usually provide at least partial control of the weevil but they are labour-intensive and costly.

## Integrated disease and pest management

A wide range of technologies for integrated disease and pest management (IDPM) has been developed and/or employed in East and Central Africa over recent years. Increasing access to and the use of healthy planting material should remain a key focus and the foundation of any IDPM package. Complementary options include the use of resistant *Musa* germplasm, chemical control, cultural control (mulching away from the plants, removal and destruction of residual pseudostems), sucker treatment, pest trapping, male bud removal, disinfection of tools, soil management and endophyte enhancement of planting material. Comparative studies that would have characterized the contribution of each IDPM component under field conditions remain limited. These methods usually provide at least partial control of banana diseases and pests, but are labour-intensive and costly. Hence their adoption by resource-poor farmers is often limited.

Management of several important pathogens and pests such as the BBTV, Xcm, Foc and nematodes, is based on two approaches: exclusion from unaffected areas and eradication of affected plants (Table 7.1). Exclusion involves a strict national quarantine programme and requires legislation, awareness and the use of pathogen-free planting material. Eradication involves the removal of infected plants, the use of insecticides and herbicides and a 1–2-year production system. These control measures are not all possible in East and Central African countries where banana is produced by resource-limited smallholders and where the plants are grown in perennial systems (Lepoint *et al.* 2011).

*Exclusion.* Banana planting material cannot be moved internationally unless it has been certified as free of pests and diseases by plant health services (Lescot and Staver 2010). Before planting material is moved internationally, an export permit is required, indicating the source, variety name, quantity, recipient institute or organization and destination. In addition, the recipient needs to obtain an import permit. Tissue culture should be the sole source of material for the international movement of plantlets to commercial plantations and elite varieties to research institutes for local evaluation and selection. Released varieties can be further multiplied using mother gardens. Mother gardens should be rejuvenated with new stocks of clean tissue-cultured plants or replenished with new varieties every 3–4 years. Domestic movement of planting material should be accompanied by a material transfer agreement (MTA) indicating its source, variety name, quantity, destination and preliminary information on agronomic characteristics. The movement of suckers should be discouraged and growers should be encouraged to buy and use planting material from a known source. Regular technical backstopping is necessary on rapid multiplication techniques to generate clean planting material from mother gardens for local distribution.

*Eradication.* Eradication involves the removal of infected plants, sanitation, the use of pesticides and biological control agents, and soil management. The management of a disease such as XW is largely based on field hygiene and includes early de-budding using a forked wooden stick, decontamination of garden tools and the use of clean planting materials (Blomme *et al.* 2005; Karamura and Tinzaara 2009). Machetes needed to prune leaves to provide more light for the annual crops should also be disinfected in regions infected with bacterial wilt, otherwise diseases such as XW and Fusarium wilt will spread. The same applies when suckers are thinned to reduce competition. Xcm is immobile, oligophagous and does not survive in decaying plant debris or in soil. Leaving land fallow or planting break crops for at least six months, therefore, will remove any risk of soil-borne infection when replanting with clean planting materials. This, however, is not the case with Fusarium wilt, as chlamydospores may survive in fields for decades (Stover 1962). All new plantings, therefore, will be affected by the pathogen.

Attempts to control pathogens and pests with pesticides have generally met with limited success and are most often not an option for smallholders. Biological control is also difficult, often because of pathogen biology and pest behaviour – for instance the banana weevil's cryptic habit (Gold *et al.* 2001). No effective parasitoids and predators of the weevil have been found; indigenous ant species in East Africa have shown little promise as biological control agents. Strains of *Beauveria bassiana* are effective against banana weevils but are beyond the means of smallholders unless economically feasible mass production and efficient delivery systems are developed. Integrating *B. bassiana* with pheromone- and kairomone-based traps which concentrates weevils at entomo-pathogen release sites has shown promise (Tinzaara *et al.* 2007).

*Protection.* Smallholders in Africa have difficulty in getting access to suitable pesticides as a consequence of unavailability or high cost. They often have poor knowledge of pesticide application and poor information on the products and consequently rely on altered field management practices to protect their crops against pathogens and pests.

*Cultural practices.* Black leaf streak was reported to be less pronounced in banana grown in backyards close to the homestead than in banana cultivated in fields (Mobambo *et al.* 1994), possibly because there are man-made soils rich in organic matter content, good drainage, rooting and microbial diversity (Jimenez *et al.* 2007; Mobambo *et al.* 2010). Mobambo *et al.* (2010) also showed that mulch from cereals had the highest negative effect on black leaf streak severity; this is presumably caused by the high silicon content (Kablan *et al.* 2012). Methods that contribute to rapid leaf emission rates (to replace infected leaves by new uninfected leaves) and reduce black leaf streak are the use of healthy planting material, the reduction of nematode and weevil infestation by paring and heat treatment (Hauser 2007), the use of varietal combinations or high plant densities (Ndabamenye *et al.,* submitted), action to reduce weeds and the practice of intercropping. Most importantly, infected leaves have to be pruned and destroyed to reduce the inoculum. Other pests and diseases can affect black leaf streak incidence, and its control therefore needs to be considered holistically.

*Biological control.* Enhancing tissue culture material with bioprotectants for biologically based protection against nematodes has received significant attention recently. Naturally occurring endophytes, such as non-pathogenic *F. oxysporum*, when introduced into tissue culture plants, have shown promise for nematode and weevil management (Dubois and Coyne 2011). The ability to establish *B. bassiana* as an endophyte in banana for the biological control of weevil larvae during feeding was also investigated. This work proved promising under screenhouse conditions (Akello *et al.* 2007, 2008, 2009).

*Genetic resistance.* High-yielding exotic and improved varieties introduced from the International Transit Centre (ITC) in Belgium, and hybrids developed at Fundación Hondureña de Investigación Agrícola (FHIA) in Honduras, by the IITA and the National Agricultural Research Organisation (NARO) in Uganda, were tested for resistance to diseases and pests across East and Central Africa. These varieties combine higher resistance with higher yields. All plantain (AAB), East African highland banana (AAA-EAHB), 'Bluggoe' (ABB cooking banana) and the dessert banana 'Sukali Ndizi' (AAB), 'Cavendish' and 'Gros Michel' (AAA) were affected by *M. fijiensis*. Only 'Kayinja' (ABB Pisang Awak) and Yangambi Km5 (AAA) were resistant. Screening trials have also demonstrated that varieties such as Yangambi Km5, Calcutta-4 and FHIA-03 are resistant to the banana weevil and could be utilized in breeding programmes. All currently tested *Musa* genotypes have succumbed to XW. However, wild *Musa balbisiana* recovered from initial symptoms (Ssekiwoko *et al.* 2006). Differences in susceptibility among cultivars to BBTV are being investigated (Niyongere *et al.* 2011). The host range of the vector is limited to *Musa* spp., and aphid populations were reported to be higher on Yangambi Km5 (AAA) and plantain (AAB). No source of resistance has been found within cultivated *Musa* spp. to date.

In the medium to long term, farmers will benefit from the deployment of resistant varieties produced by conventional breeding (Tenkouano *et al.* 2003; Tenkouano and Swennen 2004) and genetic engineering (Kovács *et al.* 2012). East African highland (AAA-EAHB) banana hybrids developed jointly by IITA and NARO were tested for black leaf streak resistance in CIALCA-organized multi-location trials on farmers' fields in Burundi and Rwanda, and North and South Kivu provinces of DRC. Farmers' evaluations in both Burundi and North Kivu identified 9750S-13 as a superior, high-yielding cooking variety and 9518S-12 as a superior, high-yielding juice variety suitable for brewing. Two new AAA-EAHB hybrid selections, 'M2' and 'M9', have shown moderate resistance to the banana weevil (Ssali *et al.* 2010). Nematode resistance remains a key target in breeding programmes (Tenkouano and Swennen 2004; Lorenzen *et al.* 2009). Sources of resistance to nematode species have been demonstrated (De Waele and Elsen 2002; Dochez *et al.* 2009) and recent breeding successes have been achieved at IITA (Lorenzen *et al.* 2009) and CIRAD (Quénéhervé *et al.* 2009).

A first generation of genetically engineered resistant lines holds some promise for the development of XW resistance (Tripathi *et al.* 2010). *Bacillus thuringiensis* (Bt) genes, including Cry3a currently being used against lepidopteron pests of genetically modified (GM) maize and cotton, have been the focal point for transgenic pest control in crops. However, these were not effective against coleopteran pests, such as the banana weevil. More recently, Bt strains have been discovered with considerable activity against coleopteran pests. One such strain, Cry6A, is being investigated for banana weevil control in a NARO–Bioversity collaborative study. Cysteine proteinase inhibitors (cystatins) from plants such as rice and papaya are also being tested against the weevil (Kiggundu *et al.* 2003) as well as parasitic nematodes (Atkinson *et al.* 2004).

## The importance of clean seed systems

A major constraint to the expansion of banana cultivation is the scarcity of healthy planting materials (Schill *et al.* 1997; Nkendah and Akyeampong 2003; Tenkouano *et al.* 2006). Farmers usually depend on the natural regeneration of suckers in their own or neighbouring fields for the supply of planting material because of the easy access and low cost compared with other sources (Table 7.1). Although suckers are easily accessible, they have a low multiplication rate and are often contaminated with pathogens and pests (Swennen 1990; Faturoti *et al.* 2002). Transplanting contaminated materials spreads diseases and reduces the longevity of plantations. To overcome this constraint, several techniques have been developed and disseminated to producers in East and Central Africa to clean the suckers (paring of corms, boiling water treatment) and rapidly multiply planting material (macro- and micro-propagation).

*Paring of corms.* Cutting off the roots and paring back the corm surface with a knife is a simple and effective method to remove infected material from sucker planting material. Ideally, this should be conducted in combination with hot or boiling water treatments, as some residual nematode and weevil infestation can remain on pared corms. However, this is an uncomplicated technique that will provide a high level, but not complete, disinfestation suitable for rapid dissemination with farmers. Removal of the roots will not affect sucker germination.

*Hot water treatment.* Hot water treatment of suckers after the removal of infected roots is an effective technique for sanitizing sucker planting material against nematodes and weevils. Suckers are immersed in water, heated to 50–55°C for a period of 20 minutes, which is suitable for killing nematodes and weevil larvae. This technique has been further adapted using a 30-second exposure period in boiling water, following poor uptake and use of the hot water technique by smallholders (Coyne *et al.* 2010). However, weevil larvae lying deep in the corm tissue are less likely to be controlled than with the hot water treatment due to the short duration of the exposure to heat.

*Macro-propagation of plants.* Macro-propagation is an alternative method that is cost-effective and affordable for rapidly multiplying healthy planting material using a germination chamber. It relies on the suppression of apical dominance to enhance the growth of numerous lateral buds. Destroying the apical meristem of pre-flowered plants constitutes one way of breaking the dormancy of the lateral buds to enhance their development into suckers. The macro-propagation technique, although genotype-dependent, can produce 8–15 new plants/corm within 15 weeks; scarification of the emerging lateral buds has the potential to further increase the number of plantlets by a factor of 2–3 within the same time period. Plantlets obtained through this method have the uniformity of micro-propagated seedlings while being less prone to post-establishment stress in the field.

The technique is user-friendly, simple and cheap, but requires a minimum investment to set up the germination chambers and weaning facilities, which makes it suitable for small- and medium-scale enterprise farmers. To prevent the multiplication of infected plantlets, the corm should first be treated in hot or boiling water to kill nematodes and weevil larvae. To sustain a rapid production and distribution of healthy planting material a three-tier (primary, secondary and tertiary) multiplication scheme is essential. Tissue culture plantlets free of viruses are distributed to development partners; macro-propagated materials are delivered by development partners to farmers' associations. Healthy suckers (pared and treated with boiling water) (Coyne *et al.* 2010) can then reach the farmers. Regular backstopping on multiplication techniques is essential, and the provision of nuclear stocks of improved varieties to rejuvenate seed production nurseries or replenish them with new varieties every 3–4 years is important. On the other hand, smallholders' knowledge and technical skills regarding the use of healthy planting material should be reinforced through field inspection as well as exchange visits with other farmers.

*Plant tissue culture.* Plants produced by tissue culture and certified as pest- and disease-free are ideal. Such material provides healthy planting material that can be transported with relative ease across international borders. For commercial production tissue-cultured planting material is now routinely used, but has yet to gain wider use by smallholders (Dubois *et al.* 2006). Fields established from tissue-cultured material are uniform and more easily managed for market timelines than sucker-derived plantations. Due to the process of production, however, tissue-cultured plantlets are sterile and thus also free of associated beneficial microbes, and initially very fragile. Recent research on identifying and reintroducing fungal endophytes that are effective in providing pest and disease protection and improving plant vigour is showing considerable promise (Dubois *et al.* 2006).

*International transfer of banana germplasm.* Established in 1985, the ITC secures the long-term conservation of the entire *Musa* gene pool, provides a service for the safe movement of *Musa* germplasm and maintains related information. The ITC is thus central to the conservation of *Musa* germplasm on a global level (Van den Houwe *et al.* 2003). The ITC currently holds nearly 1,400 accessions *in vitro* under slow growing conditions and over 900 accessions have been cryopreserved. This collection is continuously enriched with new germplasm from collecting missions and breeding programmes. Within the collection 66.8 per cent are virus-free and distributed upon demand. The ITC has distributed so far about 10,000 accessions to 103 countries free of charge and more than 75 per cent of the entire collection's genotypes have been distributed. Two-thirds of the requests were for cultivated forms, landraces, varieties and cultivars for evaluation for resistance or tolerance to biotic stresses (especially black leaf streak, Foc and nematodes), agronomic characteristics and fruit quality. The demand is high for improved materials (20 per cent of samples distributed). Most of the germplasm distributed by the ITC is intended for use by farmers, and the ITC is the only viable source of improved and clean germplasm for many countries. Consequently, the highest impact from ITC comes from increased yields and the reduction of losses due to pests and diseases, leading to higher farm income. The ITC also contributes to advances in research, such as improved knowledge about the *Musa* genome and diversity, improved virus indexing and therapy methods and capacity building. Attempts are under way to improve a mechanism whereby users' evaluation data are incorporated into the ITC's documentation systems, including the *Musa* germplasm information system (MGIS: www.musanet.org). That information would then facilitate the future selection of *Musa* germplasm.

## Improved linkages between banana research, extension, the private sector and policy-makers

Over the first half of the twentieth century, banana was regarded in East and Central Africa as a hardy native crop that required little research attention. It was consumed only in the rural regions and commercial prospects were perceived as dim, given its bulkiness, perishability and perceived low nutritional value. From the 1960s, research-led knowledge and technology improvement into East and Central African banana-based cropping systems raised public awareness of banana as a potential crop for food security. By the 1980s, massive rural-to-urban migration and the resulting high demand by urban dwellers, coupled with crop destruction by weevils and nematodes, led many governments in the region to prioritize research and development. At this point in time, technology transfer for smallholder producers was largely linear; the transfer of technologies from research to extension and on to farmers.

It was eventually realized that for the effective transfer of research information and technologies to occur, information must be shared with individuals and/or groups in the target population. This would then lead to collective action and ultimately to mutual agreement and mutual understanding. This approach ushered in participatory research in which constraints and attendant remedies were discussed by researchers, extension people and farmers, and which emphasized dialogue rather than linear, one-way communication. However, by the 1990s the limitations of participatory research were already apparent, primarily because of the assumption that all participants are equal, ignoring the huge differentials in the knowledge base among them. Those with a lot of knowledge ultimately led and guided the direction of research. This shifted the debate from persuasion and diffusion models to horizontal communication and information/ technology exchange. Moreover, the governments of the region realized that banana, which by then occupied up to 30–40 per cent of the region's total crop land, had an important role in efforts to fight poverty.

With lessons learned from the past, it was decided that value addition for banana would open more opportunities to the smallholder farmers who target markets of both fresh and processed products. This required non-traditional partners, such as the private sector and other chain actors along the production–consumption continuum, leading to the creation of groups of stakeholders with different but often complementary experiences, linked together to address a common priority problem. In Uganda, for example, a number of banana-producing districts formed farmers' organizations that linked them directly with urban traders, processors of banana figs, chips, crisps and wine, with some of the products reaching export markets. This resulted in the formation of innovation platforms whose functions included network development and the organization of producers and consumers into groups. It also facilitated access to technologies, expertise, markets, credit, inputs, training on new approaches and the strengthening of linkages with service providers. These innovation platforms were very useful during epidemic outbreaks of diseases, such as XW, which required widespread mobilization to address the problem and prevent food insecurity/income losses in the affected banana-dependent regions of East and Central Africa. Innovation platforms at both the national and regional levels developed a regional XW mitigation strategy that formed the basis for the development and deployment of national action plans (NAPs) for mobilizing action, negotiating and advocating for resources to address the problem.

## Conclusion

Banana, a perennial crop that can be harvested throughout the year, is of major importance to the income and food security of the rural populations in East and Central Africa. Banana not only provides an important socio-economic buffer but also an important biophysical buffer – the permanent canopy and soil cover (i.e. mulch) that prevents soil erosion and conserves soils in hilly landscapes. Unfortunately, this important buffer crop is vulnerable to pest and disease outbreaks due to its narrow genetic base. In addition, poor quarantine regulations and a considerable movement of planting material between fields, districts and countries facilitate the continued spread of diseases and pests. New biotic constraints, e.g. Foc TR4, could be introduced into Africa with infected planting material. Pests and diseases can only partly be counteracted with good field management practices. The big challenge, therefore, is to find inexpensive solutions to improve pest and disease resistance that smallholder farmers can easily obtain and adopt. Conventional breeding using existing genetic variability has produced several hybrids with host plant resistance and high yields. Some of the hybrids warrant further multiplication and distribution to farmers. The use of biotechnology to obtain *Musa* genotypes resistant to pests and diseases (e.g. XW) has also moved higher on the agenda.

Markets in East and Central Africa are very inelastic. A rapid increase of production would strongly decrease prices and, vice versa, a decrease in supply will increase prices. Although the post-harvest range of products has been widened (e.g. drinks, chips), this means that intensification should happen smartly and cannot focus solely on increasing total production. A primary challenge is to maintain and develop healthy production and seed systems. In addition, investments have to be carefully targeted. The costs of interventions (e.g. technical requirements and the cost of seed system options) and returns to (and incentives for) these investments need to be balanced. For example, relatively costly investments in tissue-cultured plants or inorganic fertilizers are profitable near urban markets but may not be very useful far from the markets when transport costs for fresh produce are high and farm-gate prices are low. As such, banana may not follow the path of other crops (e.g. maize) where intensification and increased production can automatically be absorbed by the regional/international market. Hence, intensification options for banana production makes greatest sense (1) near large (urban) fresh produce markets, (2) when improving the system buffer functions (e.g. healthy planting material) and (3) when efficiency is improved (e.g. yield/unit land/labour/money invested) without necessarily increasing total national/regional production.

## References

Adriko, J., Aritua, V., Mortensen, C. N., Tushemereirwe, W. K., Kubiriba, J. and Lund, O. S. (2011) 'Multiplex PCR for specific and robust detection of *Xanthomonas campestris* pv. *musacearum* in pure culture and infected plant material', *Plant Pathology*, vol. 10, pp. 1–9.

Akello, J., Dubois, T., Gold, C. S., Coyne, D. and Nakavuma, J. (2007) '*Beauveria bassiana* (Balsamo) Vuillemin as an endophyte in tissue culture banana (*Musa* spp.)', *Journal of Invertebrate Pathology*, vol. 96, pp. 34–42.

Akello, J., Dubois, T., Coyne, D. and Kyamanywa, S. (2008) 'Effect of endophytic *Beauveria bassiana* on populations of the banana weevil, *Cosmopolites sordidus*, and their damage in tissue-cultured banana plants', *Entomologia Experimentalis et Applicata*, vol. 129, pp. 157–165.

Akello, J., Dubois, T., Coyne, D. and Kyamanywa, S. (2009) 'Effects of *Beauveria bassiana* dose and exposure duration on colonization and growth of tissue cultured banana (*Musa* spp.) plants', *Biological Control*, vol. 49, pp. 6–10.

Allen, R. N. (1978) 'Spread of bunchy top disease in established banana plantations', *Australian Journal of Agricultural Research*, vol. 29, pp. 1223–1233.

Allen, R. N. and Barnier, N. C. (1977) 'The spread of bunchy top disease between banana plantations in the Tweed River District during 1975–1976', *NSW Department of Agriculture, Biology Branch Plant Disease Survey (1975–1976)*, pp. 27–28.

Anhalt, M. D. and Almeida, R. P. P. (2008) 'Effect of temperature, vector life stage, and plant access period on transmission of *Banana bunchy top virus* to banana', *Phytopathology*, vol. 98, pp. 743–748.

Atkinson, H. J., Grimwood, S., Johnston, K. and Green, J. (2004) 'Prototype demonstration of transgenic resistance to the nematode *Radopholus similis* conferred on banana by a cystatin', *Transgenic Research*, vol. 13, pp. 135–142.

Blomme, G., Mukasa, H., Ssekiwoko, F. and Eden-Green, S. (2005) 'On-farm assessment of banana bacterial wilt control options', *African Crop Science Society Conference Proceedings*, Entebbe, Uganda, 5–9 December 2005, pp. 317–320.

Carter, B. A., Reeder, R., Mgenzi, S. R., Kinyua, Z. M., Mbaka, N., Doyle, K., Nakato, V., Mwangi, M., Beed, F., Aritua, V., Lewis Ivey, M. L., Miller, S. L. and Smith. J. J. (2010) 'Identification of *Xanthomonas vasicola* (formerly *X. campestris* pv. *musacearum*), causative organism of banana Xanthomonas wilt, in Tanzania, Kenya and Burundi', *Plant Pathology*, vol. 59, p. 403.

Coyne, D. L., Wasukira, A., Dusabe, J., Rotifa, I. and Dubois, T. (2010) 'Boiling water treatment: a simple, rapid and effective technique for producing healthy banana and plantain (*Musa* spp.) planting material', *Crop Protection*, vol. 29, pp. 1478–1482.

Dale, J. L. (1987) 'Banana bunchy top: an economically important tropical plant virus disease', *Advances in Virus Research*, vol. 33, pp. 301–325.

de Lapeyre de Bellaire, L., Fouré, E., Abadie, C. and Carlier, J. (2010) 'Black leaf streak disease is challenging the banana industry', *Fruits*, vol. 65, pp. 327–342.

De Waele, D. and Elsen, A. (2002) 'Migratory endoparasites: *Pratylenchus* and *Radopholus* species', in J. J. Starr, R. Cook and J. Bridge (eds), *Plant Resistance to Parasitic Nematodes*, CAB International, Wallingford, pp. 175–206.

Dochez, C., Tenkouano, A., Oritz, R., Whyte, J. and De Waele, D. (2009) 'Host plant resistance to *Radopholus similis* in a diploid banana hybrid population', *Nematology*, vol. 11, no. 3, pp. 329–335.

Dubois, T. and Coyne, D. (2011) 'Integrated pest management of banana', in M. Pillay and A. Tenkouano (eds), *Banana Breeding: Constraints and Progress*, CRC Press, Boca Raton, FL, pp. 121–144.

Dubois, T., Coyne, D. L., Kahangi, E., Turoop, L. and Nsubuga, E. N. (2006) 'Endophyte-enhanced banana tissue culture: technology transfer through public–private partnerships in Kenya and Uganda', *African Technology Development Forum Journal*, vol. 3, pp. 18–23.

Eden-Green, S. (2004) 'How can the advance of banana Xanthomonas wilt be halted?', *InfoMusa*, vol. 13, no. 2, pp. 38–41.

Faturoti, B., Tenkouano, A., Lemchi, J. and Nnaji, N. (2002) *Rapid Multiplication of Plantain and Banana, Macropropagation Techniques: A Pictorial Guide*, International Institute of Tropical Agriculture, Ibadan, Nigeria.

Gaidashova, S. V., Van Asten, P., De Waele, D. and Delvaux, B. (2009) 'Relationship between soil properties, crop management, plant growth and vigour, nematode occurrence and root damage in East African Highland banana-cropping systems: a case study in Rwanda', *Nematology*, vol. 11, pp. 883–894.

Gold, C. S., Karamura, E. B., Kiggundu, A., Bagamba, F. and Abera, A. M. K. (1999) 'Geographic shifts in highland cooking banana (*Musa* spp., group AAA-EA) production in Uganda', *International Journal of Sustainable Development and World Ecology*, vol. 6, pp. 45–59.

Gold, C. S., Pena, J. E. and Karamura, E. B. (2001) 'Biology and integrated pest management for the banana weevil, *Cosmopolites sordidus* (Germar) (Coleoptera: Curculionidae)', *Integrated Pest Management Reviews*, vol. 6, pp. 79–155.

Gowen, S. C., Quénéhervé, P. and Fogain, R. (2005) 'Nematode parasites of bananas and plantains', in M. Luc, R. A. Sikora and J. Bridge (eds), *Plant Parasitic Nematodes in Subtropical and Tropical Agriculture. Second Edition*, CAB International, Wallingford, pp. 611–643.

Hauser, S. (2007) 'Plantain (*Musa* spp. AAB) bunch yield and root health response to combinations of physical, thermal and chemical sucker sanitation measures', *African Plant Protection*, vol. 13, pp 1–15.

Jimenez, M. I., Van der Veken, L., Neirynck, H., Rodriguez, H., Ruiz, O. and Swennen, R. (2007) 'Organic banana production in Ecuador: its implications on black Sigatoka development and plant-soil nutritional status', *Renewable Agriculture & Food Systems*, vol. 22, no. 4, pp. 297–306.

Jones, D. R. (2000) *Diseases of Banana, Abaca and Ensete*, CAB International, Wallingford.

Jones, D. R. (2009) 'Diseases and pest constraints to banana production', *Acta Horticulturae*, vol. 828, pp. 21–36.

Kablan, L., Lagauche, A., Delvaux, B. and Legrève, A. (2012) 'Silicon reduces black sigatoka development in banana', *Plant Disease*, vol. 96, pp. 273–278.

Karamura, E. B. (1993) 'The strategic importance of bananas/plantains in Uganda', in C. S. Gold and B. Gemmill (eds), *Biological and Integrated Control of Highland Banana and Plantain Pests and Diseases*, IITA, Cotonou, Bénin, pp. 384–387.

Karamura, E. B. and Tinzaara, W. (2009) 'Management of Banana Xanthomonas wilt in East and Central Africa', *Proceedings of the Workshop on Review of the Strategy for the Management of Banana Xanthomonas Wilt*, Hotel la Palisse, Kigali, Rwanda, 23–27 July 2007, Bioversity International, Uganda, p. 88.

Karamura, E. B., Frison, E. A., Karamura, D. A. and Sharrock, S. (1999) 'Banana production systems in eastern and southern Africa', in C. Picq, E. Fouré and E. A. Frison (eds), *Bananas and Food Security*, INIBAP, Montpellier, pp. 401–412.

Kashaija, I. N., Speijer, P. R., Gold, C. S. and Gowen, S. R. (1994) 'Occurrence, distribution and abundance of plant parasitic nematodes on bananas in Uganda', *African Crop Science Journal*, vol. 2, pp. 99–104.

Kiggundu, A., Pillay, M., Viljoen, A., Gold, C., Tushemereirwe, W. and Kunert, K. (2003) 'Enhancing banana weevil (*Cosmopolites sordidus*) by genetic modification: a perspective', *African Journal of Biotechnology*, vol. 2, pp. 563–569.

Kovács, G., Sági, L., Jacon, G., Arinaitwe, G., Busogoro, J. P., Thiry, E., Strosse, H., Swennen, R. and Remy, S. (2012) 'Expression of a rice chitinase gene in transgenic banana ("Gros Michel", AAA genome group) confers resistance to black leaf streak disease', *Transgenic Research*, doi: 10.1007/s11248-012-9631-1.

Lepoint, P., Sibomana, R., Blomme, G. and Niyongere, C. (2011) 'Good cultural practices for banana bunchy top disease management: a sustainable option for Burundian smallholders?' *Bananas and Plantains: Toward Sustainable Global Production and Improved Uses*, Salvador, Bahia, Brazil, 10–14 October 2011, pp. 46–47.

Lescot, T. and Staver, C. (2010) 'Bananas, plantains and other species of Musaceae', in J. Fajardo, L. NeBambi, M. Larinde, C. Rosell, I. Barker, W. Roca and E. Chujoy (eds), *Quality Declared Plant Material,* FAO plant production and protection paper 195, pp. 126.

Lewis Ivey, L. M., Tusiime, G. and Miller, A. S. (2010) 'A polymerase chain reaction assay for the detection of *Xanthomonas campestris* pv. *musacearum* in banana', *Plant Disease*, vol. 94, pp. 109–114.

Lorenzen, J., Tenkouano, A., Bandyopadhyay, R., Vroh, B. I., Coyne, D. L. and Tripathi, L. (2009) 'The role of crop improvement in pest and disease management', *Acta Horticulturae*, vol. 828, pp. 305–314.

Magee, C. J. P. (1953) 'Some aspects of the bunchy top disease of banana and other *Musa* spp.', *Journal and Proceedings of the Royal Society of New South Wales*, vol. 87, pp. 3–18.

Mbaka, J., Ndungu, V. and Mwangi, M. (2007) 'Outbreak of Xanthomonas wilt (*Xanthomonas campestris* pv. *musacearum*) on banana in Kenya', *ISHS/ProMusa symposium: Recent Advances in Banana Crop Protection for Sustainable Production and Improved Livelihoods*, Greenway Woods Resort, White River, 10–14 September 2007, Bioversity International, Montpellier.

Mbwana, A. S. S. and Rukazambuga, N. D. T. M. (1999) 'Banana IPM in Tanzania', in E. Frison, C. S. Gold, E. B. Karamura and R. A. Sikora (eds) *Mobilizing IPM for Sustainable Banana Production in Africa: Proceedings of a Workshop on Banana IPM*, Nelspruit, South Africa, 23–28 November 1998, INIBAP, Montpellier, France, pp. 237–245.

Mobambo, K. N., Gauhl, F., Vuylsteke, D., Ortiz, R., Pasberg-Gauhl, C. and Swennen, R. (1993) 'Yield loss in plantain from Black Sigatoka leaf spot and field performance of resistant hybrids', *Field Crops Research*, vol. 35, pp. 35–42.

Mobambo, K. N., Zuofa, K., Gauhl, F., Adeniji, M. O. and Pasberg-Gauhl, C. (1994) 'Effect of soil fertility on host response to black leaf streak of plantain (*Musa* spp., AAB group) under traditional farming systems in southeastern Nigeria', *International Journal of Pest Management*, vol. 40, no. 1, pp. 75–80.

Mobambo, K. N., Gauhl, F., Swennen, R. and Pasberg-Gauhl, C. (1996) 'Assessment of the cropping cycle effects on black leaf streak severity and yield decline of plantain and plantain hybrids', *International Journal of Pest Management*, vol. 42, no. 1, pp. 1–7.

Mobambo, K. N., Gauhl, F., Pasberg-Gauhl, C., Swennen, R. and Staver, C. (2010) 'Factors influencing the development of black streak disease and the resulting yield loss in plantain in the humid forests of West and Central Africa', *Tree and Forestry Science and Biotechnology*, vol. 4 (Special Issue 1), pp. 47–51.

Ndabamenye, T., Van Asten, P. J. A., Vanhoudt, N., Blomme, G., Swennen, R., Annandale, J. G., and Barnard, R. O. (2012) 'Ecological characteristics influence farmer selection of on-farm plant density and bunch mass of low input East African Highland banana (*Musa* spp.) cropping systems, *Field Crops Research*, vol. 135, pp. 126–136.

Niyongere, C., Ateka, E. M., Losenge, T., Blomme, G. and Lepoint, P. (2011) 'Screening *Musa* genotypes for banana bunchy top disease resistance in Burundi', *Acta Horticulturae*, vol. 897, pp. 438–448.

Niyongere, C., Losenge, T., Ateka, E. M., Nkezabahizi, D., Blomme, G. and Lepoint, P. (2012) 'Occurrence and distribution of banana bunchy top disease in the Great Lakes Region of Africa', *Tree and Forestry Science and Biotechnology*, vol. 6, pp. 102–107.

Nkendah, R. and Akyeampong, E. (2003) 'Socioeconomic data on the plantain commodity chain in West and Central Africa', *InfoMusa*, vol. 12, pp. 8–13.

Okumu, M. O., van Asten, P. J. A., Kahangi, E., Okech, S. H., Jefwa, J. and Vanlauwe, B. (2011) 'Production gradients in smallholder banana (cv. Giant Cavendish) farms in Central Kenya', *Scientia Horticulturae*, vol. 127, pp. 475–481.

Ploetz, R. C., Herbert, J., Sebasigari, K., Hernandez, J. H., Pegg, K. G., Ventura, J. A. and Mayato, L. S. (1990) 'Importance of Fusarium wilt in different banana growing regions', in R. C. Ploetz (ed.), *Fusarium Wilt of Banana*, APS Press, St Paul, MN, pp. 13–21.

Plowright, R., Dusabe, J., Coyne, D. and Speijer, P. (2013) 'Analysis of the pathogenic variability and genetic diversity of the plant-parasitic nematode *Radopholus similis* on bananas', *Nematology*, 15, 41–56.

Price, N. S. (2006) 'The *Musa* burrowing nematode, *Radopholus similis* (Cobb) Thorne in the Lake Victoria region of East Africa: its introduction, spread and impact', *Nematology*, vol. 8, pp. 801–817.

Price, N. S. and Bridge, J. (1995) '*Pratylenchus goodeyi* (Nematoda: Pratylenchidae): a plant-parasitic nematode of the montane highlands of Africa', *Journal of African Zoology*, vol. 109, pp. 435–442.

Quénéhervé, P., Salmon, F., Topart, P. and Horry, J. P. (2009) 'Nematode resistance in bananas: screening results on some new *Mycosphaerella* resistant banana hybrids', *Euphytica*, vol. 165, pp. 123–136.

Reeder, R. H., Muhinyuza, J. B., Opolot, O., Aritua, V., Crozier, J. and Smith, J. (2007) 'Presence of banana bacterial wilt (*Xanthomonas campestris* pv. *musacearum*) in Rwanda: new disease reports', *Plant Pathology*, vol. 56, p. 1038.

Rukazambuga, N. D. T. M., Gold, C. S. and Gowen, S. R. (1998) 'Yield loss in East African highland banana (*Musa* spp., AAA-EA group) caused by the banana weevil, *Cosmopolites sordidus* Germar', *Crop Protection*, vol. 17, pp. 581–589.

Rutherford, M. A. (2001) 'Fusarium wilt of banana in East and Central Africa', in A. B. Molina, N. H. Masdek and K. W. Liew (eds), *Banana Fusarium Wilt Management: Towards Sustainable Cultivation*, INIBAP-ASPNET, Los Banos, p. 87.

Sarah, J. L. (2000) 'Burrowing nematode', in D. R. Jones (ed.), *Diseases of Banana, Abacá and Enset*, CAB International, Wallingford, pp. 295–303.

Schill, P., Afreh-Nuamah, K., Gold, C. S., Ulzen-Aprah, F., Paa Kwesi, E., Peprah, S. A. and Twumasi, J. K. (1997) *Farmers' Perception of Constraints in Plantain Production in Ghana*, International Institute of Tropical Agriculture, Ibadan, Nigeria.

Seshu Reddy, K. V., Prasad, J. S., Speijer, P. R., Sikora, R. A. and Coyne, D. L. (2007) 'Distribution of plant-parasitic nematodes on *Musa* in Kenya', *InfoMusa*, vol. 16, no. 1–2, pp. 18–23.

Simmonds, N. W. (1966) *Bananas. Second Edition*, Longman, London.

Smith, J. J., Jones, D. R., Karamura, E., Blomme, G. and Turyagyenda, F. L. (2008) 'Analysis of the risk from *Xanthomonas campestris* pv. *musacearum* to banana cultivation in Eastern, Central and Southern Africa', www.promusa.org/images/stories/infomusa/features/pdf/pra.pdf

Ssali, R. T., Nowakunda, K., Barekye, A., Erima, R., Batte, M. and Tushemereirwe, W. K. (2010) 'On-farm participatory evaluation of East African Highland banana "matooke" hybrids (*Musa* spp)', *Acta Horticulturae*, vol. 879, pp. 585–603.

Ssekiwoko, F., Taligoola, H. K. and Tushemereirwe, W. K. (2006) '*Xanthomonas campestris* pv *musacearum* host range in Uganda', *African Crop Science Journal*, vol. 14, pp. 111–120.

Stover, R. H. (1962) *Fusarial Wilt (Panama Disease) of Bananas and other Musa Species*, CAB International, Wallingford.

Stover, R. H. (2000) 'Diseases and other banana health problems in tropical Africa', *Acta Horticulturae*, vol. 540, pp. 311–317.

Su, H. J., Tsao, L. Y., Wu, M. L. and Hung, T. H. (2003) 'Biological and molecular categorisation of strains of banana bunchy top virus', *Journal of Phytopathology*, vol. 151, pp. 290–296.

Swennen, R. (1990) *Plantain Cultivation under West African Conditions: A Reference Manual*, IITA, Ibadan, Nigeria.

Tenkouano, A. and Swennen, R. (2004) 'Progress in breeding and delivering improved plantain and banana to African farmers', *Chronica Horticulturae*, vol. 44, no. 1, pp. 9–15.

Tenkouano, A., Vuylsteke, D., Okoro, J., Makumbi, D., Swennen, R. and Ortiz, R. (2003) 'Diploid banana hybrids TMB2x5105-1 and TMB2x9128-3 with good combining ability, resistance to Black Sigatoka and nematodes', *HortScience*, vol. 38, no. 3, pp. 468–472.

Tenkouano, A., Hauser, S., Coyne, D. and Coulibaly, O. (2006) 'Clean planting materials and management practices for sustained production of banana and plantain in Africa', *Chronica Horticulturae*, vol. 46, pp. 14–18.

Thwaites, R., Eden-Green, S. J. and Black, R. (2000) 'Diseases caused by bacteria', in D. R. Jones (ed.), *Diseases of Banana, Abaca and Enset*, CABI Publishing, Wallingford, pp. 213–239.

Tinzaara, W., Gold, C. S., Dicke, M., van Huis, A., Nankinga, C. M. and Kagezi, G. H. (2007) 'Pheromone-baited traps enhance dissemination of entomopathogenic fungi to control the banana weevil in Uganda', *Biocontrol Science and Technology*, vol. 17, pp. 111–124.

Tripathi, L., Mwangi, M., Abele, S., Aritua, V., Tushemereirwe, W. K. and Bandyopadhyay, R. (2009) 'Xanthomonas wilt: a threat to banana production in East and Central Africa', *Plant Disease*, vol. 93, no. 5, pp. 440–451.

Tripathi, L., Mwaka, H., Tripathi, J. N. and Tushemereirwe, W. K. (2010) 'Expression of sweet pepper *Hrap* gene in banana enhances resistance to *Xanthomonas campestris* pv. *Musacearum*', *Molecular Plant Pathology*, vol. 11, pp. 721–846.

Tushemereirwe, W., Kangire, A., Ssekiwoko, F., Offord, L. C., Crozier, J., Boa, E., Rutherford, M. and Smith, J. J. (2004) 'First report of *Xanthomonas campestris* pv. *musacearum* on banana in Uganda', *Plant Pathology*, vol. 53, p. 802.

van Asten, P. J. A., Gold, C. S., Okech, S. H., Gaidashova, S. V., Tushemereirwe, W. K. and De Waele, D. (2004) 'Actual and potential soil quality constraints in East African Highland banana systems and their relation with other yield loss factors', *Infomusa*, vol. 13, no. 2, pp. 20–25.

van Asten, P. J. A., Fermont, A. M. and Taulya, G. (2011) 'Drought is a major yield loss factor for rainfed East African highland banana', *Agricultural Water Management*, vol. 98, pp. 541–552.

Van den Berg, N., Berger, D. K., Hein, I., Birch, P. R. J., Wingfield, M. J. and Viljoen, A. (2007) 'Tolerance in banana to Fusarium wilt is associated with early up-regulation of cell wall-strengthening genes in the roots', *Molecular Plant Pathology*, vol. 8, pp. 333–341.

Van den Houwe, I., Lepoivre, P., Swennen, R., Frison, E. and Sharrock, S. (2003) 'The world banana heritage conserved in Belgium for the benefit of small-scale farmers in the tropics', *Plant Genetic Resources Newsletter*, vol. 135, pp. 18–23.

Wairegi, L. W. I., van Asten, P. J. A., Tenywa, M. M. and Bekunda, M. A. (2010) 'Abiotic constraints override biotic constraints in East African highland banana systems', *Field Crops Research*, vol. 117, pp. 146–153.

Yirgou, D. and Bradbury, J. F. (1974) 'A note on wilt of banana caused by the enset wilt organism Xanthomonas musacearum', *East African Agricultural and Forestry Journal*, vol. 40, pp. 111–114.

# 8 Challenges for the improvement of seed systems for vegetatively propagated crops in Eastern Africa

*J. J. Smith,*[1] *D. Coyne*[2] *and*
*E. Schulte-Geldermann*[3]

[1] *Food and Environment Research Agency*
[2] *International Institute of Tropical Agriculture*
[3] *International Potato Center*

## Overview of the relative values of vegetatively propagated crops

Vegetatively propagated crops (VPCs) include some of those most commonly cultivated in the region of Eastern Africa and are considered vital for food security. Most notable are *Musa* (banana and plantain), potato, cassava, sweet potato and, to a lesser extent, yam and *taro*. However, in review of this subject it is worth stating that while it may be useful to cluster these crops with respect of commonalities of propagation and in that they constitute key staple crops, substantial differences exist between them in terms of their optimum climatic needs, resilience to adverse factors, seasonality of production and added-value market potential. To appreciate the nature of these differences is as important as commenting on their similarities in trying to bring together a coherent analysis for progressing vegetative seed systems. Figure 8.1 provides an analysis of production and area data for these crops for Eastern African countries over the past 50 years.

These data are often used to infer importance and changes therein, with an increase normally taken as a positive trend. Similarly, references in the literature for consumption per capita and by other metrics attest to the importance of VPCs and the rising demand for these crops. Indeed, this is a strong thread to the justification for a Consortium Research Programme on Roots, Tubers and Banana (Consortium of International Agricultural Research Centers 2011). Table 8.1 draws comparisons of VPCs with the primary grain staples, wheat and maize, and estimates the percentage increase in production a single percentage increase in land leverages.

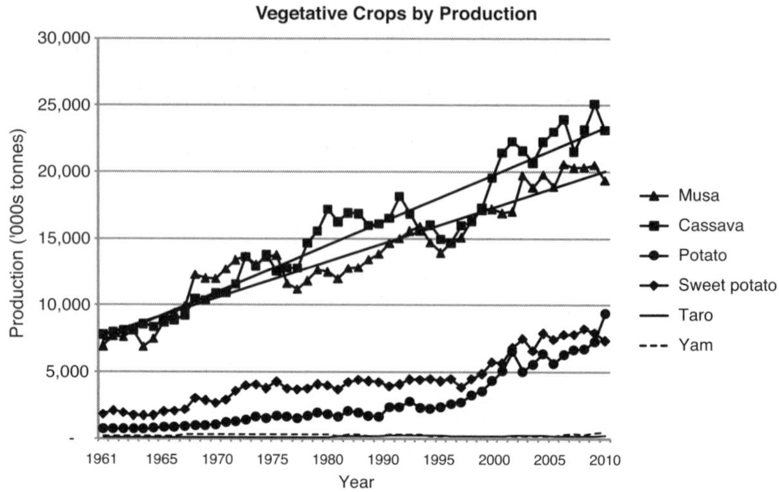


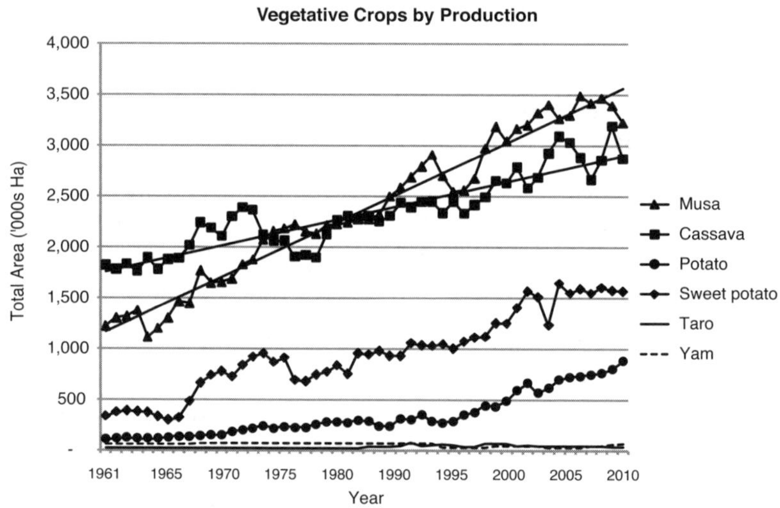


*Figure 8.1* Comparison of VPCs of Eastern Africa (source: FAO crop production data, 2009; listed East African countries only, excluding non-mainland countries).

Based on the data presented in Figure 8.1 and Table 8.1, some general observations are possible, the most striking being as follows:

- The level of cultivation of *taro* and yam is low, compared with other VPCs.
- Cassava and *Musa* have exchanged positions with respect to area; *Musa* has 'overtaken' cassava. However, this is not reflected in production data where cassava production exceeds that of *Musa* and there is evidence of the 'gap' widening.

*Table 8.1* Comparison of cassava and *Musa* as primary vegetative staples with maize and wheat

| *Crop* | *Area ('000s ha)** | | | *Production ('000 of tonnes)** | | | *Ratio %P:%A* |
|---|---|---|---|---|---|---|---|
| | *1960* | *2010* | *% increase* | *1960* | *2010* | *% increase* | |
| Maize | 5,250 | 12,250 | 133 | 5,000 | 18,000 | 260 | 1.95 |
| *Musa* | 1,100 | 3,800 | 245 | 500 | 2,500 | 400 | 1.63 |
| Cassava | 1,900 | 3,000 | 58 | 7,500 | 23,000 | 207 | 3.57 |
| Wheat | 1,000 | 1,700 | 70 | 8,000 | 20,000 | 150 | 2.14 |

*Note*: *Values estimated from trend lines generated from FAO crop production data (Figure 8.1).

- Potato and sweet potato production have increased markedly, disproportionately to area, over the past 10–15 years.
- Per unit area of cultivation, cassava has increased productivity more than wheat, *Musa* and maize over the past 50 years.

Moreover, it is equally important to note that crops such as potato and cassava occupy very different agricultural environments. Thus, it makes little sense to compare the productivity and value of these crops without taking into account the cropping system they occupy, the market opportunity that may exist and alternative crops farmers compare in deciding what to grow. Likewise, whereas yam and *taro* may not provide as much as banana and cassava, in the context of where these crops are cultivated they are frequently of an equal or greater value to the local community. In this type of analysis, we begin to question what is important and how success may look; how we measure a change in the cultivation of a crop as an indicator of a healthier, or less healthy, cropping system and an improved, or more vulnerable, position for the dependent community. Cassava provides an interesting example of the contrasting ways in which its increased cultivation and productivity may be viewed. The traditional values of cassava are as a food security crop, suitable for cultivation in the least favourable soils and climates. Accordingly, any increase in the area under cassava would reflect a 'declining' cropping system health position, unless countered by evidence (e.g. due to high market value) of a positive choice being made by farmers. From the data presented (Figure 8.1, Table 8.1), the latter scenario appears the more probable, with farmers increasingly cultivating cassava and realizing higher productivity, at the expense of *Musa*, which is experiencing a stagnant or declining level of productivity. Of late, cassava has received considerable institutional investment in the area of plant breeding and improved processing for added-value products. A further factor contributing towards cassava becoming more favoured may be associated with climate change, with less rain and higher temperatures being experienced in the region, as cassava is less sensitive to drought (Heath 2010).

## Characterization of VPC seed systems

For all VPCs, especially within Eastern Africa, the main seed requirement is met by non-commercial, informal seed systems. Strictly, the sale of seed must be by a seed producer, registered with the government, which submits to, and meets, a certification standard.

Non-certified seed producers effectively are farmers, farmers' cooperatives and non-governmental organizations (NGOs). These entities, in a phytosanitary policy sense, are producing material that is intended as suitable for planting, but cannot constitute the true meaning of the term 'seed'. But probably the main source of seed is from a farmer's practice of home-saving or obtained from a 'nearest-neighbour', where farmers hold back planting material from the current season's production. Inherent in all informal seed sources are issues of quality measured in terms of cultivar authenticity, pests, tuber size and readiness for germination. The cycling of VPCs over seasons and build-up of infection in planting material are frequently cited as the largest constraints to VPCs and the association with seed the main failing of the seed system. As an example, the impact of seed degeneration was shown in a recent study in Kenya, where seed potatoes sourced from two field multiplication generations increased yields by 2.6 times compared with farmers' seed and where every further field multiplication generation significantly reduced this yield advantage as pests accumulated (Schulte-Geldermann 2010). However, to some extent, farmers are successful at 'self-policing' in that they will seek seed material from outside the farm during a poor year and know of farmers that have experienced good yields from which to source. For farmers' cooperatives and NGOs, seed production may be expected to be more expansive, with periodic access to certified seed and greater vigilance displayed towards maintaining basic standards for bulking in servicing a larger catchment area with material suitable for planting.

Figure 8.2 depicts some phases in seed bulking, from breeder seed to the farm, that are broadly aligned with most VPC seed multiplication systems. Experience would suggest it is not as linear as Figure 8.2 suggests and a major frustration is that basic seed often fails to realize sufficient cycles of certified multiplication before entering the informal system.

### *What motivates a seed sector?*

At its simplest, a farmer will see the need to buy new seed from a seed sector only if that seed delivers something different and is otherwise unavailable, and a commercial seed sector will be sustained only if that demand for seed is of a scale that makes production a viable business. Thus, to secure this demand, farmers must be positively disposed to accessing:

- new cultivars that provide something different by way of yield or choice of market to exploit;

*Figure 8.2* Schematic of a breeder to farm VPC seed production pipeline, a seed system.

- healthy seed to replace the home-saved seed, which has become unproductive.

In considering these requirements both VPCs and true seed crops have been bred for many new releases, and there is nothing intrinsically different in offering these to the market. However, the need for renewal between true seed and vegetative seed is a more complex consideration. Two main factors contribute to the need for renewal of seed.

First, both vegetative and true seed are associated with pests; however, VPCs are more predisposed to high levels of pest burden due to the vegetative nature of propagation. The greater decline associated with pest carry-over in vegetative seed than true seed is well documented. Second, implicit in the terms of vegetative and true seed is the understanding that VPCs are genetically stable over generations, whereas true seed will experience genetic change over generations due to sexual segregation that is associated with phenotypic and performance change. This factor potentially supports a higher demand for renewal of true seed, although this may be a simplification when a farmers' preference to maintain landraces is considered. Farmers' decision-making for home-saving true seed will be influenced by many factors, including the desire for homogeneity or heterogeneity, level of mechanization in production, resilience and taste preferences. These disparities between renewal and genetics also overlook the issue that true seed is equally an option for VPC crops, yet vegetative propagation provides for greater establishment, genetic stability, vigour and higher yield than their true seed equivalent.

## Challenges facing vegetative seed

From the above analysis the main drivers for a VPC seed system are as follows.

- Vegetative seed becomes infected with pests over cycles of home-saving, or otherwise becomes unhealthy.
- Healthy vegetative seed outperforms true seed with respect to establishment and vigour, and also has the advantage of maintaining genetic uniformity.

Both of these factors should point to VPC seed being more vital than true seed. Some of the primary impediments to a vibrant vegetative seed sector in Africa, however, include the following.

- The traditional view of VPCs as a staple for home consumption and of low-value market foods. Consequently, the low value does not facilitate start-up costs such as for seed.
- Vegetative seed, compared with true seed, is bulky (e.g. potato) and often of short shelf life (e.g. sweet potato and yam vines); this creates challenges and expense over the timing of production and distribution.
- Multiplication rates are low compared with true seed, necessitating greater land and labour requirements.
- While the need for healthy seed is a primary driver for renewal, the practices around ensuring healthy seed production require comparatively sophisticated inspection and testing. Nascent seed sectors have frequently failed to secure a reliable level of quality associated with their seed.
- Except for a little success with potato, VPCs have yet to attract significant commercial investment as compared with true seed. Yet in many cases the commercial true seed distribution is an outreach of large companies that originate from outside Africa.

None of the above factors are mutually exclusive, and in general provide a bleak outlook for private sector investment for VPCs. Likewise, no single innovation or critical change is likely to lead directly to success, with a number of factors needing to come together.

The most evident feature is that VPC seed is considered expensive in relation to the value of the crop product, i.e. there is insufficient value associated with vegetative staples to support the price of seed. This has been true for starchy staple crops such as cassava and increasingly less so for potato, where the higher value is associated through processing to potato chips, for example, which command a better market. Potato has traditionally commanded a better price than other VPCs, reflecting to some degree a history of potato seed production. However, this cannot be entirely disentangled from the colonial influences that would have been familiar with either the value or culture of using seed potatoes. The critical changes needed are therefore based on the following:

- building a market value for the crop that better supports a seed cost;
- reducing the cost of seed;
- providing robust assurances over quality;
- ensuring local availability at planting time.

## Building markets for added-value products of vegetative crops

To review market influences on VPC seed aspects exceeds the remit of this chapter.

However, some brief observations are made.

- Increasingly, there is value associated with processing VPCs: e.g. chips/crisps, flour, starch, ethanol, etc.
- The cultivars best suited to such added-value products are not necessarily those preferred for home consumption. This identifies a role for breeders. It also identifies with an absolute requirement that the market for added-value products is real and low risk, as the option to use excess production for home use or local markets is by deduction unattractive.
- Added-value products need to attract a price premium for quality, which will be best achieved by planting healthy seed. This is true for crisps/chips, but also for other processed food products such as flour and, importantly, in relation to mycotoxins.

## Reducing costs of seed production

Innovation technology for basic seed is evident in many forms, with some of the more notable including aeroponics for mini-potato tubers and tissue culture (TC) for all VPCs. In priming the seed system pipeline, the opportunity these technologies present is well demonstrated. To cite a few examples:

- For potato, the International Potato Center (CIP) and its partners tested the use of aeroponics as a rapid multiplication technology that both dramatically lowered the cost of basic seed potato and increased availability six-fold in Kenya (Schulte-Geldermann 2010).
- TC was recently successfully used for cassava under the Great Lakes Cassava Initiative, whereby in 2011 approximately 40,000 plants of breeder clones were commercially multiplied using TC techniques over a 3–4-month period and distributed with the oversight of plant health authorities across six East African countries (Smith 2012).
- Yam multiplication has been increased by the use of vine cuttings (*R4D Review* 2011) instead of the traditional method of planting tubers; a method that has also been linked with lowering pest loads.

With all of these methods for rapid multiplication, the next stage is for planting out and bulking-up, ideally through certification and in meeting phytosanitary standards.

## Ensuring VPC seed quality and the policy of standards

The role of phytosanitary policy is critical for developing and maintaining sustainable healthy seed systems. However, the need for working at scale and meeting set standards provides a significant conundrum, especially if the standards are initially set too high or require sophisticated procedures such as laboratory diagnostics.

Leading the way for policy-based standards is potato, where the harmonization of standards is well advanced among members of the East African Community (EAC) with basic, breeder and certified seed classes (Generation 1, 2 and 3) being recognized. The EAC has identified similar efforts for harmonization with planting materials for cassava and banana among standards for true seed crops. These standards are comprehensive and address issues of authenticity, pest status and size, and the needs for labelling and traceability, all of which are critical to maintaining consumers' confidence. Each of these quality metrics presents specific challenges, some technical, as with pest detection, some more logistical, as with size and traceability, but all remain critical for their value realized.

The technical challenge regarding pest detection is perhaps the most steep, especially the need for laboratory testing. For viruses and bacteria it is evident that the standards set in plant health seed certification policies are somewhat aspirational. The major challenge in moving from breeder to basic to certified seed (see Figure 8.2), and in scaling up the volume of material associated with these levels, is to maintain pest levels within acceptable levels, i.e. below the set threshold. When these pest thresholds are set at zero, as with pathogens such as *Ralstonia solanacearum* for potato or cassava brown streak viruses for cassava, and the multiplication is, by virtue of the scale required, on farmers' land that may be adjacent to, or on, land with a history of these pests, then the implementation of testing is vital and the likelihood of rejection can be assumed to be high. In reality, most vegetative seed is not laboratory tested, or certified, undermining the authority of the legislation as the requirements are over-burdensome and the risk of rejection is too high. Most importantly, the setting of pest threshold levels that are too exacting is also a disincentive to investment in a seed sector since achieving a certification standard and a market premium comes with too high a risk. This is not to question the value of plant health policies, but to note that these must be evidence-based for the receiving environment they address. Policies can subsequently evolve to gradually tighten standards with successive achievements and increased progress.

Quality assurance for germplasm, as held by the Consultative Group for International Agricultural Research (CGIAR), presents a different situation, but is especially critical with respect to ensuring pest-free material when this is shared worldwide for breeding. The ideals for these standards are set out among the International Standards for Phytosanitary Measures of the International Plant Protection Convention and it is thus incumbent on UN treaty organizations, such as the CGIAR, to conform. Accordingly, across the CGIAR, assurances are provided that germplasm held is free from named pathogens. A particularly good example lies with CIP, which has received ISO 17025 accreditation for the germplasm held at Lima as free from named pathogens. The consequences of failing in this trust, in sharing diseased plant material with a partner in another country that was free of the pathogen, have obvious significance.

## Plant breeding investment profiles and valuing the role of landraces in the environment

There are many examples of success to support the importance of breeding. For the purpose of this review, only two areas are considered:

1 the post-release phase of a cultivar that contrasts the life cycle of a public good and a commercial product;
2 an appreciation of landraces and the need to maintain resilience within a landscape.

For VPCs, breeding is typically an undertaking of public funding in the delivery of public goods. With the exception of potato, there are no VPCs cultivated in Eastern Africa for which breeding rights pertain to a commercial entity, and even with potato these breeder rights are not collected. This reflects the particular commercial realities around VPC seed production, as discussed, but it also identifies an additional consequence which is significant in realizing the market potential of a cultivar developed as a public good as opposed to one developed by a commercial sector. The below sets out a principal difference in the investment cycle of a public good and a private sector entity, and presents an admittedly provocative assessment as to why a private sector approach provides a more durable outcome.

Investment for a public good is typically undertaken externally (by a donor) with a public sector entity (university, CGIAR, etc.) which pays for the development of the product (a cultivar). On delivery the product is made available and the contract is in most cases fulfilled; there has been no agreement for further investment to be ventured and both parties are content. The donor may subsequently decide to invest in the dissemination of the cultivar, and likewise the public sector entity may look for support to do the same. Often the partnership between the donor and the public sector entity is maintained. However, this investment is not commercial in the sense of realizing value, in maximizing the services the product may carry, but centred on establishing access to the cultivar by farmers, with the primary metric being numbers of farming households reached.

With the private sector, the situation is very different, with the commercial release of the product (e.g. a cultivar) realizing the point at which the company can begin to recover its outlaid investment. The business imperative thus dictates that the company is fully aligned in maximizing its warranty over the product, i.e. the cultivar should be purchased by farmers to an optimum level and be associated with a package of the company's services, akin to extension, thus making further profit. Intrinsic in this is the private sector's imperative to maintain the cultivar in the market for as long as possible.

A final comment on breeding and the intention of seed systems to replace existing cultivars is to recognize the value of landraces. Landrace is a term rarely used in developed countries. However, in the context of a developing country much greater significance is placed on specific characteristics missing from bred cultivars (but very much liked by consumers) and for the added resilience these lineages tend to possess to local edaphic and biotic factors. Despite these traits, landraces have not been commercialized, and their exploitation presents significant challenges over intellectual property, and consequently they do not form part of the rhetoric of a commercial sector. The risk is that a commercial seed sector will have no reason to value this resource. It is already evident with the success of cassava cultivars resistant to mosaic disease that a successful seed dissemination programme can lead to the displacement of landraces, leading to a more genetically uniform and vulnerable cropping system (Kizito *et al.* 2005; Smith and Tomlinson 2012). In our designs of breeding and seed system delivery we need to value landraces and the resilience they build. We also need to recognize that farmers will choose reliability in yield year-on-year, over the promise of a bumper yield when conditions are ideal. It is not sufficient to lament at the yield gap between high productivity trials on-station and the realities of the farm situation where a breeder's elite material struggles to perform. On-farm productivity has to be the determining factor.

## References

Consortium of International Agricultural Research Centers (2011) *Consortium Research Programme on Roots, Tubers, and Banana for food Security and Income, CRP RTB 3.4.* www.cgiarfund.org/sites/cgiarfund.org/files/Documents/PDF/fc5_CRP3_4_final_proposal_main_vol.pdf

Heath, T. (2010) *Kenya Climate Change Briefing*, Cranfield University Kenya. www.wsup.com/sharing/documents/Kenyaclimatechangesummary2010.pdf

Kizito, E. B., Bua, A., Fregene, M., Egwang, T., Gullberg, U. and Westerbergh, A. (2005) 'The effect of cassava mosaic disease on the genetic diversity of cassava in Uganda', *Euphytica*, vol. 146, pp. 45–54.

*R4D Review* (2011) 'Yam-growing technique'. http://r4dreview.org/2009/03/yam-growing-technique

Schulte-Geldermann, E. (2010) 'Effect of field multiplication generation on seed potato (*Solanum tuberosum*) quality in Kenya', Conference proceedings, 8th African Potato Association Conference, 5–9 December 2010, Cape Town, South Africa.

Smith, J. (2012) *Great Lakes Cassava Initiative: Building a Community of Practice in Mitigating Cassava Brown Streak Disease, GLCI: Final Report of the Food and Environment Research Agency*, Food and Environment Research Agency.

Smith, J. and Tomlinson, D. (2012) *A Review on Cassava Brown Streak Disease and Movement of Planting Material in the Great Lakes Region of East Africa: A Report for the Great Lakes Cassava Initiative*, Food and Environment Research Agency.

# Part II

# System integration

# 9 CIALCA's efforts on integrating farming system components and exploring related trade-offs

*P. van Asten,*[1] *B. Vanlauwe,*[2] *E. Ouma,*[1] *P. Pypers,*[2] *J. Van Damme,*[3] *G. Blomme,*[4] *P. Lepoint,*[5] *J. Ntamwira,*[6] *H. Bouwmeester,*[7] *E. Birachi,*[8] *L. Jassogne,*[3] *T. Muliele,*[9] *S. Bizimana,*[10] *A. Nibasumba,*[10] *S. Delstanche,*[3] *P. V. Baret,*[3] *J. Sanginga,*[11] *F. Bafunyembaka*[11] *and M. Manzekele*[6]

[1] *International Institute of Tropical Agriculture, Uganda;* [2] *Tropical Soil Biology and Fertility – International Center for Tropical Agriculture, Kenya;* [3] *Université Catholique de Louvain, Earth & Life Institute, Louvain-la-Neuve, Belgium;* [4] *Bioversity International, Uganda;* [5] *Bioversity International, Burundi;* [6] *Institut de l'Environnement et Recherches Agricoles/Consortium for Improving Agriculture-based Livelihoods in Central Africa, DRC;* [7] *International Institute of Tropical Agriculture, Tanzania;* [8] *International Center for Tropical Agriculture, Rwanda;* [9] *Institut de l'Environnement et Recherches Agricoles/Consortium for Improving Agriculture-based Livelihoods in Central Africa/UCL, DRC;* [10] *Institut de l'Environnement et Recherches Agricoles/Consortium for Improving Agriculture-based Livelihoods in Central Africa/UCL, Burundi, Belgium;* [11] *Tropical Soil Biology and Fertility – International Center for Tropical Agriculture, DRC*

## Introduction

The densely populated humid highlands of eastern DRC, Rwanda and Burundi are characterized by small farms (<1 ha), large families (seven people), few livestock (0.4 tropical livestock unit (TLU)) and a large dependence on a few staple food crops, such as cassava, banana and beans. Farmers wish to improve their food security and income, but have limited resources to achieve this. CIALCA has made an effort to increase resource use efficiency by improving interactions between major crop components. The aim is to achieve triple-wins of improved (1) productivity/profitability, (2) sustainability and resilience and (3) adoptability by smallholders, taking into account resource constraints and gender dimensions. For the annual crops, new planting arrangements have been developed for cassava–legume and maize–legume systems. When combined with judicious applications of mineral fertilizers and organic matter inputs, production and income are often doubled. However, the 'best' combination depends on the agro-ecological conditions and production objectives. For the perennially based systems, banana–legume and banana–coffee, intercropping can provide large agronomic and economic benefits to smallholders while minimizing risks and improving system resilience. Soil and water conservation measures often do not offer short-term benefit, making adoption unattractive for resource-poor and risk-averse farmers. Several integrated technologies are highly beneficial but knowledge-intensive and context-specific. The 'Green Revolution' in the humid highlands does not need to follow a 'traditional' mechanized, large-scale, high-input approach. However, achieving impact will require a more favourable environment in policy, extension and market that encourages the development of locally owned solutions adapted to the constraints and opportunities of the farming systems.

The East African highlands are some of the most densely populated areas in Africa, exceeding 500 people per square kilometre in central Rwanda. The high population density probably stems from the rather favourable ecological conditions. The medium to high altitude (generally 800–2,300 masl) and position around the equator provide a moderate average annual temperature (16–24°C) and fairly well distributed (i.e. bi-modal) rainfall (800–2,000 mm/yr). Soils near the Albertine rift are often relatively fertile when derived from volcanic or young metamorphic parent rock. The cooler temperatures reduce the mineralization of soil organic matter, leading to better soil fertility compared with that in the more hot and arid lowlands. However, soil fertility remains poor in large areas of central Rwanda, central Burundi and eastern DRC due to the silica-rich parent rock (i.e. quartzite and granite) that provides few minerals upon weathering. High population density leads to very small farms (average <1 ha) on which large families live (seven people) with few livestock (0.4 TLU – the exception is Rwanda with >0.8 TLU per family, on average,) and high illiteracy rates of up to 60 per cent (Ouma and Birachu 2011). Limited (land) resources are likely to have fuelled the civil conflicts that have had a serious impact on much of the region over the past decades. The combination of small farms, large families and civil conflict has led to high levels of food insecurity, particularly in areas such as Sud Kivu and much of Burundi, where over 60 per cent of the families interviewed during the CIALCA baseline survey in 2006 (Ouma and Birachu 2011) indicated that food quantities were insufficient at least for part of the year. Despite the high pressure on land, the majority of the population (>70 per cent) still live in the rural areas. These areas are increasingly required to produce surplus food to feed the rapidly growing urban populations. Climate change will put further pressure on the traditional smallholder systems. The farming systems in this area are classified as highland perennial systems (Dixon *et al.* 2001) with banana and coffee as important food and cash crops, in combination with roots and tubers (cassava, sweet potato and Irish potato at higher altitudes), cereals (maize and sorghum) and legumes (bush and climbing beans) in mixed systems. Livestock are mostly limited to zero-grazing and are dominated by small ruminants. Yields of most food and cash crops are only one-third or less of what is achieved elsewhere in the world in similar ecological conditions, largely due to the non-use of external inputs. To encourage farmers to produce and market surplus food, public and private sector actors have increasingly taken a commodity chain approach. However, farmers often do not consider profit maximization or resource use efficiency as a primary target. Their production systems are composed of diverse crop and livestock components to minimize risks and satisfy a broad set of needs (i.e. social, economic) within their resource constraints. Additional resources acquired are often not reinvested into agriculture. Farmers prefer to invest these in the education of their children or off-farm income opportunities. CIALCA tried to understand the drivers of technology adoption for smallholders by looking at livelihood needs and resource constraints (capital, land, knowledge, among others). The objectives of this chapter are to explain and illustrate CIALCA's farming systems research approach by (1) characterizing and understanding the current (diversity in) farming systems in the East African highlands, (2) illustrating some important interactions of key farming system components and exploring how technological efficiencies can be improved within the existing and future resource constraints and (3) discussing the key lessons learned from our experiences in farming systems research and how this research can be taken forward in the future.

## Materials and methods

We broadly explain the tools used to characterize the farming systems in the CIALCA highland region. We will also explain how we evaluate technological opportunities from a farming systems perspective.

### *Characterizing farming systems in the East African highlands*

A large baseline survey (>2,500 farms) was conducted in 27 CIALCA action sites located in the highland areas of eastern DRC (i.e. Nord and Sud Kivu), Rwanda and Burundi in 2006. To distinguish characteristic patterns of livelihood strategies existing among the surveyed households, a clustering method was applied to some of the primary variables following Gockowski and Baker (1996). This allowed us to distinguish major farm typologies. Details are available in the report of Ouma and Birachu (2011). Besides the baseline survey, a series of 'qualitative' studies were conducted on understanding the diversity and dynamics of the farming system in the region – details are available in the theses of Julie Van Damme, Damien de Bouver and Paul Cox, available on the CIALCA website (www.cialca.org).

### *Exploring opportunities to improve farming systems performance*

Great diversity in crop performance and cropping systems exists among regions, sites and farms. We've tried to use the diversity in farming system practices to pick out integrated management practices that appear promising in terms of productivity, profitability and sustainability. Rather than evaluating single enterprise technologies (e.g. improved germplasm, integrated pest management (IPM), fertilizer), we've tried to look at the components of integrated crop and resource management that are traditionally not considered by the agricultural research and extension bodies. The latter are generally dominantly organized along single commodities and value chains. In this chapter, we wish to deviate from this single commodity approach and want to explore options that require a broader 'farming systems' approach. Entry points that CIALCA worked on are (1) mixed legume–maize/cassava cropping systems, (2) banana–legume systems, (3) coffee–banana systems and (4) erosion control systems, integrating bunding, forage and annual crop components. Rating the suitability of these 'novel' integrated technologies was done by evaluating the potential of technologies to (1) *increase efficiencies*, such as returns to (a) land, (b) labour and (c) external input investments (e.g. fertilizers), (2) *improve system resilience and sustainability*, including (a) maintaining and improving the natural resource base, (b) adapting to and mitigating the effects of climate change, (c) reducing pest and disease pressure in the system, and (3) *responding to the needs and constraints of smallholder livelihoods*, including the ability of the 'technology' to (a) increase income and food, (b) reduce risks and (c) generate (more) equitable benefits across gender, age and social class. As we integrate farm system components, the ability of the novel technology packages tested to have a positive impact on the three criteria listed above may not always be universally positive or negative. Trade-offs and synergies will be explored and described.

## Results and discussion

### *Farming systems characterization*

#### *Land, livestock and crops*

Households' land holdings varied greatly across the region. Sud Kivu, Gitega and Kirundo had the lowest land sizes with a median of 0.4 ha, whereas eastern Rwanda had the largest median value of 1.6 ha. The land cultivated is generally owned, with less than 20 per cent being rented for annual crop production. Only in Sud Kivu, over 30 per cent of the fields were rented from the traditional chief. Soil protection measures such as erosion control and the planting of perennials are not practised on rented land. Livestock numbers were very low (0.2–0.5 TLU) and these were local breeds of small ruminants. The exception was Rwanda, where averages ranged from 0.8 to 1.7 TLU following government efforts to have at least one cow at every farm. Whereas livestock in the past were often free grazing, this practice has become rare and most livestock are kept within the farm and fed with locally produced forage and crop residues. This has reduced nutrient flows from communal grazing land to the cropped land. Nonetheless, fertilizer use was virtually absent, with the exception of Gitega (53 per cent of farmers), Gitarama (21 per cent) and the Rusizi plain (17 per cent). The dominant food crops were banana, beans and cassava, followed by sorghum, sweet potato and maize. At the higher altitudes (>1,800 masl) Irish potato, climbing beans and wheat were more important. Traditional cash crops, such as coffee, were widespread, but generally considered a 'government crop' with few benefits for farmers.

### *Technology adoption and access to capital*

Access to improved technologies was generally poor, though this varied by crop and region. Less than 20 per cent of the farmers had improved banana varieties. For bush and climbing beans, over 60 per cent of the farmers in Burundi but fewer than 15 per cent of the farmers in Rwanda had improved germplasm. Poor adoption may be partly related to the weak extension systems in the region and the relatively low social capital of the farmers; i.e. fewer than 45 per cent are part of a farmers' group. Technology adoption is often driven by capital availability and risks profile. The major sources of financial capital were off-farm income and financial credit. Estimated off-farm income/household varied between \$38 and \$849, while small trade generated \$57 to \$308 annually. Formal credit systems including banks and micro-finance institutions were virtually absent with the exception of Rwanda (10–22 per cent of households).

### *Farm typologies*

We observed that there are regional differences in farm characteristics and there are generally also large differences within the sites, following patterns that are found throughout the region. Using principal component analysis (PCA) and clustering tools, three major farm typologies (i.e. clusters) were identified. Cluster 1 could be classified as the 'resource-rich entrepreneurs', representing only 4 per cent of the households. Households in this cluster had access to capital through (informal) credit and/or from a relatively high farm income. They also had at least 1 TLU, but their farm size was often just below average. Their success could be attributable to their relatively close proximity to markets and basic amenities, allowing the commercialization of agricultural produce and engagement in off-farm activities. Due to their status, general household food sufficiency was highest in this cluster. Proportionally more farmers belonged to this cluster in Gitarama (Rwanda) and Cibitoke (Burundi). Due to their access to capital and markets, these farmers are generally more inclined to adopt new technologies that increase productivity. Cluster 2 could be classified as the 'resource constrained' and contained the majority of households (72 per cent). These farmers had the least physical, natural, social and financial capital and household food sufficiency is most problematic, despite their engagement in selling fresh food produce. Livelihood improvements of this cluster require a multi-faceted approach aimed at the development of capital and the improvement of agricultural productivity and household welfare. Most households in Burundi and the DRC belonged to this cluster. Farmers may often be risk-averse and their weak capital status makes it difficult for them to adopt new technologies. Cluster 3 contained 24 per cent of the households. They could be classified as the 'naturally resource-rich' with relatively large land holdings, TLU and physical capital. However, they are often remotely located, being far from basic amenities. The reasons why these land-abundant and livestock-owning households may not be focused on production for the market may be market inaccessibility as well as a lack of access to productivity-enhancing technologies. Strategies aimed at reducing the associated transaction costs, such as the formation of cooperatives and collective marketing, may build their entrepreneurial ability. A relatively high proportion of households from Nord Kivu and eastern Rwanda belonged to this cluster. We observed no clear differences in terms of human capital across the three clusters as most household members received only primary education.

### *Farm management and productivity*

Various crop and livestock units have different requirements in terms of land, labour and nutrient investments. Farmers often try to optimize the interaction (i.e. find synergies) between the various crop and livestock units through (intuitive) spatial and temporal planning of their activities. Homestead fields or gardens receive more nutrient inputs from the nearby livestock and from kitchen residues. At a medium distance, annual crops, such as beans and maize, are cultivated. In the far-out (often rented) fields in the upland, cassava is often cultivated whereas the lowlands are often used for sweet potato, vegetables, rice or cash crops such as sugar cane. The relative importance and management of the various crop and livestock units vary strongly but a few trends are common throughout the region:

1. Perennial crops are generally dominant close to the homestead where soil fertility is higher. Soil fertility declines sharply after the first 50 m away from the homestead (see Figure 9.1).
2. Livestock is dominantly zero-grazing and located close to the homestead, particularly the small ruminants. In some sites, limited grazing areas for livestock are available outside the 'village'.
3. The relative importance of livestock has decreased over the past decades due to reduced farm sizes and civil conflict (Cox 2012). Farmers no longer consider cattle to be an engine for the management of soil fertility in cropped fields.

4 Medium-sized farms of around 0.8 ha generally produced more value per unit area (i.e. annual produce per hectare converted to monetary value using local farm-gate prices) than the very small farms (<0.5 ha) or relatively larger farms (>2.0 ha) (Figure 9.2). We hypothesize that this is likely to be due to the fact that small farms 'sell' their labour and crop residues (as fodder) to more wealthy neighbours. Hence, their labour and nutrient investments per unit area may turn out to be smaller than that achieved by medium-sized farmers who spend most of their labour on the farm and recycle nutrients as much as possible. The larger farms generally had more TLU and labour available, but in proportion to the amount of land this may be less. Similar observations were made by Tittonell *et al.* (2010) in the highlands of western Kenya and by Okumu *et al.* (2011) on banana systems in central Kenya. Average yields of most crops are <30 per cent of the best on-farm yields in the region.

### *Evaluating opportunities to improve farming systems performance*

In this section we briefly explore and evaluate opportunities for improved farming systems performance based on the 'smarter' integration of various farming system components. These options will then be compared on the basis

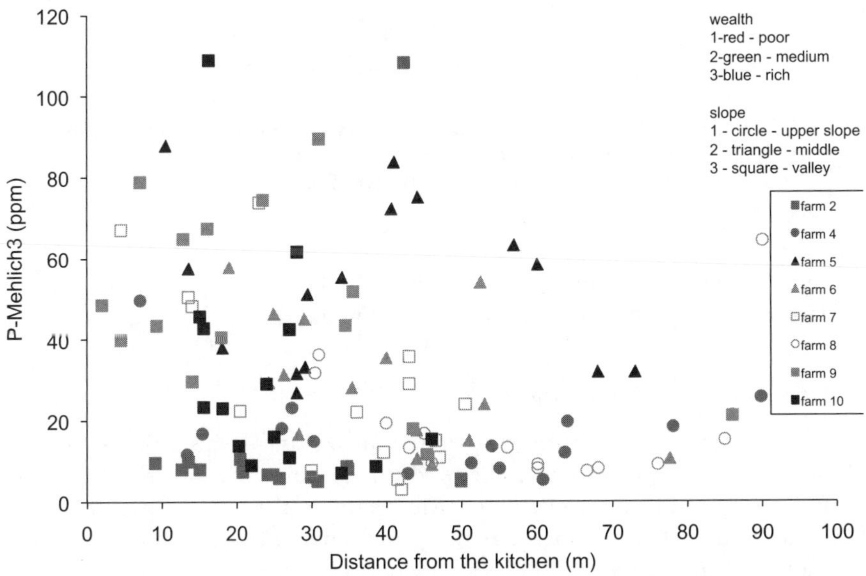


*Figure 9.1* Typical soil fertility gradient from a smallholder farm in Burhale (east DRC), with kitchen and livestock residue input being relatively large near the homestead compared with further away. The graph depicts real nutrient (P-Melich3) gradients as measured in ten farms in Ntungamo (southwest Uganda) (van Asten, unpublished).

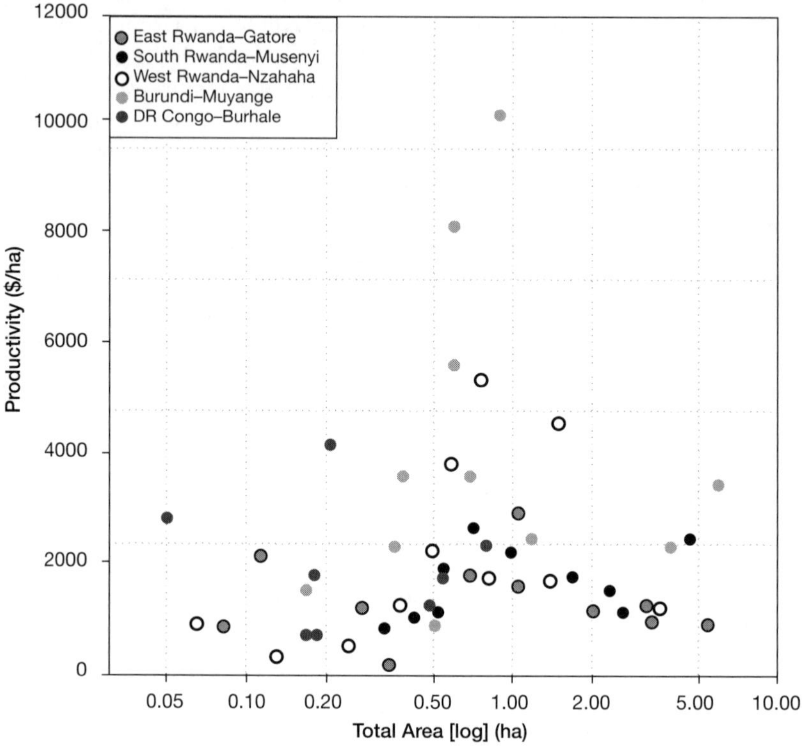


*Figure 9.2* The value of farm produce/ha/year (y-axis) is highest in the medium-sized farms (0.8 ha – x-axis is farm area) as opposed to small (<0.5 ha) and 'large' farms (2.0 ha) (source: De Bouver 2011).

of their technical efficiency, system sustainability/resilience and ability to respond to farmers' needs and resource constraints.

### *Mixed legume–maize/cassava crop systems*

Farmers across the region often intercrop maize and beans (*Zea mays* with *Phaseolus vulgaris*), as this offers them advantages in terms of risk avoidance, crop diversity and land use efficiency. Most farmers plant the beans in between their maize lines (75 cm apart). In western Kenya, a system called *mbili* was developed (Woomer *et al.* 2004) in which a staggered arrangement was used with two rows of maize closer together (0.5 m) and a larger distance (1.0 m) between the double rows of maize, which was then used to plant two rows of beans. When this spacing was tested in Sud Kivu, yields of beans were increased by up to 10 per cent and of maize by up to 20 per cent when compared with yields from the traditional planting arrangements. When combined with

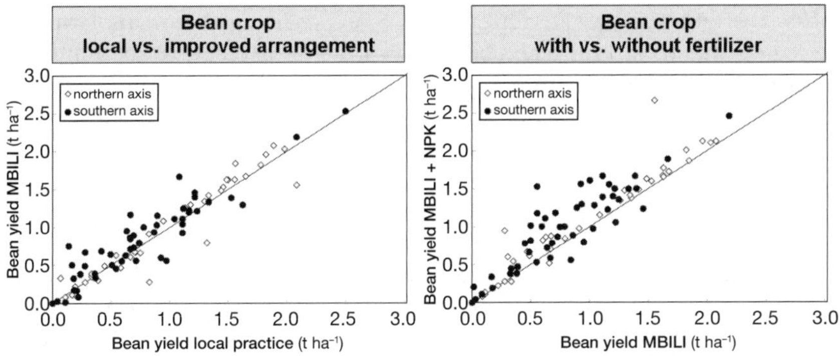


**% yield increase relative to the traditional practice (averaged across all participants)**

| | Legume | Maize |
|---|---|---|
| MBILI | 0-10% | 5-20% |
| MBILI + Fertilizer | 15-50% | 15-60% |

*Figure 9.3* Improved spatial arrangement of bean and maize mixed systems (*mbili*) increases yields (y-axis) marginally (<20 per cent) compared with yields from the traditional planting arrangement (x-axis). When combined with fertilizer, yield increments for both legume and maize can be achieved (table) as found in on-farm trials in Sud Kivu (DRC).

modest amounts of NPK fertilizer application (100 kg/ha of NPK-17-17-17), yields were further increased by 15–60 per cent (Figure 9.3). Similarly, an improved spatial arrangement for the cassava–legume intercrop was tested in combination with modest fertilizer applications (150 kg/ha of NPK-17-17-17). Traditional 'random' planting arrangements were compared with cassava planting arrangements of 1.0 × 1.0 m and 0.5 × 2.0 m. In between the lines, four lines of bush beans or groundnut at 0.4 m distance were planted during the first rainy season, and during the second season two lines of soybean were planted. The 1.0 × 1.0 m arrangement did not improve yields of either crop much (<10 per cent), but the addition of fertilizer did improve yields (15–40 per cent) for both the cassava and legume intercrop. In the 0.5 × 2.0 m cassava arrangement, yields were increased by 40–60 per cent for the legume intercrop and by 20–35 per cent for the cassava, as compared with the traditional practice. Marginal rates of returns to fertilizer investments varied between 1.6 and 2.7 (Pypers *et al.* 2011). Poor soils did not respond very well to fertilizer input. Generally, we observe that substantial productivity improvements can be achieved when combining improved spatial arrangements with modest fertilizer inputs. Unfortunately, the response is not homogeneous across soil types, and vulnerable farmers on poor soils are particularly exposed to poor returns on investment. Although mixed cropping systems improve system resilience, the enhanced productivity also puts more pressure on soil nutrient stocks. Last but not least, fertilizer use is a real challenge for many farmers in the region due to lack of technical know-how, poor access to the inputs and relatively high prices.

*Banana–legume intercropping*

Bush beans are the legume most commonly grown under banana in the region. They are a means for the family (particularly women) to provide the much-needed proteins. Most farmers till the soil at the start of the wet season to prepare bean intercropping, thereby damaging the superficial banana root system. In southwestern Uganda, farmers often practise a zero-tillage system with mulch application to suppress weeds and enhance banana growth. Legume intercrops are then planted in single planting holes (made by pushing a stick into the ground) in the mulched non-tilled plots. Within CIALCA we wished to (1) explore if this practice would result in a better system performance and (2) find out to what extent the banana canopy could be trimmed to reduce the shade levels and allow more light-demanding legumes, such as soybean and climbing beans, to be intercropped. Trials in Sud Kivu (DRC) and Burundi revealed that zero-tillage and mulch systems indeed increased the performance of banana substantially (see Chapter 4). However, whereas zero-tillage and

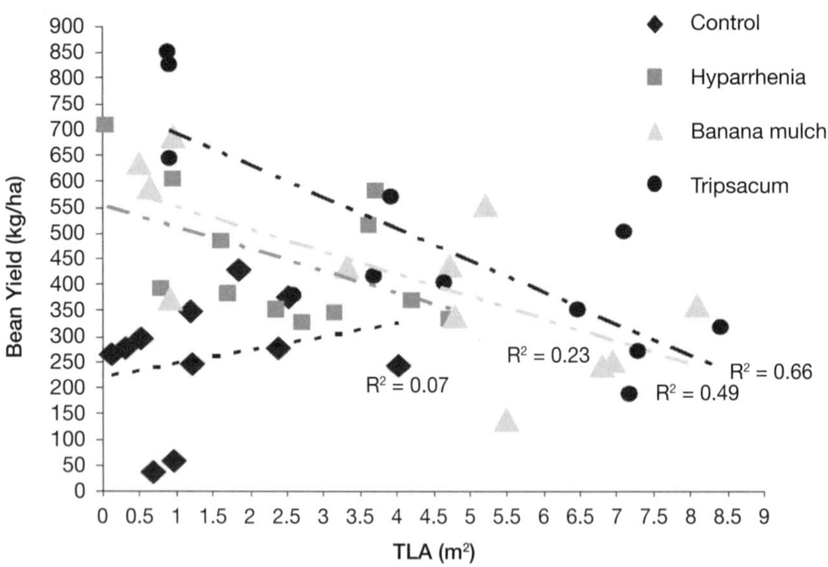


*Figure 9.4* Bush bean yields (*Phaseolus vulgaris*) initially increase with the application of mulch, but at high banana canopy levels (total leaf area/plant > 4 m$^2$), bean yields tend to be negatively affected by the enhanced performance of banana. Data presented from trials in Gitega (Burundi).

mulching initially improved yields, enhanced performance by banana increased the total leaf area (TLA) to >4 $m^2$/plant (leaf area index (LAI) > 0.67), which had a negative impact on bush bean yields (see Figure 9.4). Pruning the banana canopy to seven, or even four leaves strongly increased yields of the legume intercrops. Yields of soybean and climbing beans increased over 300 per cent from yields when planted to unpruned banana. Yields of bush beans increased by only 100 per cent at the maximum pruning levels. Despite their low average yields (<200 kg/ha) farmers prefer this variety since its yields are relatively little affected by shading by banana. Intercrop system improvements will require more investments in knowledge and labour. Since intercropped legumes are generally managed by women and banana by men, it's imperative to take the gender dimensions into account when recommending these systems to farmers.

### *Coffee–banana systems*

When Arabica coffee was first introduced in Rwanda and Burundi during the colonial era (1940s–1950s) farmers were obliged to mono-crop a few hundred coffee plants. In Tanzania farmers were encouraged to plant coffee in their banana gardens and in Uganda mono-cropping was recommended but not enforced. CIALCA explored to what extent the coffee–banana intercropping system provides benefits to farmers in the region. Research in Uganda revealed that the intercropped field generated much more revenue (>50 per cent) than mono-cropped banana or coffee (van Asten *et al.* 2011). Follow-up surveys in Burundi and Rwanda appeared to confirm these trends. In Burundi, Nibasumba *et al.* (2011) showed that coffee shaded by banana at the border of the coffee plot did not yield less than mono-cropped coffee in the middle of the plot. Likewise, a recent unpublished study in Rwanda in the framework of CIALCA seems to confirm that coffee intercropped with banana in eastern and southern Rwanda is performing equally as well as mono-cropped coffee. Jassogne *et al.* (2012) asked coffee stakeholders along the value chain in Uganda about the perceived benefits and constraints of the intercropping system. All actors agreed that the intercrop systems were beneficial for the farmers in that they provided more total food and income from limited land. In addition, the shaded coffee appeared less sensitive to climate shocks (e.g. drought, hail storms) and produced larger 'higher quality' cherries. Male farmers indicated that banana intercropping (for food) encouraged their women to help maintain the coffee fields, possibly putting additional pressure on women when this practice is promoted. Soil fertility was perceived as a key constraint, as it was difficult to maintain a productive intercrop system when soil quality was poor. On the positive side, leaf rust appeared to have a much lower (50 per cent) incidence in intercropped coffee than when mono-cropped. Generally, the intercrop practice seems to provide a real opportunity to improve the productivity and resilience of the coffee systems, but recommendations need to be fine-tuned based on ecological conditions and gender implications.

### *Erosion control systems, integrating bunding, forage and annual crop components*

The hilly landscape of the East African highlands is prone to erosion, and this is the primary reason why erosion control measures are encouraged (and sometimes enforced) in Rwanda and Burundi. Various erosion control techniques are recommended, including contour bunding, the planting of hedgerows along the contour lines and zero-tillage practices. CIALCA executed a long-term trial to compare these practices in terms of their ability to control erosion and provide farmers with the benefits of improved food and forage production. A full factorial design was tested with and without bunding, tillage and the presence of *Calliandra* hedgerows. In terms of their capacity to reduce erosion, the embankments (i.e. contour bunding) appeared very effective, with erosion dropping from 30 to 10 kg/m$^2$ after four years. Tillage had virtually no impact on erosion. *Calliandra* hedgerows decreased erosion by roughly one-third only when combined with embankments. Unfortunately, the embankments had a large negative impact on crop yields, with soybean yields in the first year dropping from over 600 kg/ha to less than 400 kg/ha with embankments. After seven seasons, the soybean yields in the embanked treatments recovered to some degree, but still did not match the treatments without embankments (Figure 9.5). The no-tillage and *Calliandra* practice consistently performed less well in terms of crop yield than treatments with tillage or without *Calliandra*. The findings suggest that not all of the erosion control practices recommended are effective in reducing erosion (e.g. zero-tillage). Moreover, none of the practices led to yield improvements in the short to medium term, making it difficult to convince farmers to invest their precious labour in these practices.

## Conclusion

The farming systems in the East African highlands are diverse, but are generally characterized by their small farm size (<2 ha), low yields (<30 per cent of those attainable) and subsistence orientation with pervasive food insufficiency particularly present in Sud Kivu and Burundi (>60 per cent in 2006). Although soil fertility is a major production constraint, external inputs such as fertilizers and improved seeds are rarely used. Nutrients recycled by the few livestock are minimal and communal grazing land has largely disappeared. Agricultural intensification is required to feed the rapidly growing population in this land-limited region. This will require the use of external nutrient inputs and improved germplasm. However, our findings suggest that we do not have to copy the model of the Green Revolution as used elsewhere in the world, which was largely based on mono-crop practices, high external input use and generally an increase in scale through land consolidation and mechanization. In fact, the evidence collected thus far seems to suggest that the medium-sized smallholder farm (±1 ha) can be quite productive through the optimal use of land, nutrients

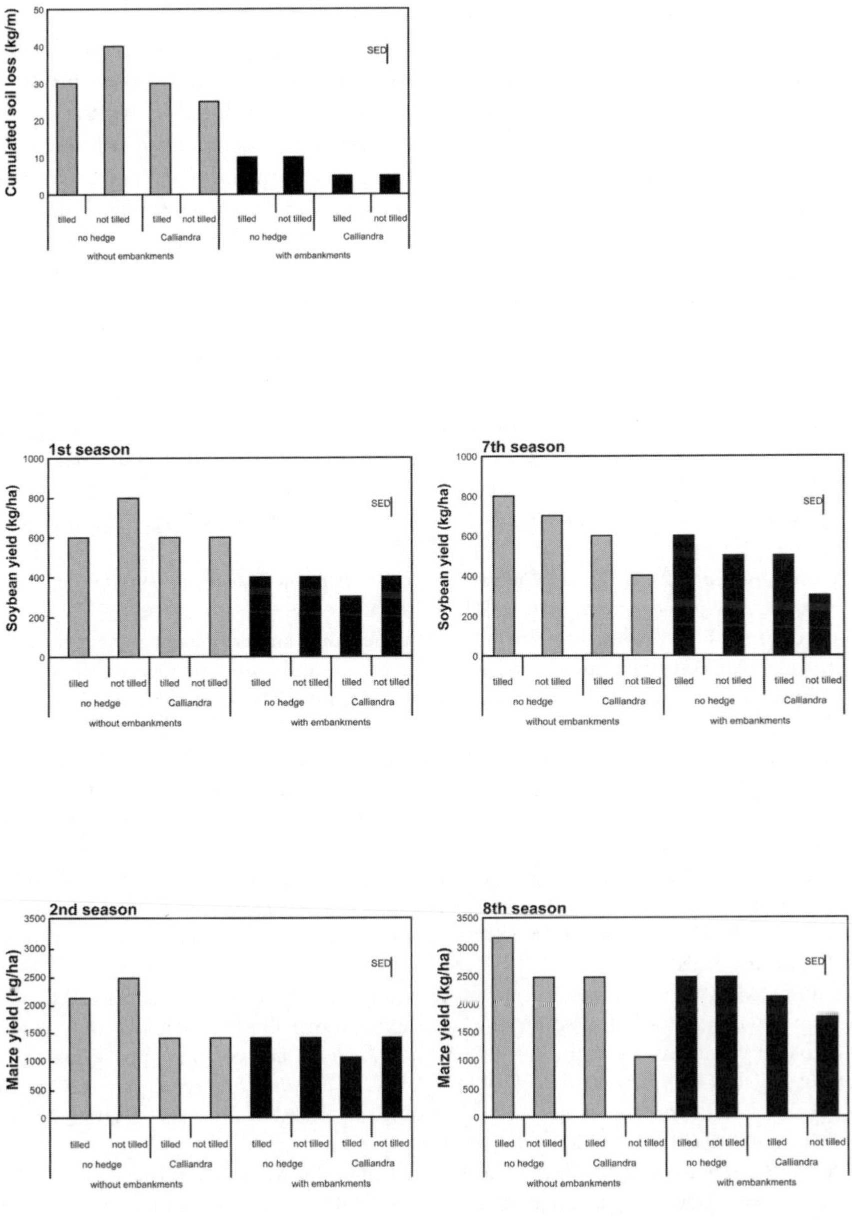


*Figure 9.5* Erosion control trials in Sud Kivu (DRC) reveal that contour embankments, *Calliandra* contour hedgerows and zero-tillage practices do not have a positive impact on yields of soybean and maize in the short (first/second season) or medium term (seventh/eighth season), despite a marked decrease in culminated soil loss over eight seasons when embankments are used.

and labour. These systems can further be improved by optimizing the spatial arrangement of mixed cropping systems, integrated with improved soil fertility management, where possible. Top-down encouragement of erosion control may need some further thoughts on how to offset short-term losses in productivity with long-term soil sustainability. To develop suitable technologies, farming systems analysis is required, including an understanding of resource constraints, risk behaviour and opportunities and constraints to input/output markets. Translating this knowledge into extension messages for farmers may prove to be a challenge due to the complex nature of the interactions. Co-generation of knowledge with farmers is a key means to create ownership in the decision-making on the adoption of new technologies.

## References

Cox, T. P. (2012) 'Farming the battlefield: the meanings of war, cattle and soil in South Kivu, Democratic Republic of the Congo', *Disasters,* vol. 36, pp. 233–248.

De Bouver, D. (2011) 'Diagnostic des systèmes agraires s'appuyant sur la culture de la banana dans la région des Grands Lacs Africaines', MSc thesis, UCL, Louvain-la Neuve, Belgium.

Dixon, J. A., Gibbon, D. P. and Gulliver, A. (2001) *Farming Systems and Poverty: Improving Farmers' Livelihoods in a Changing World*, FAO, Rome.

Gockowski, J. J. and Baker, D. (1996) 'Targeting resource and crop management research through an eco-regional methodology for the humid forest of Central and West Africa', Paper presented at 1996 Biennial Meeting of Rockefeller Social Science Research Fellows, 15 August 1996, Nairobi, Kenya.

Jassogne, L., van Asten, P. J. A., Wanyama, I. and Baret, P. V. (2012) 'Perceptions and outlook on intercropping coffee with banana as an opportunity for smallholder coffee farmers in Uganda', *International Journal of Agricultural Sustainability*, doi: 10.1080/14735903.2012.714576

Nibasumba, A., Baret, P., Jassogne, L. and van Asten, P. (2011) 'Plant and soil interactions between adjunct coffee and banana plots in Burundi: preliminary assessment in farmer field conditions', Poster presentation at CIALCA conference, October 2011, Kigali, Rwanda.

Okumu, M. O., van Asten, P. J. A., Kahangi, E., Okech, S. H., Jefwa, J. and Vanlauwe, B. (2011) 'Production gradients in smallholder banana (cv. Giant Cavendish) farms in Central Kenya', *Scientia Horticulturae*, vol. 127, pp. 475–481.

Ouma, E. A. and Birachu, E. (2011) *CIALCA Baseline Survey Report.* CIALCA Technical Report series no. 17. Available at: www.cialca.org/files/files/cialca%20baseline%20survey%20report_print.pdf

Pypers, P., Sanginga, J. M., Kasereka, B., Walangululu, M. and Vanlauwe, B. (2011) 'Increased productivity through integrated soil fertility management in cassava–legume intercropping systems in the highlands of Sud-Kivu, DR Congo', *Field Crops Research*, vol. 120, pp. 76–85.

Tittonell, P., Muriuki, A., Shepherd, K. D., Mugendif, D., Kaizzi, K. C., Okeyo, J., Verchot, L., Coe, R. and Vanlauwe, B. (2010) 'The diversity of rural livelihoods and their influence on soil fertility in agricultural systems of East Africa: a typology of smallholder farms', *Agricultural Systems*, vol. 103, pp. 93–97.

van Asten, P. J. A., Wairegi, L. W. I., Mukasa, D. and Urungi, N. (2011) 'Agronomic and economic benefits of coffee–banana intercropping in Uganda's smallholder farming systems', *Agricultural Systems*, vol. 104, no. 4, pp. 326–334.

Woomer, P., Lan'gat, M. and Tungani, J. O. (2004) 'Innovative maize–legume intercropping results in above- and below-ground competitive advantages for understory legumes', *West African Journal of Applied Ecology*, vol. 6, pp. 85–94.

# 10 Towards ecologically intensive smallholder farming systems

## Design, scales and trade-offs evaluation

*Pablo Tittonell*

*Farming Systems Ecology Group, Department of Plant Sciences, Wageningen University*

## Introduction

Smallholder farming systems are diverse, spatially heterogeneous and dynamic. They operate in uncertain and changing environments to which they need to adapt constantly. Resources and investments are often limited, and their strategic allocation in space and time has an impact on system attributes such as efficiency, vulnerability and resilience. Any technological strategy aiming to improve system performance and sustainability through ecological intensification should be designed considering the integrated nature of smallholder farming systems. This is particularly the case for mixed crop–livestock farming systems, which differ in terms of intensity and integration across the agro-ecological gradients of East and Central Africa.

Biophysical interactions at the scale of field plots or livestock units are central to agro-ecosystems; biophysical responses and affordability are key determinants of the adoption of management practices proposed to smallholders. Yet to be realistic and disposed to achieve an efficient use of resources, the design of ecologically intensive management should consider resource interactions across scales, from farms, to landscapes or territories. This calls for integrated approaches that allow the evaluation of the effectiveness and fitness of the technologies proposed in the light of local socio-economic contexts and livelihood systems. Recent examples of such approaches can be found, for example, in Groot and Rossing (2011), Giller *et al.* (2011) and Tittonell *et al.* (2012).

In the humid highlands of East and Central Africa, dense human population coupled with insufficient resources and sometimes inadequate agricultural practices has often resulted in resource degradation at farm and landscape levels, and conflict over the control and utilization of communally owned resources. When resources are scarce, decisions on their allocation to competing activities within the farming system entail trade-offs of different natures. Such trade-offs must be quantified to guide the design of strategies for integrated resource management. Although smallholder farmers are system managers by nature, system integration remains a major challenge in the field of agricultural research. Disciplinary standpoints, institutional interests and different scales of analysis often lead to competing research efforts, disaggregated results, replication of old experiences, and/or impractical recommendations. But one fundamental problem is the stronger emphasis placed on research to the detriment of the attention paid to design. Whereas research operates through systems analysis to arrive at conclusions, design operates through knowledge integration and synthesis to arrive at decisions.

Several integrative approaches have been proposed in the field of agronomy over the last two decades: integrated pest management (IPM), integrated soil fertility management (ISFM) or integrated crop management (ICM). All of these are examples of knowledge synthesis at the service of design. Their concepts and acronyms, however, have been frequently misused to describe disciplinary research that unfortunately ignores the integrated nature of these practices and of the smallholder farming systems they target. This chapter presents examples that illustrate the need for system integration in agricultural research, for linking analysis to design, and for the assessment of re-design options through the evaluation of trade-offs at different scales. The first half deals with scale effects on the design of ecologically intensive smallholder systems, with examples from ISFM and IPM. As moving up in scale leads to resource constraints and to possibly competing allocations of land, labour and capital, the second half deals with trade-offs and effective methods for their assessment.

## Moving through scales in agro-ecosystems design

Most statements in this section rest on the hypothesis that agro-ecosystems are cybernetic systems. Unlike natural ecosystems, which have no apparent goal per se, agro-ecosystems can be considered cybernetic systems because they are steered towards the fulfilment of one or more human objectives, and thus their functionality can be influenced by design. In cropping systems, for example, the association of species in space and time represents a case of design of a specific *structure*, with the objective of promoting desirable ecological *functions* that serve a number of social, cultural or economic *purposes*. Such a chain of causalities from structure to function to purpose – and vice versa – is central to systems design (Goewie 1997). Design starts from identifying the purpose (e.g. productivity, food security, ecological services) to then identifying the functions that are relevant to that purpose, and the structures that sustain such functions in space and time. Biological research, on the other hand, aims at analysing natural systems by unravelling the relationship between observed structures, the functions they support and the purpose they may serve. In an agro-ecosystem, which is both a socio-economic and a biophysical entity, a large number of processes that govern their functioning may remain unknown, uncontrollable or regulated by external forces. The rules of research in biology apply largely to agro-ecosystems; their analysis is thus a prerequisite for their design (unlike the design of a purely cybernetic system, such as a computer), and because in most cases we do not start from zero we often speak of agro-ecosystem re-design.

Agro-ecosystems are often defined as a nested hierarchy of systems (e.g. Fresco and Westphal 1988), in which the *i* – 1 integration level represents a sub-system or system component and the *i* + 1 the supra-system or immediate context. The number of stakeholders involved and their complexity increase as we move up in scale. At the lower end, probably the smallest scale at which an agro-ecosystem could be considered is the cultivated field (cf. Conway 1987). While scaling issues in agro-ecosystem analysis occupy an important place in the scientific literature, the objective here is to examine and discuss to what extent the context, in the broadest sense of the term, affects or should influence decisions concerning design. The chain of causalities from structure to function to purpose may be conspicuously nested within a certain scale, although in most cases it will involve processes that operate at one or two levels above and/or below the reference scale. The two examples below offer insights into the choice of scales at which agro-ecosystem design needs to be addressed.

### *Towards whole-farm, truly integrated soil fertility management approaches*

ISFM is often associated with the use of organic nutrient resources. An operational definition of ISFM has been offered by Vanlauwe *et al.* (2010) which considers the diverse responsiveness of soils that underwent different intensities of degradation, and recognizes the knowledge-intensive nature of this practice. Emphasis is placed on improving nutrient use efficiencies by crops through the strategic targeting of nutrient resources. Organically held nutrients from a diversity of sources are applied alone or in combination with inorganic fertilizers, often to improve total efficiency in the use of both resources, especially when the proper crop genotypes are used. When looking at opportunities for the implementation of IFSM practices on-farm, however, a common challenge faces smallholder farmers: where do these much-needed organic resources come from? Beyond incipiently adopted practices relying on transfers of green manure biomass or agroforestry, a major source of organic matter and nutrients *ex-situ* is still animal manure, particularly in smallholder crop–livestock systems. Yet the efficiency of livestock-mediated nutrient transfers through feeding and manure handling is seldom the target of ISFM research.

Livestock densities vary widely across sub-Saharan Africa (SSA). The greatest number of animals per area, and thus per unit of cultivated land, is found in the densely populated highlands of East and Central Africa. In these locations, cattle densities range from ten to more than 50 head per square kilometre (www.earthtrends.wri.org). If we consider half of the land area to be under cultivation, the maximum cattle densities would be in the order of 100 head per square kilometre, or one head per hectare of cultivated land. Mass measurements on-farm across Kenya indicated that 1–2 t of manure dry matter can be recovered from a well-fed cross-bred dairy cow kept in a hard-floored zero grazing (i.e. cut-and-carry) unit (Lekasi 2000). These rough calculations put together indicate that the potential average application rates would be in the order of 1–2 t/ha/yr for manure if all farmers had access to this resource.

Manure samples from five crop–livestock farms within a radius of 10 km in western Kenya showed wide diversity in terms of carbon and nutrient concentrations (Table 10.1). Samples were taken from manure that was ready for application. This illustrates the extent to which livestock feeding, housing and manure handling affect the efficiency of nutrient retention and recycling within the farming system. Due to limitations imposed by the quantities available and the need to homogenize treatment effects, most ISFM experiments are conducted using manure from an experimental or commercial dairy farm, which differs widely in quality from smallholders' manure. If manure from Farm E in Table 10.1 is applied at the rates of 1–2 t/ha calculated above, this would represent an addition of 5–10 kg N and 1–2 kg P per hectare per year, which would be rather ineffective in terms of both immediate crop responses and soil carbon build-up. But farmers often concentrate manure resources in certain fields of their farms. Concentration in one-tenth of the surface would allow application rates of 10 t/ha/year of manure, which would represent as much as 120 kg N per hectare and 32 kg P per hectare if the manure comes from Farm A.

Thus, the concentration of manure seems to make sense for smallholder farmers, and this leads to important farmer-induced soil heterogeneity. Soil samples from 250 farms in six highland districts of Kenya and Uganda showed large inter-field (within-farm) variability in topsoil C and available P, which tended to decrease in farms having a large number of cattle (Figure 10.1a). An important part of the variability in manure quality in smallholder systems is due to differences in nutrient losses from manure throughout collection and

*Table 10.1* Dry matter, carbon and nutrient content in manure samples from an experimental dairy farm and from five smallholder farms in western Kenya

| *Manure origin* | *Content (%)* | | | | |
|---|---|---|---|---|---|
| | *Dry matter* | *C* | *N* | *P* | *K* |
| Experimental farm* | 82 | 39 | 2.1 | 0.22 | 4.0 |
| Farm A | 56 | 30 | 1.2 | 0.32 | 2.0 |
| Farm B | 59 | 29 | 1.0 | 0.30 | 1.6 |
| Farm C | 77 | 25 | 1.0 | 0.10 | 0.6 |
| Farm D | 43 | 35 | 1.5 | 0.12 | 3.3 |
| Farm E | 41 | 23 | 0.5 | 0.10 | 0.6 |

Source: Castellanos-Navarrete 2007.

*Note*: * Maseno University Farm.

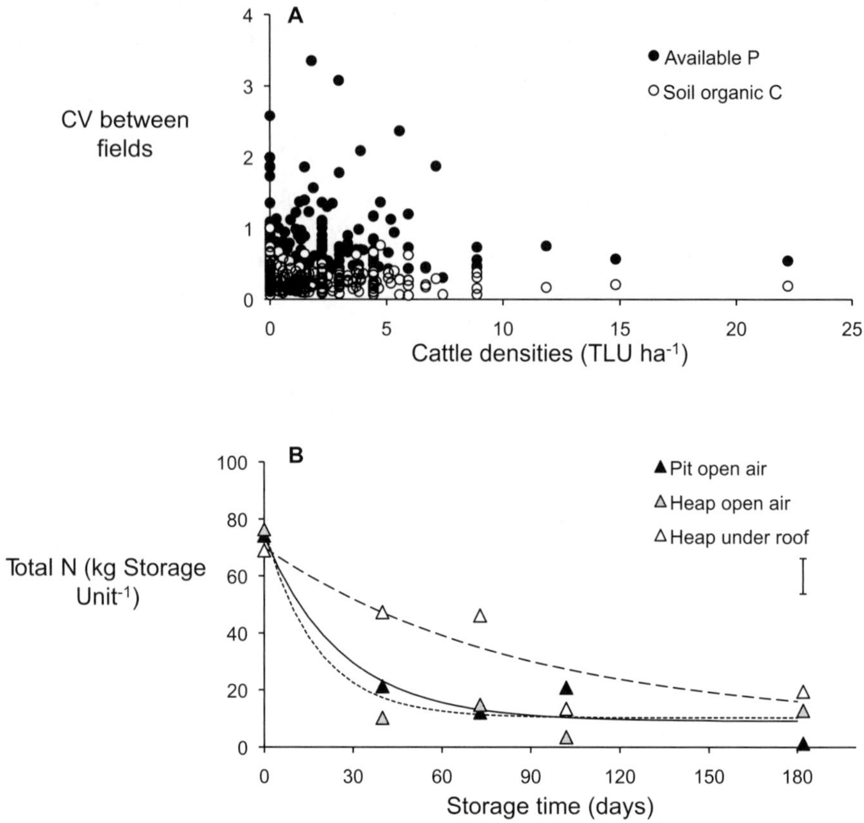


*Figure 10.1* (a) coefficient of variation (CV) between the field of each individual farm for topsoil carbon and available P in 250 farms from six districts in Kenya and Uganda, as a function of the number of cattle per unit area on each farm; (b) changes in total N content in manure stored in pits or heaped in open air or under a roof in western Kenya (source: Tittonell *et al.* 2010a, 2010b).

Notes: TLU: tropical livestock unit; SU: storage unit, a heap of 300 kg; FW: fresh weight.

storage. An experiment to mimic farmers' storage systems and to test simple improvements such as roofing indicated important differences in nutrient losses between storage systems when a mix of manure and maize stover was stored for less than two months (Figure 10.1b). After six months of storage, however, most nitrogen had been lost from all storage treatments without significant differences among them.

It is obvious that for ISFM approaches to be truly integrated there is a need to consider the farming system as a whole to design management strategies that improve its total resource use efficiency. This includes the efficiency of nutrient retention and cycling throughout the soil–plant–animal continuum, and thus livestock feeding and/or manure handling are as much an essential part of ISFM as fertilizer use efficiency. This holistic perspective is also necessary to place researchers within the realm of what is quantitatively realistic on-farm or within a territory: a recent study in the highlands of northeast Zimbabwe (Zingore *et al.* 2011) indicates that about 30 ha of grassland would be needed to sustain crop productivity through manure application in one hectare of land on sandy soils; a clear case in which limitations to ecological intensification are imposed by the context.

### *Designing ecologically intensive farming landscapes for pest control*

A conspicuous example of agro-ecosystem design that aims at influencing agro-ecological processes in the benefit of crop production is the 'push–pull' system. The push–pull is a stimulus–deterrent diversionary strategy that consists of repelling pest insects from a harvestable crop, and simultaneously attracting them to a 'trap' crop. The push effect may be obtained through masking host attraction, using repellents, anti-feedants, oviposition deterrents or attractants for predators and parasitoids. Semio-chemicals mediating interactions between organisms of the same species (pheromones) or from different species (allelo-chemicals) push undesirable insects away from the crop and attract predators or parasitoids into the area. The pull effect may be achieved through host attractants, aggregation, sex or oviposition pheromones or visual cues. Although the chemical ecology of many plant defence mechanisms behind the push–pull effect is still unknown, evidence shows that the design of the cropping system can have an important impact in reducing pest infestation on crops (e.g. Khan *et al.* 1997).

When designing cropping systems it is possible to achieve some of the push–pull principles (e.g. host masking, visual cues, etc.) through the proper choice of companion plants and surrounding trap crops in intercropping systems. A large body of evidence and practical examples are available for East African cropping systems as developed by the International Centre for Insect Physiology and Ecology (ICIPE) in their experimental station in Kenya (see Kahn *et al.* 2010 and references therein). However, in anisotropic agricultural landscapes, such as those dominating the western Kenya highlands, a large diversity of plant species, cultivated or not, is likely to have push–pull effects. A recent study in the area presents evidence on the functionality of this biodiversity (André 2011): infestation rates of the aphid species *Aphis fabae* and *A. craccivora* in fields of maize associated with common bean (*Phaseolus vulgaris*) and cowpea (*Vigna unguiculata*), and of their natural antagonists (e.g. predators and fungi), were significantly affected by landscape structure, notably by the presence, morphology, density and species composition of the hedgerows surrounding those fields, by the crops grown in adjacent fields and by the degree of fragmentation of space along a gradient of patch (land use) diversity. Some examples are shown in Table 10.2.

*Table 10.2* Effect of presence and species composition of hedgerows on levels of aphid infestation and predator abundance in beans and cowpea fields in western Kenya

| | *Plant infestation index* | *No. of predators per field* |
|---|---|---|
| Fields with hedgerows | 2.2 ± 0.6 | 3.7 ± 0.5 |
| Fields without hedgerows | 3.4 ± 0.7 | 6.5 ± 2.0 |
| Dominant hedgerow species | | |
| *Euphorbia* | 0.6 ± 0.3 | 2.1 ± 1.1 |
| *Lantana* | 2.4 ± 0.9 | 4.0 ± 0.9 |
| *Eucalyptus* | 1.5 ± 1.0 | 3.2 ± 1.4 |

Source: André 2011.

These and other similar results suggest that: (1) landscape structure and species composition may interfere in the dissemination of stimulus–deterrent cues and therefore with the performance of crop associations designed for push–pull effects; and (2) that the proper scale for agro-ecosystem design is not the field plot but the landscape and its temporal dynamics. In this case, for example, the challenge is to search for push–pull elements already existing in the landscape to study their impact on the distribution of pests and antagonists. This is a key research area that remains largely ignored in the context of African smallholder farming, and that consists of the study of patterns of landscape anisotropy (niches), their relation with ecological functions that sustain agro-ecological services of interest and the perception and utilization of such structure-to-function causalities by local farming communities. Current work in western Kenya aims precisely at establishing such links between landscape structure, local knowledge and ecological services (Birman *et al.* 2010).

## Trade-offs analysis

The two examples above indicate the need to move across scales in agro-ecosystem design. The design of such multi-functional agricultural landscapes aims at fulfilling various objectives simultaneously. Often two or more of such objectives may be in conflict when they are mutually exclusive or in competition when the fulfilment of one of them results in a partial detriment for the other. In analysing trade-offs between two objectives, such an inverse relationship can be described in the simplest of cases as being proportional and constant. The slope of the resulting line indicates the elasticity of competition between the two objectives. In most cases, however, trade-offs are more complex than that and may exhibit cases of strong substitution, in which fulfilling an objective makes it almost impossible to fulfil a second one, or cases in which complementarities or compromise solutions are possible (Figure 10.2). The first derivative of the trade-offs curve indicates the marginal rate of substitution between two objectives at any given level of the fulfilment of one of them.

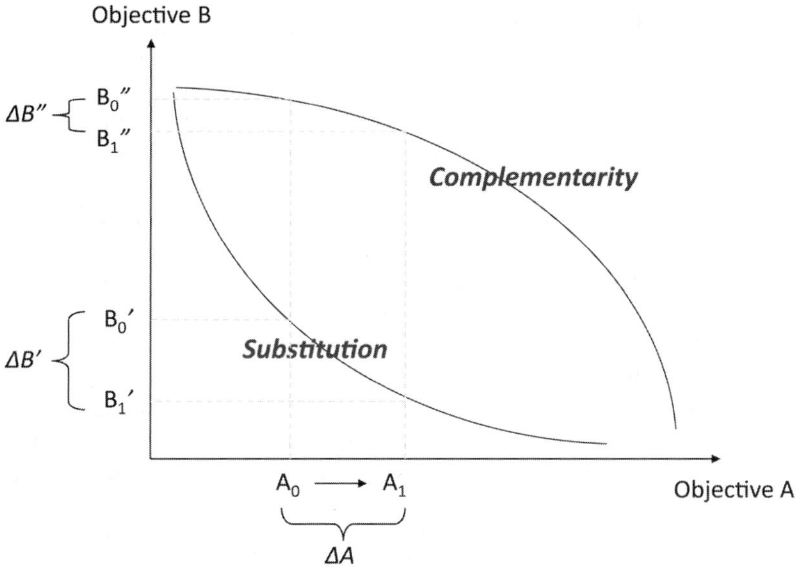


*Figure 10.2* Schematic representation of possible trade-offs between two objectives (A and B). A change of a certain magnitude in the value achieved for objective A (ΔA, from A0 to A1) may imply a relatively large (ΔB′ – substitution) or a relatively small (ΔB″ – complementarity) 'sacrifice' in the value achieved for objective B, respectively.

This information is very valuable for decision-making and justifies the need to map trade-offs.

Trade-offs are central to rational decision-making when resources are scarce. Trade-offs often exist between objectives that are prioritized by different stakeholders, considering different spatial and temporal scales, and responding to diverging concerns. For example, trade-offs are made between today's productivity and long-term sustainability, between farm economic efficiency and regional employment, between food security and negative externalities in the form of environmental impact, etc. Trade-offs may be analysed quantitatively through the simple calculation of the elasticity of substitution between objectives (i.e. absolute or relative changes in A with respect to absolute or relative changes in B). Shadow prices, opportunity costs, payment for environmental services, etc., are all examples of such ways of quantifying trade-offs.

To capture the complexity of the interactions between the socio-economic and biophysical dimensions of agro-ecosystems, trade-offs can be better assessed through model-based explorations or the *ex-ante* evaluation of the consequences of (management or strategic) decisions on the fulfilment of two or more objectives under different scenarios. The models used for this vary in their complexity, and their implementation requires an initial identification of trade-offs formulated as hypotheses. Multiple goal optimization models, referred to nowadays as bio-economic models (Janssen and van Ittersum 2007), are frequently used to analyse trade-offs. This approach has an economic bias and the limitation of being 'static' (i.e. the model runs for a single season, or a single sequence of seasons) and thus unable to capture either relevant biophysical feedbacks or temporal variability in the system, particularly in long-term analyses. An alternative is to use inverse modelling techniques (global search algorithms, e.g. Groot and Rossing 2011), which allows a number of objectives to be optimized by running dynamic models with a huge number of different combinations of parameters that represent farm management decisions.

Let us examine an example of a farm-scale trade-offs map obtained through inverse modelling (Tittonell *et al.* 2007). A single farm, one decision-maker, a number of fields with a contrasting history of past use and current fertility level, a limited amount of cash to invest in hired labour for land preparation, weeding or mineral fertilizers, plus the possibility of allocating these to the various fields of the farm lead to the overwhelming universe of possible model solutions (i.e. each point is a solution obtained through a particular combination of management parameters) that is depicted in Figure 10.3a. This is because the various components of the farm and all their daily interactions are considered in the model; for instance, allocating less cash to hiring labour would result in late planting, poorer yields, probably more weed pressure and greater losses of fertilizer N by leaching and of soil by erosion. The points that are relevant in trade-offs analysis are those corresponding to the optimization of the objectives considered, in this case reducing N losses while increasing whole-farm maize production. Each point along the outer trade-offs curve, or Pareto efficient frontier, represents a combination of management parameters (dates, investments, application rates, etc.) at farm scale that complies with both objectives.

Trade-offs in this example were quantified for three scenarios of cash availability at planting – which is a more relevant indicator of investment capacity by smallholders than annual revenues. If we focus on the lower right corner of the cloud of possible solutions, which represents for each scenario the best compromise between grain production and whole-farm N-use efficiency, we see that a large number of parameter combinations are possible (Figure 10.3b). Note that to allow the comparison of scenarios the investments in either hired labour or fertilizer are expressed in relative terms, as a fraction of the total cash available. The fraction of total cash available that is allocated to buying fertilizer vs. to hiring labour for weeding varies under the different scenarios, but tends to favour the latter when less cash is available. In other words, investment in fertilizers was favoured only when sufficient cash was available to hire labour and secure proper weeding (see the 10,000 KSh scenario).

Although this may be a powerful tool, its weaknesses reside in the identification of relevant indicators that represent farmers' objectives and rationale, and in the risk of over-interpretation of the results obtained. Identifying indicators – and their meaningful thresholds – is an important step in agro-ecosystem design. Such indicators may be expressed as objectives in multiple

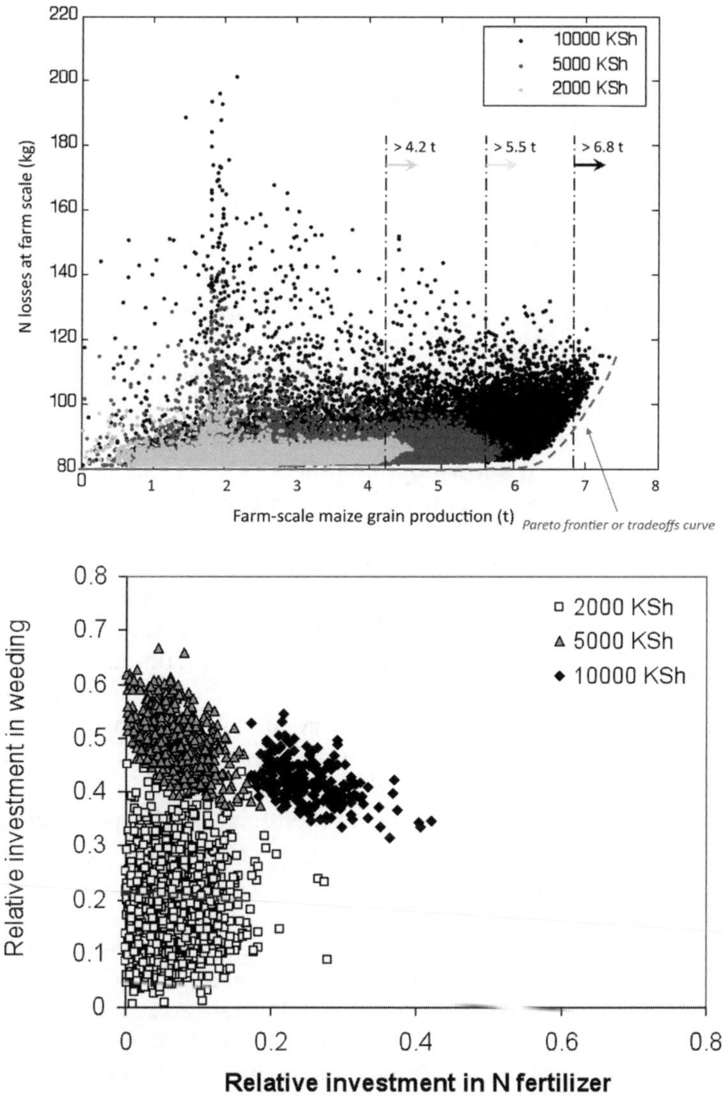


*Figure 10.3* (a) Results of the optimization of the objectives maximizing maize production at farm scale and minimizing nitrogen losses at farm scale for three scenarios of liquidity at planting (2,000, 5,000 and 10,000 KSh/ha; 1 KSh = $0.95) to invest in extra labour and agricultural inputs in a heterogeneous smallholder farm (2.2 ha) of Kakamega, western Kenya. To the right of the vertical lines is the subset of solutions that satisfy high maize production while keeping N losses close to their minimum. (b) Fraction of cash available that is invested in labour for weeding and in N fertilizer in the best-yielding cases. Simulations were done using the crop–soil dynamic model DYNBAL and the global search engine MOSCEM.

criteria and multi-scale analyses, representing the goals of different stakeholders. The definition of meaningful indicators and thresholds also allows the negotiation and participatory prototyping of desired agro-ecosystem functions. An example from western Kenya shows how the feasibility and the transition towards prototype farms designed by local farmers could be evaluated by means of a simulation model (Tittonell *et al.* 2009). Several approaches using models as *discussion* support tools to enhance participatory learning and action were developed worldwide, using different types of models. The Companion Modelling Approach (Bousquet *et al.* 1999) is one of them, which combines participatory tools such as role-playing games with the simulation of the interaction between actors in a community using agent-based modelling techniques.

To minimize the risk of over-interpretation of modelling outcomes, it is important to situate the study in its specific context of causality and uncertainty. Future-oriented or *ex-ante* studies used in agro-ecosystem design and evaluation may be classified based on these two criteria (van Ittersum *et al.* 1998): the degree of causality that is known on the processes being studied and the level of uncertainty associated with the scenarios analysed. Uncertainty is often associated with the time horizon of the scenario (e.g. 'what is likely to happen next year' vs. 'what if the world oil market collapses in 20 years'). According to this classification, when uncertainty is low, but little is known about the relevant causes and effects governing key phenomena (e.g. when an empirical function is used), the study is defined as a projection. Still under low uncertainty, but when causality increases (e.g. when using a mechanistic, process-based model), the study is defined as a prediction. When using a mechanistic, high-causality approach to study highly uncertain scenarios, the study is defined as an exploration. Finally, when analysing highly uncertain scenarios using empirical, descriptive tools we may be in the presence of a speculation.

## Concluding remarks

The entry point of agricultural research is still largely the cultivated field plot, the plant, the cell or the gene. The benefits expected from the progress made on innovative cropping systems (e.g. push–pull crop associations), new germplasm or genetic engineering (e.g. IR maize) are hardly attained under smallholder conditions, where such technologies are either non- or sub-performing, overridden by larger-scale processes, or simply perceived as unsuitable by local farmers. All this indicates that agricultural research needs to move up some steps in scale to embrace the diversity of situations that characterize smallholder farming landscapes, which are the result of biophysical and socio-economic processes that shape them through their history. Moving up in scale implies, among other things, that resource trade-offs will be encountered. Currently, a diversity of tools is there to support trade-offs analysis and the evaluation of the multiple consequences of management interventions, across scales and in consultation with the relevant stakeholders. A number of advantages can be drawn from using inverse modelling techniques as one possible method, as both biophysical and socio-economic interactions may be captured. But the result of model-based studies should not be over-generalized and considered only within the context of their parameterization.

Designing ecologically intensive farming systems imposes on agronomists the need to move beyond optimizing current processes, to the design of the structures that support such processes (and new ones) in space and time. The diversity, spatial heterogeneity and temporal dynamics of smallholder farming systems in the humid highlands of Africa call for the flexible, adaptive targeting of policies, management practices and technologies. Integrative management strategies, such as ISFM or IPM, suit the holistic nature of resource allocation by smallholders, who are naturally system managers. However, agricultural research remains largely disciplinary, often concentrating huge expectations around a single set of processes or technologies as the solution to smallholders' problems, whereas in many cases the mismatch between proposed technical solutions and farmers' realities may be simply the result of mismatches of scale.

## References

André, L. (2011) 'Les déterminants de la variabilité spatiale et temporelle de la pression des pucerons et de leurs ennemis naturels dans une région agricole du Kenya', MSc thesis, Ecole d'Ingénieurs en Agro-Développement International, ISTOM, Paris, France.

Birman, D., Moraine, M., Tittonell, P., Martin, P. and Clouvel, P. (2010) 'Ecosystem services assessment in complex agricultural landscapes using farmers' perceptions', in J. Wery, I. Shili-Touzi and A. Perrin A. (eds), *Proceedings of Agro 2010: The XIth ESA Congress, August 29th–September 3rd, 2010*, Montpellier, France.

Bousquet, F., Barreteau, O., Le Page, C., Mullon, C. and Weber, J. (1999) 'An environmental modelling approach: the use of multi-agent simulations', in F. Blasco and A. Weill (eds), *Advances in Environmental and Ecological Modelling*, Elsevier, Paris.

Castellanos-Navarrete, A. (2007) 'Cattle feeding strategies and manure management in Western Kenya', MSc thesis, Wageningen University, the Netherlands.

Conway, G. R. (1987) 'The properties of agroecosystems', *Agricultural Systems,* vol. 24, pp. 95–117

Fresco, L. O. and Westphal, E. (1988) 'A hierarchical classification of farm systems', *Experimental Agriculture*, vol. 24, pp. 399–419.

Giller, K. E., Tittonell, P., Rufino, M. C., van Wijk, M. T., Zingore, S., Mapfumo, P., Adjei-Nsiah, S., Herrero, M., Chikowo, R., Corbeels, M., Rowe, E. C., Baijukya, F., Mwijage, A., Smith, J., Yeboah, E., van de Burg, W. J., Sanogo, O. M., Misiko, M., de Ridder, N., Karanja, S., Kaizzi, C., K'ungu, J., Mwale, M., Nwaga, D., Pacini, C. and Vanlauwe, B. (2011) 'Communicating complexity: integrated assessment of trade-offs within African farming systems to support innovation and development', *Agricultural Systems*, vol. 104, pp. 191–203.

Goewie, E. A. (1997) 'Designing methodologies for prototyping ecological production systems', Course reader MSc Ecological Agriculture (F800-204), Department of Ecological Agriculture, Wageningen Agricultural University, Wageningen.

Groot, J. C. J. and Rossing, W. A. H. (2011) 'Model-aided learning for adaptive management of natural resources: an evolutionary design perspective', *Methods in Ecology and Evolution*, vol. 2, pp. 643–650.

Janssen, S. and van Ittersum, M. (2007) 'Assessing farm innovations and responses to policies: a review of bio-economic farm models', *Agricultural Systems*, vol. 94, pp. 622–636.

Khan, Z. R., Ampong-Nyarko, K., Chiliswa, P., Hassanali, A., Kimani, S., Lwande, W., Overholt, W. A., Pickett, J. A., Smart, L. E., Wadhams, L. J. and Woodcock, C.M. (1997) 'Intercropping increases parasitism of pests', *Nature*, vol. 388, pp. 631–632.

Khan, Z. R., Midega, C. A. O., Bruce, T. J. A., Hooper, A. M. and Pickett, J. A. (2010) 'Exploiting phytochemicals for developing a "push–pull" crop protection strategy for cereal farmers in Africa', *Journal of Experimental Botany*, vol. 61, pp. 4185–4196.

Lekasi, J. K. (2000) 'Manure management in the Kenya Highlands: collection, storage and utilisation to enhance fertiliser quantity and quality', PhD thesis, Coventry University.

Tittonell, P., van Wijk, M. T., Rufino, M. C., Vrugt, J. A. and Giller, K. E. (2007) 'Analysing trade-offs in resource and labour allocation by smallholder farmers using inverse modelling techniques: a case-study from Kakamega district, western Kenya', *Agricultural Systems*, vol. 95, pp. 76–95.

Tittonell, P., van Wijk, M. T., Herrero, M., Rufino, M. C., de Ridder, N. and Giller, K. E. (2009) 'Beyond resource constraints: exploring the physical feasibility of options for the intensification of smallholder crop–livestock systems in Vihiga district, Kenya', *Agricultural Systems*, vol. 101, pp. 1–19.

Tittonell, P., Rufino, M. C., Janssen, B. and Giller, K. E. (2010a) 'Carbon and nutrient losses from manure stored under traditional and improved practices in smallholder crop–livestock systems: evidence from Kenya', *Plant and Soil*, vol. 328, pp. 253–269.

Tittonell, P., Muriuki, A. W., Shepherd, K. D., Mugendi, D., Kaizzi, K. C., Okeyo, J., Verchot, L., Coe, R. and Vanlauwe, B. (2010b) 'The diversity of rural livelihoods and their influence on soil fertility in agricultural systems of East Africa: a typology of smallholder farms', *Agricultural Systems*, vol. 103, pp. 83–97.

Tittonell, P., Scopel, E., Andrieu, N., Posthumus, H., Mapfumo, P., Corbeels, M., van Halsema, G. E., Lahmar, R., Lugandu, S., Rakotoarisoa, J., Mtambanengwe, F., Pound, B., Chikowo, R., Naudin, K., Triomphe, B. and Mkomwa, S. (2012) 'Agroecology-based aggradation-conservation agriculture (ABACO): targeting innovations to combat soil degradation and food insecurity in semi-arid Africa', *Field Crops Research*, vol. 101, pp. 1–7.

van Ittersum, M. K., Rabbinge, R. and van Latesteijn, H. C. (1998) 'Exploratory land use studies and their role in strategic policy making', *Agricultural Systems*, vol. 58, 309–330.

Vanlauwe, B., Bationo, A., Chianu, J., Giller, K. E., Merckx, R., Mokwunye, U., Ohiokpehai, O., Pypers, P., Tabo, R., Shepherd, K., Smaling, E. M. A. and Woomer, P. L. (2010) 'Integrated soil fertility management: operational definition and consequences for implementation and dissemination', *Outlook on Agriculture*, vol. 39, pp. 17–24.

Zingore, S., Tittonell, P., Corbeels, M., van Wijk, M. T. and Giller, K. E. (2011) 'Managing soil fertility diversity to enhance resource use efficiencies in smallholder farming systems: a case from Murewa District, Zimbabwe', *Nutrient Cycling in Agroecosystems*, vol. 90, pp. 87–103.

# 11 Using the 'livestock ladder' as a means for poor crop–livestock farmers to exit poverty in Sud Kivu province, eastern DR Congo

*Brigitte L. Maass,*[1] *Wanjiku L. Chiuri,*[2] *Rachel Zozo,*[3] *Dieudonné Katunga-Musale,*[3,5] *Thierry K. Metre*[4] *and Eliud Birachi*[2]

[1] *International Center for Tropical Agriculture (CIAT), Kenya*
[2] *International Center for Tropical Agriculture (CIAT), Rwanda*
[3] *International Center for Tropical Agriculture (CIAT), DRC*
[4] *Université de Liège-Gembloux Agro-Bio Tech, Belgium*
[5] *Institut National pour l'Etude et la Recherche Agronomiques (INERA), DRC*

## Introduction

Livestock in general are central to the livelihoods of a large percentage of rural households living in poverty, especially in sub-Saharan Africa (SSA) (Sumberg 2002). Small-stock play a critical role in conflict and post-conflict situations in the region. In Sud Kivu province of the Democratic Republic of the Congo (DRC), for example, cavies[1] provide incomes, manure and animal protein, especially for women and teenagers in small-scale farming households. After decades of war and unrest, livestock numbers have substantially declined, dropping most families lower on the livestock ladder. Small animals have become prominent in the agriculture of Sud Kivu (Cox 2012; Maass *et al.* 2012) as well as the adjacent Great Lakes region of Rwanda and Burundi (Ouma and Birachi 2011). Agriculture in this region has traditionally been managed with crop–livestock integration (Cox 2012; Ouma and Birachi 2011). Livestock, especially cattle, are a sign of wealth and enhance social status (Dolberg 2001). During the diagnosis of a Participatory Rural Appraisal (PRA) in Sud Kivu, households consequently associated the general lack of livestock with absolute poverty (Zozo *et al.* 2010) when self-assessing the classes of wealth. Cattle and large livestock in general, however, represent a major risk for most farmers in Sud Kivu, particularly through theft and diseases. They are substantial assets that mostly serve as sources for savings and insurance or for manure production, but are not intended for the regular provision of food. Small-stock, such as chickens, rabbits or cavies, are lower risks (Lammers *et al.* 2009); they supply high-quality nutrition and create an access to market, particularly for women farmers. Teenagers from the communities also raise small-stock to pay their school fees and cover other personal needs (Maass *et al.* 2012). The significance of cavies being raised by a considerable proportion of livestock keepers in Eastern and Central Africa has not been investigated in detail. Perry *et al.* (2002) recognize the role for the poor of diverse 'unconventional' livestock species, but do not include them in their study due to lack of information. Sporadic media reports (e.g. Bii 2007; Mwangi 2011), however, indicate that cavies, for example, may play a discernible role in the livelihoods of poor rural people, including their potential, similar to that of keeping chickens, to assist these people in climbing up the livestock ladder.

Various authors (Todd 1998; Dolberg 2001; Udo *et al.* 2011) have used the term livestock ladder to indicate the potential for poor rural people to acquire assets and, hence, get out of poverty (Figure 11.1). This chapter aims to depict the potential of the livestock ladder for poor livestock keepers in Sud Kivu and adjacent regions, with particular emphasis on the role of cavies as small-stock.

## Approach

This chapter is based on synthesizing data obtained from our research in Sud Kivu, revising the findings in the frame of a literature review. The existing

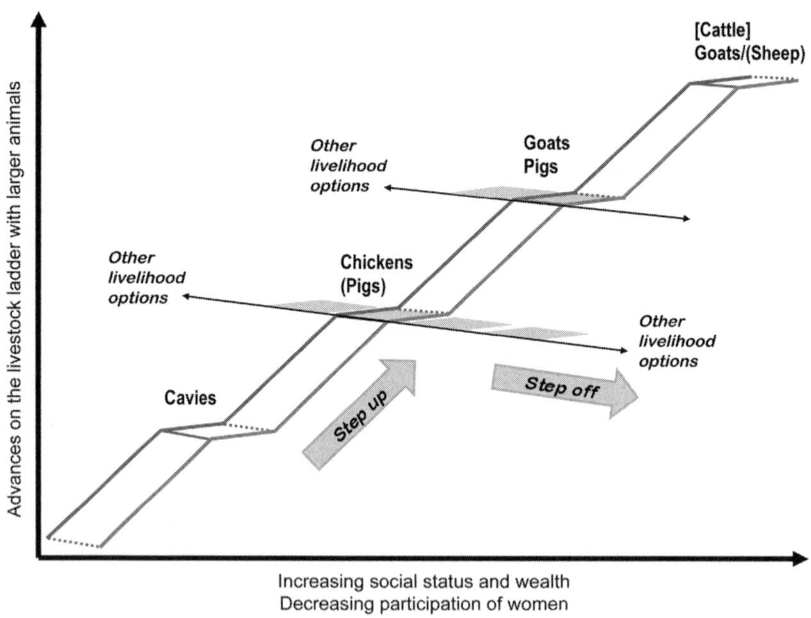


*Figure 11.1* The 'livestock ladder' showing the progression of animal ownership, modified from Todd (1998) and adapted to Sud-Kivu conditions.

situation for livestock production in Sud Kivu was assessed by applying a number of surveys, such as a diagnostic survey (Maass *et al.* 2010, 2012) and PRA (Zozo *et al.* 2010) with producers, and value chain analysis (VCA) including producers, retailers and urban consumers (Chiuri *et al.*, unpublished) around the provincial capital of Bukavu with an estimated 700,000 inhabitants. Agronomic evaluation and participatory variety selection (PVS) were conducted on small forage plots (Katunga-Musale *et al.* 2011) in four representative locations from the conducted studies. Descriptions of details can be found in each of the documents cited above. Livestock species reported by survey participants were converted into tropical livestock units (TLUs) (Maass *et al.* 2010). We attempted to measure the strength of association by calculating odds ratios from the livestock species reported (presence/absence), applying Fisher's exact test (Pezzullo 2011). The lack of a clear pattern of species' associations in the different data sets (data not shown) indicates that most people rear a diverse combination of livestock. This strengthens the general argument that livestock are an integral part of agriculture in the region.

## Findings

### *Livestock holdings and feeding systems*

The livestock ladder in Sud Kivu begins with cavies and ends with goats and, rarely, with cattle (Figure 11.1). In all surveys, livestock holdings expressed in TLU showed strongly skewed distributions, with 65–75 per cent of the livestock being owned by the first quartile of respondents, whereas the fourth quartile only had 2–5 per cent (data not shown). Species composition is clearly distinct when households with low TLU values are compared with those with high values (Figure 11.2). For the less well-endowed households, goats and pigs provide 75 per cent of their – generally low – livestock assets, while chickens and cavies together with other small-stock account for almost 25 per cent of their TLU. Cattle, on the other hand, were the livestock assets of the wealthier households, while small-stock played an insignificant role (less than 5 per cent).

Small-stock were usually produced in flexible smallholder backyard systems, with chickens, cavies and/or pigs. Feed for these animals was typically kitchen scraps and others sourced by scavenging around the homesteads. In addition, 45 per cent of chickens and 70 per cent of pigs received small complements of maize or palm oil cake. This feeding system normally leads to monogastric animals being underfed, which is partly due to lack of knowledge about zero-grazed animal husbandry. The second most important issue raised by most respondents after diseases was a lack of dry season feed. Farmers welcomed forage research on-farm, although they do not have a history of cultivating forages. By applying PVS, they chose forages with visible dry season-tolerance that were palatable for their small-stock, such as the herbaceous legumes *Canavalia brasiliensis* and *Stylosanthes guianensis*.

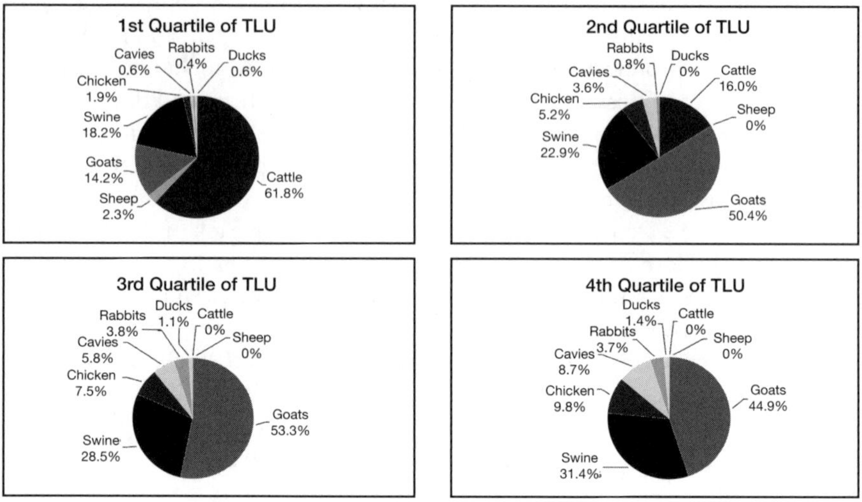


*Figure 11.2* Relative contribution of different species to total livestock assets in households with many compared to those with fewer livestock assets (in tropical livestock units, TLU), by quartiles. Under first quartile TLU = 5.43, under 2nd quartile TLU = 1.11, under 3rd quartile TLU = 0.54, and under 4th quartile TLU = 0.36.

## *Livestock production, trading and marketing*

The peasant majority produces too few livestock throughout the year for regular sales of a surplus. For example, not even half of 112 people interviewed had sold animals during the last year (Maass *et al.* 2010). Their production is too small even to satisfy subsistence dietary needs; rural producers stated an average household (mean of 7.7 persons) would consume annually five broilers, five hens, four cavies/rabbits, 10 kg of pork and 54 litres of milk used in the form of the locally made cheese known as *masanja*. The largest portion of animal products consumed appears to originate from fish (Chiuri *et al.*, unpublished). Consequently, farmers sell only when the need arises. Despite the obvious market demand and potential market size, sales rarely reach Bukavu, at a distance of 20–80 km from the production areas surveyed. Poor road infrastructure within the region and the accompanying insecurity are additional reasons. Hence, value chains are simple and provide only a little for urban consumers. Nevertheless, sizeable margins can be obtained along the value chain. Producer sales prices, for instance, vary between $4.9 and $6.2 per hen in the village, while the sales price at town markets may rise up to between $6.7 and $8.2 per hen, depending on the season. Similarly, a producer will sell one cavy for $1.1–1.3, while in town a roast cavy is sold for $3.5 (Maass *et al.*, unpublished). Trans-border trade has recently become significant in Sud Kivu. The Bukavu market for animal products is supplied from nearby Rwanda, Uganda and Kenya. For example, large quantities of eggs are regularly brought across the border between Rwanda and Bukavu since it has become much smoother to pass between the two countries.

## Discussion and outlook

### *Species composition and roles of livestock*

Holding diverse livestock species may serve as insurance for poor smallholder farmers. However, including larger animals is almost impossible for them because of the efforts and resources required for maintaining such species (Pica-Ciamarra *et al.* 2011). Our data corroborate this fact that poorer livestock keepers had a lower diversity in terms of the number of different species reared (Figure 11.2). Species composition varied across livestock asset classes (quartiles of TLU) and also the relative importance of a species in TLU. This indicates how dependent results are on both the chosen sample of respondents and sample size, in addition to any traditional, cultural or personal preferences.

Relative species composition reflects the diverse roles that livestock perform in a farm household, which differ depending on the species and also vary according to intensification levels (Udo *et al.* 2011). General roles attached to distinct livestock species have been derived from Todd (1998), Dolberg (2001), Perry *et al.* (2002), Sumberg (2002) and Ampaire and Rothschild (2010), adjusting the importance of functions to conditions in Sud Kivu (Table 11.1) based on experiences from our research. In fact, livestock functions should be assessed in their entire complexity, including their effects on gender. Perry *et al.* (2002) distinguished financial, social, physical, natural and human roles, whereas Sumberg (2002) recognized multiple objectives towards biological performance as well as economic, social and cultural outcomes of keeping livestock. Most authors implicitly associate cattle with the term livestock. The relative roles of small-stock, however, may be quite distinct from those of cattle. Small-stock such as cavies are an essential source of security and provide a small but frequent income for the very poor (Udo *et al.* 2011), and especially for women. Despite its assumed relatively low productivity, small-stock production provides a steady source of high-quality animal protein for household consumption, thus enhancing the security of both food and nutrition (Matthiesen *et al.* 2011; Metre 2011). Importantly, cavies can be consumed at any time, as they are fully under the control of women or the youth in Sud Kivu, whereas a chicken is slaughtered only for extraordinary events, and this decision must be made by the male household head (Metre, unpublished; Chiuri, unpublished).

When observing the livestock situation in Sud Kivu, we therefore consider that small-stock, especially cavies, have a particular role regarding the empowerment of women and the youth (Table 11.1) as a non-monetary service and an indirect benefit. Consequently, husbanding small-stock in Sud Kivu may not translate into larger livestock assets for all, but indirectly improve livelihoods through the empowerment of women. Notably, it will help to meet

*Table 11.1* The complex functions of smallholder-held livestock in a household, adapted to the situation in Sud Kivu

| *Livestock species* | *Human nutrition* | | *Social and cultural* | | | *Economic and financial* | | *Farming system* | | |
|---|---|---|---|---|---|---|---|---|---|---|
| | *Home consumption* | *Purchased food* | *Hospitality* | *Women's benefits* | *Social status* | *Savings/ insurance* | *Regular cash income* | *Manure production* | *Resource use (residues)* | *Resource use (labour)* |
| Cavies | +++ | + | + | +++ | + | ++ | ++ | ++ | ++ | + |
| Chicken – eggs | ++ | + | + | ++ | + | + | ++ | + | + | + |
| Chicken – meat | + | + | +++ | + | + | + | + | + | + | + |
| Pigs | + | + | + | ++ | + | ++ | ++ | + | ++ | ++ |
| Goats/[sheep] | + | ++ | +++ | + | ++ | ++ | + | ++ | ++ | ++ |
| [Cattle – beef] | – | ++ | – | – | +++ | [+++] | + | [+++] | [++] | [+++] |
| [Cattle – dairy] | [+++] | + | [+] | [++] | +++ | [+++] | [+++] | [+++] | [++] | [+++] |

*Notes*: Higher importance of the respective function for the individual livestock species expressed by more '+'; cattle are in squared brackets as most of their functions are currently of little relevance in the region.

some of the requirements of household nutrition and contribute to the educational cost of many children.

### *Cavies as small-stock*

Although cavies occur widely in smallholder households in the rural Sud Kivu region, they are rarely marketed in towns such as Bukavu (Chiuri *et al.*, unpublished). However, considerable numbers are usually sold in some rural markets and also prepared for on-site consumption (Metre 2011). This corroborates the statement (Todd 1998) that the significance of returns from small-stock is more in their frequency than in their magnitude.

In the post-conflict environment, the peasant population in Sud Kivu in general is relatively passive and does not take advantage of the rapidly rising urban demand for animal products. Humanitarian assistance during the past decades apparently discouraged peasants from taking initiatives, preferring to rely on donations. This is understandable as there are also high risks involved for farmers if they restock by increasing their livestock numbers. Large animals are still targets for looting by armed gangs (Cox 2012; Maass *et al.* 2012), thereby increasing the risk for poor farmers to the extent that certain activities are economically non-viable. Smaller animals such as pigs and chickens are comparatively susceptible to certain lethal diseases; consequently their rearing also involves a high risk in the absence of veterinary services and products (Maass *et al.* 2012). This situation has apparently made the rearing of cavies a very useful technology and a better option for keeping livestock in the circumstance of the region, although essentially no R&D support has been given to improving cavy culture from the mere keeping of the animal to its production. One of the general advantages over other livestock species is that very little capital is needed to start raising cavies. On the other hand, the animal responds well to improved husbandry (Lammers *et al.* 2009), thus offering potential for the use of microcredit programmes for those who want to seriously engage in small-stock production.

### *Opportunities and risks of the livestock ladder*

Women own and control small-stock; this means that by promoting small-stock women can be empowered. Todd (1998) and Dolberg (2001) demonstrate how women benefit along the livestock ladder both through increased cash income and improved family nutrition, when they improve their village poultry rearing. The same authors demonstrate, however, that benefits from small-stock need not necessarily go into the acquisition of larger animals in order to climb up the assets ladder. Frequently, women lease or purchase pieces of land that assure their family's food security (Todd 1998); they also improve housing or move into alternative livelihood activities such as petty trade. Hence, those who seriously engage in small-stock production have opportunities for both stepping up the ladder and for stepping off it (Figure 11.1). Analyses by

Pica-Ciamarra *et al.* (2011; for 12 developing countries) and by Udo *et al.* (2011; in six case studies) seem to challenge this concept as they demonstrate that poor households keeping livestock would not benefit much from investments into small-stock; however, both studies acknowledge that this country-level research may not capture non-monetary services and the indirect benefits from livestock.

In his survey, Metre (2011) found that some of the experienced cavy breeders in Sud Kivu purchase other animals, such as goats and pigs, in exchange for a certain number of cavies; for example, a young she-goat's cost was equivalent to 15–20 cavies, which is the approximate annual offspring attained from one cavy held under good conditions; or a piglet was bartered for 10–14 cavies. While ascending the livestock ladder, there is a risk that women may lose control over the larger species acquired (Todd 1998), which may prompt them to step off the ladder. But this can be countered by ensuring that the spouses are also participating in other small-stock activities such as chickens, which are currently a man's livestock in the area. Research is needed to identify those households that will respond to an R&D focus on improving small-stock husbandry for the enhancement of their livelihoods.

### *The potential role of forages*

Undoubtedly, feeding small-stock better by utilizing improved forages could play a role in enhancing livestock production. Adult cavies under an improved feeding regime, for instance, reach live weights of 700/900 g (female/male) compared with 500/600 g otherwise (Metre 2012). Improved forages also sustain soil fertility by preventing soil erosion and improving nutrient cycling. Hence, forages can have an impact on livestock productivity directly by providing feed and indirectly through enhanced soil fertility for crop production. Millar and Connell (2010) show that improved forages can induce a beneficial change of mixed production systems in the highlands of Laos that also target monogastric (pigs and poultry) animals and fish. Lisson *et al.* (2010) illustrate that introducing improved forages can be a chief component of a strategy to enhance livelihoods in cattle-based crop–livestock systems of Indonesia when integrating both farming systems and participatory research. In our studies, PVS has helped farmers to identify suitable forages; however, when they were offered seeds, farmers planted only food/feed legumes (e.g. cowpea or lablab) and not the forages selected earlier, because rural people are hungry and have restricted access to food. Improved feeding and forages can be an entry point, however, and their potential impact depends strongly on the importance given to livestock due to its dominant role(s) for a farming household (Table 11.1). Livestock keeping is based on multiple objectives, and livestock is only one component of the farming system. Sumberg (2002) argues that, despite their undeniable biophysical merit, cultivated forages may turn out to be insignificant in the context of the larger farming system and additional livelihood options, such as trading and seasonal labour migration. Therefore, it is yet to be demonstrated for the post-conflict environment of the Great Lakes region that a similar R&D approach (Lisson *et al.* 2010; Millar and Connell 2010) through enhanced livestock feeding of small-stock can lead to comparable livelihood improvements.

## Conclusions

The livestock ladder depicts a system that poor smallholders can use to ascend from keeping small-stock to acquiring larger animals. Investment, therefore, is suggested in the lowest rung of the livestock ladder and investigation of ways to improve small-stock systems, emphasizing the provision of dry season feed. This was raised as a major issue by most respondents in various surveys in Sud Kivu, DRC. Women particularly would greatly benefit from the increased cash income generated through enhanced small-stock production, especially that of cavies. Family and especially children's nutrition would additionally benefit, as small-stock is owned and controlled by women. Nevertheless, smallholders will adopt livestock production innovations only if they match household priorities and resources (Udo *et al.* 2011). This demands an inclusive and participatory R&D approach, underlining also the need for food/feed plants when addressing improved livestock feeding.

The current low livestock density in the Great Lakes region is a concern when crop–livestock integration is regarded as a major support for agricultural production. Small-stock such as cavies are under-researched. The role of small-stock for the livelihood of people in the region, including non-monetary services and indirect benefits, is undocumented. Livestock feeding and forage research have been neglected compared with crop research in the recent past. The role that forages can play in nutrient cycling for the larger farming system requires attention.

## Acknowledgements

We acknowledge the unreserved collaboration by all survey respondents and enumerators. Funding by the Federal Ministry for Economic Cooperation and Development (BMZ), Germany through the project 'More Chicken and Pork in the Pot, and Money in Pocket: Improving Forages for Monogastric Animals with Low-income Farmers' is recognized.

## Note

1 The term 'cavy' is used throughout the text instead of the colloquial 'Guinea pig' because the latter provides a wrong impression of a domestic animal that neither originates from Guinea nor is a pig. We follow the scientific literature, where 'cavy' is considered the appropriate naming if neither the laboratory animal nor the pet is concerned.

## References

Ampaire, A. and Rothschild, M. F. (2010) 'Pigs, goats and chickens for rural development: smallholder farmers' experience in Uganda', *Livestock Research for Rural Development,* vol. 22. www.lrrd.org/lrrd22/6/ampa22102.htm (accessed 4 January 2012).

Bii, B. (2007) 'When the hungry refuse to be used like guinea pigs: one farmer's dream to get rich while ending food shortages proves elusive', *Daily Nation (Nairobi, Kenya),* 18 April, p. 3, http://allafrica.com/stories/200704171135.html (accessed 9 October 2012).

Cox, T. P. (2012) 'Farming in the battlefield: the meanings of war, cattle, and soil in South Kivu, Democratic Republic of the Congo', *Disasters*, vol. 36, no. 2, pp. 233–248.

Dolberg, F. (2001) 'A livestock development approach that contributes to poverty alleviation and widespread improvement of nutrition among the poor', *Livestock Research for Rural Development*, vol. 13. http://lrrd.cipav.org.co/lrrd13/5/dolb135.htm (accessed 9 October 2012 ).

Katunga-Musale, D., Ngabo, T., Bacigale-Bashizi, S., Muhimuzi-Lwaboshi, F. and Maass, B. L. (2011) 'Testing agro-ecological adaptation and participatory acceptability of ten herbaceous legumes in South Kivu, DR Congo', presented at Development at the Margin, Tropentag, 5–7 October 2011, University of Bonn, Germany. www.tropentag.de/2011/abstracts/links/Maass_H4BtL3bR.pdf (accessed 9 October 2012).

Lammers, P. J., Carlson, S. L., Zdorkowski, G. A. and Honeyman, M. S. (2009) 'Reducing food insecurity in developing countries through meat production: the potential of the guinea pig (*Cavia porcellus*)', *Renewable Agriculture and Food Systems*, vol. 24, pp. 155–162.

Lisson, S., MacLeod, N., McDonald, C., Corfield, J., Pengelly, B., Wirajaswadi, L., Rahman, R., Bahar, S., Padjung, R., Razak, N., Puspadi, K., Dahlanuddin, Sutaryono, Y., Saenong, S., Panjaitan, T., Hadiawati, L., Ash, A. and Brennan, L. (2010) 'A participatory, farming systems approach to improving Bali cattle production in the smallholder crop–livestock systems of Eastern Indonesia', *Agricultural Systems*, vol. 103, pp. 486–497.

Maass, B. L., Katunga-Musale, D., Chiuri, W. L. and Peters, M. (2010) *Diagnostic Survey of Livestock Production in South Kivu/DR Congo*, Working Document No. 210, Centro Internacional de Agricultura Tropical (CIAT), Nairobi, Kenya. www.ciat.cgiar.org/ourprograms/Agrobiodiversity/forages/Pages/Publications.aspx (accessed 9 October 2012).

Maass, B. L., Katunga-Musale, D., Chiuri, W. L. and Peters, M. (2012) 'Challenges and opportunities for smallholder livestock production in post-conflict South Kivu, eastern DR Congo', *Tropical Animal Health and Production*, vol. 44, no. 6, pp. 1221–1232.

Matthiesen, T., Nyamete, F., Msuya, J. M. and Maass, B. L. (2011) 'Importance of guinea pig husbandry for the livelihood of rural people in Tanzania – a case study in Iringa Region', in Development at the Margin, Tropentag, 5–7 October 2011, University of Bonn, Germany, www.tropentag.de/2011/abstracts/links/Matthiesen_lIDdf2DY.pdf (accessed 9 October 2012).

Metre, T. K. (2011) 'Small, healthy, high-yielding', *Rural21: The International Journal for Rural Development*, vol. 45, no. 1, pp. 40–42. www.rural21.com/uploads/media/Small_healthy_highyielding_01.pdf (accessed 9 October 2012).

Metre, T. K. (2012) 'Possibilités d'amélioration de l'élevage de cobaye (*Cavia porcellus* L.) au Sud Kivu, à l'est de la République Democratique du Congo', MSc thesis, Université de Liège-Gembloux, Belgium.

Millar, J. and Connell, J. (2010) 'Strategies for scaling out impacts from agricultural systems change: the case of forages and livestock production in Laos', *Agriculture and Human Values*, vol. 27, pp. 213–225.

Mwangi, B. (2011) 'Granny is fed and clothed by rodents', *Daily Nation (Nairobi, Kenya)*, 4 August, p. 6. http://allafrica.com/stories/201108040014.html (accessed 9 October 2012).

Ouma, E. and Birachi, E. (eds) (2011) *CIALCA Baseline Survey, Cialca Technical Report 17*. CIAT, IITA and Bioversity International, Nairobi, Kenya and Ibadan, Nigeria, www.cialca.org/files/files/cialca%20baseline%20survey%20report_print.pdf (accessed 9 October 2012).

Perry, B. D., Randolph, T. F., McDermott, J. J., Sones, K. R. and Thornton, P. K. (2002) *Investing in Animal Health Research to Alleviate Poverty*, International Livestock Research Institute (ILRI), Nairobi, Kenya, ch. 4. www.ilri.org/InfoServ/Webpub/fulldocs/investinginAnimal/Book1/media/PDF_chapters/B1_4.pdf (accessed 9 October 2012).

Pezzullo, J. C. (2011) 'Web pages that perform statistical calculations!'. http://statpages.org/ctab2x2.html (accessed 9 October 2012).

Pica-Ciamarra, U., Tasciotti, L., Otte, J. and Zezza, A. (2011) 'Livestock assets, livestock income and rural households: cross-country evidence from household surveys', Joint Paper of the World Bank, FAO, AU-IBAR, ILRI with support from the Gates Foundation. www.africalivestockdata.org/afrlivestock/sites/africalivestock data.org/files/PAP_Livestock_HHSurveys.pdf (accessed 9 October 2012).

Sumberg, J. (2002) 'Livestock nutrition and foodstuff research in Africa: when is a nutritional constraint not a priority research problem?', *Animal Science*, vol. 75, pp. 332–338.

Todd, H. (1998) 'Women climbing out of poverty through credit; or what do cows have to do with it?', *Livestock Research for Rural Development*, vol. 10. http://lrrd.cipav.org.co/lrrd10/3/todd103.htm (accessed 9 October 2012).

Udo, H. M. J., Aklilu, H. A., Phong, L. T., Bosma, R. H., Budisatria, I. G. S., Patil, B. R., Samdup, T. and Bebe, B. O. (2011) 'Impact of intensification of different types of livestock production in smallholder crop–livestock systems', *Livestock Science*, vol. 139, pp. 22–29.

Zozo, R., Chiuri, W. L., Katunga-Musale, D. and Maass, B. L. (2010) *Report of a Participatory Rural Appraisal (PRA) in the* Groupements *of Miti-Mulungu and Tubimbi, South Kivu/DR Congo*, Working Document no. 211, Centro Internacional de Agricultura Tropical (CIAT), Nairobi, Kenya. www.ciat.cgiar.org/ourprograms/Agrobiodiversity/forages/Pages/Publications.aspx (accessed 9 October 2012).

# 12 N2Africa

## Putting nitrogen fixation to work for smallholder farmers in Africa

*Ken E. Giller,*[1] *Angelinus C. Franke,*[1] *Robert Abaidoo,*[2] *Freddy Baijukya,*[3] *Abdullahi Bala,*[4] *Steve Boahen,*[5] *Kenton Dashiell,*[6] *Speciose Kantengwa,*[7] *Jean-Marie Sanginga,*[8] *Nteranya Sanginga,*[6] *Alastair Simmons,*[9] *Anne Turner,*[10] *Judith de Wolf,*[11] *Paul Woomer*[12] *and Bernard Vanlauwe*[13]

[1] *Wageningen University, Wagenigen, Netherlands;* [2] *International Institute of Tropical Agriculture, Accra, Ghana;* [3] *International Center for Tropical Agriculture, Maseno, Kenya;* [4] *Federal University of Technology, Minna, Nigeria;* [5] *International Institute of Tropical Agriculture, Mozambique;* [6] *International Institute of Tropical Agriculture, Headquarters Ibadan, Nigeria;* [7] *International Center for Tropical Agriculture, Kigali, Rwanda;* [8] *International Center for Tropical Agriculture, Bukavu, DRC;* [9] *Tiscali, United Kingdom;* [10] *International Institute of Tropical Agriculture, Lilongwe, Malawi;* [11] *International Center for Tropical Agriculture, Harare, Zimbabwe;* [12] *International Center for Tropical Agriculture, Nairobi, Kenya;* [13] *International Institute of Tropical Agriculture, Nairobi, Kenya*

To grow beans you have to own a cow.[1]

## Introduction

Putting Nitrogen Fixation to Work for Smallholder Farmers in Africa (N2Africa – www.n2africa.org) is a large-scale project focused on increasing the production of grain legumes and the inputs from $N_2$-fixation in smallholder farming systems in sub-Saharan Africa. We focus on four major grain legumes: common beans (*Phaseolus vulgaris* L.), cowpea (*Vigna unguiculata* (L.) Walp.), groundnut (*Arachis hypogaea* L.) and soybean (*Glycine max* (L.) Merrill) and also on legume forages. The project works with many partners in the Democratic Republic of Congo (DRC), Ghana, Kenya, Malawi, Mozambique, Nigeria, Rwanda and Zimbabwe, and has recently expanded activities in Ethiopia, Uganda, Tanzania, Liberia and Sierra Leone. Building on the results of previous research conducted across many countries, a multi-pronged approach is used to:

1 increase the farm area cropped with legumes;
2 enhance productivity through the use of phosphorus fertilizer and good agronomic practices;
3 introduce improved legume varieties with good pest and disease resistance, grain and nutritional quality and yield;
4 inoculate with rhizobia where needed and select better rhizobial strains; and
5 link farmers to markets and add value through local processing.

N2Africa is designed around a central hypothesis that $N_2$-fixation by legumes, grain and biomass yield depend on:

$$(G_L \times G_R) \times E \times M$$

where $G_L$ = the legume genotype, $G_R$ = the genotype(s) of rhizobia nodulating the legume, E = the environment, including climate (temperature, rainfall, day length, etc., to encompass the length of the growing season) and soils (acidity, aluminium toxicity, limiting nutrients, etc.) and M = management, including agronomic practices (rhizobial inoculation, use of mineral fertilizers, sowing dates, plant density, weeding). The project seeks to identify socio-ecological niches for legumes within the farming systems, moving from a 'best bet' to a 'best fit' approach (Ojiem *et al.* 2006; Giller *et al.* 2011b).

In this chapter we sketch how the project came about. This is followed by a justification for our approach, based on past research conducted in Africa and elsewhere in the world. We conclude by reflecting on lessons learned during the first two years of the project, with particular emphasis on Central Africa.

## How N2Africa came about

### *A brief history*

The initial idea arose from a two-day meeting on $N_2$-fixation held in Seattle by the Bill & Melinda Gates Foundation in April 2008 to which about 25 experts were invited from around the world.[2] Subsequently, Dr Prem Warrior of the foundation approached the first author with the question, 'What approaches can we deploy now to use biological $N_2$-fixation to improve the lives of smallholder farmers in Africa?'.[3] To be offered the opportunity of putting this into practice seemed like a dream come true. At the same time, we faced a daunting challenge. Luckily, we could draw on a rich body of research and experience on the potentials and limitations of $N_2$-fixation to guide us. A series of discussions was initiated and led to an initial concept (Table 12.1) that was presented at the thirteenth meeting of the African Association of Biological Nitrogen-Fixation (AABNF) in Hammamet, Tunisia. The president of AABNF, Dr Ridha Mhamdi, devoted an afternoon of the conference to allow participants to discuss the proposed approach. The concept note was further

*Table 12.1* From first contact to project launch: steps taken in the development of N2Africa

| *Activity* | *Who* | *When* | *Where* |
|---|---|---|---|
| Convening on opportunities for nitrogen fixation research | Prem Warrior, Yvonne Pinto, Ken Giller, Mariangela Hungria and many experts from around the world | April 2008 | Seattle |
| Development of research priorities and objectives | Beatrice Anyango, Abdullahi Bala, Jo Fening, Mariangela Hungria, Didier Lesueur, Prem Warrior, Ken Giller | June–December 2008 | Wageningen |
| Plenary presentation and group discussions | Participants of the thirteenth AABNF meeting | December 2008 | Hammamet, Tunisia |
| Proposal consultation workshop | Some 30 experts from Africa, Australia, Brazil, Europe and the United States | February 2009 | Mombasa, Kenya |
| Detailed proposal writing | Abdullahi Bala, Nteranya Sanginga, Bernard Vanlauwe, Paul Woomer, Ken Giller | March 2009 | Mombasa, Kenya |
| Full proposal evaluation and reformulation | Nteranya Sanginga, Ken Giller | June 2009 | Seattle |
| Project approval and contract signing | | September 2009 | |
| Project launching and inception workshop | | February 2010 | Nairobi, Kenya |

refined at a planning workshop in Mombasa, Kenya by about 25 researchers from around the world. This formed the basis for a full proposal for submission to the foundation. After numerous discussions and adjustments, pigeonpea (*Cajanus cajan* (L.) Millsp.), chickpea (*Cicer arietinum* L.) and $N_2$-fixing trees in the savannas of West Africa were removed from the project activities to match the strategy of the foundation. Finally, the project was approved and started in September 2009.

### *The scientific background*

Maximal rates of $N_2$-fixation in the tropics reach an astonishing 5 kg of N per hectare per day with the green manure *Sesbania rostrata* (Giller 2001). More than 250 kg of N per hectare of fixed $N_2$ has been measured in soybean in Southern Africa, with associated grain yields of more than 3.5 t/ha (Giller *et al.* 2011a). Thus, potential rates of $N_2$-fixation in legumes are not limited by the efficiency of $N_2$-fixation in the legume–rhizobial symbiosis. The largest amounts of $N_2$ are fixed and added to cropping systems through green manures and agroforestry trees (Giller and Cadisch 1995). But participatory evaluations in Africa found that smallholders generally prefer grain legumes, followed by herbaceous and woody fodder legumes, and agroforestry legumes specifically for wood and fencing (Chikowo *et al.* 2004; Baijukya 2006; Kerr *et al.* 2007; Ebanyat 2009). Smallholders rarely invest in legumes that do not give direct benefits of food or fodder. Farmers widely recognize the improvement of yields of subsequent crops due to rotation with legumes and the contribution of legume $N_2$-fixation to soil fertility, but as secondary benefits. Despite claims of the widespread adoption of legume green manures and 'fertilizer trees', this has often been 'pseudo-adoption', driven by artificial markets for the seed of the legumes created by development projects (Giller 2001; Ojiem *et al.* 2006; de Wolf 2010). In several areas, the green manures and fertilizer trees vanished soon after the withdrawal of the projects that promoted them. N2Africa, therefore, focuses on increasing production and $N_2$-fixation in grain legumes and in legume forages in those areas where the production of ruminant livestock forms a major component of the farming system.

## N2Africa explained

The analysis of farming systems in East and Southern Africa had indicated that often less than 10 per cent of the farm area is planted to sole crops of legumes (Giller *et al.* 2006; Ojiem *et al.* 2006, 2007). Even at the current rates of $N_2$-fixation which are generally poor, a modest increase to 20 per cent of the area planted with legumes would automatically double the amount of N input into the farming systems. Successful interventions that have led to enhanced contributions of $N_2$-fixation from legumes in smallholder agriculture are invariably those that gave multiple benefits, including immediate food, fodder and/or income. Key entry points to expand the area planted with legumes are to focus on multi-purpose grain legumes (food, fodder and fertility) and forages (fodder, fertility and fuelwood).

Where markets already exist for the target crop, improved yields resulting from better management practices could lead to a spontaneous increase in cultivation. For instance, beans are widely cultivated and consumed in East and Southern Africa (see Chapter 15). However, low yields due to the devastating effect of root rot disease forced farmers to change to other crops. The introduction of disease-resistant climbing varieties led to large increases in yield and rapid diffusion (Sperling and Muyaneza 1995). Cowpea is a major pulse crop in the West African savanna, traditionally intercropped with cereals such as sorghum and millet. The introduction in the 1990s of a range of improved grain and dual-purpose short- and medium-duration varieties led to rapid adoption by farmers in northern Nigeria (Sanginga *et al.* 2003; Singh *et al.* 2003). The short duration varieties allow two crops of cowpea to be grown in a single season, or one crop, producing grain in the 'hunger period', followed by a cereal crop. If a more indeterminate dual-purpose variety is planted as the second crop, it will produce pods until drought finishes the crop at the end of the growing season and valuable fodder for storage and use in the dry season. By 2001, more than 8,000 farmers were planting the short-duration varieties and there is potential for this technology to reach millions of smallholder farmers in the dry savannas of West Africa (Kristjanson *et al.* 2005). In Nigeria, farmers readily accepted the promiscuous soybean varieties introduced by the International Institute of Tropical Agriculture (IITA) because, in addition to grain, the plants produced large amounts of leafy biomass which contribute N to the soil (Sanginga *et al.* 1999) and assist in the control of the parasitic weed *Striga hermonthica* in cereals. A concerted campaign was needed to promote soybean in Zimbabwe to overcome misconceptions that the crop was not suited to smallholder farming (Mpepereki *et al.* 2000; Giller *et al.* 2011a). The multiple advantages of soybean as a versatile crop for cash and household nutrition and the soil fertility benefits to subsequent maize crops were the main reasons for the rapid uptake by smallholders.

Uptake of forage legumes has been patchier (Sumberg 2002). Substantial research across many countries has provided a wide diversity of well-adapted forage legumes. For example, a number of herbaceous forage legumes were identified as alternatives to *Stylosanthes hamata* for use in legume fodder banks and pastures in West Africa (Tarawali *et al.* 1999). Demand for high-quality fodder for dairy cattle resulted in large numbers of farmers planting fodder tree legumes such as *Calliandra calothyrsus* in the central highlands of Kenya (Franzel *et al.* 2003). N2Africa focuses on forage legumes only where there is a clear demand from farmers for milk or meat production, or when the legumes have additional benefits, such as providing stakes for climbing beans. Research into self-regenerating (weedy) legumes that can improve soil fertility (indifallows – Mapfumo *et al.* 2005) shows promise while incurring little cost or trade-off in terms of labour or land. It may be worth further investing in such legumes and their rhizobia as a low-cost soil fertility management strategy for resource-poor farmers.

A 'socio-ecological niche' approach is used to identify the legume/rhizobium combination that is appropriate for a given type of farm within a given agro-ecology or farming system depending on market opportunities (Ojiem *et al.* 2006). Matching a genotype with the right socio-ecological niche is necessary for improved productivity and yield and enhanced income for farmers (Ojiem *et al.* 2007).

## $G_L$ *in the* $(G_L \times G_R) \times E \times M$ *equation*

Wide genetic variability for nodulation and $N_2$-fixation exists in all the major grain legumes. The ability to fix substantial amounts of $N_2$ is often associated with a large total production of biomass – due to creeping or indeterminate ability – and with long duration (Giller 2001). Contrary to the widely held idea that legumes suffer a yield-restricting penalty due to the energy demands for $N_2$-fixation, there is strong evidence that legumes compensate by increasing the rate of photosynthesis (Kaschuk *et al.* 2009). This is known as sink-stimulation of photosynthesis, which appears both for rhizobial and mycorrhizal symbioses in legumes. A meta-analysis suggests that dependence on $N_2$-fixation increases the legume harvest index compared with reliance on soil-N or fertilizer-N (Kaschuk *et al.* 2010).

*Phaseolus vulgaris* (common bean) is widely regarded as a poor $N_2$-fixer, but in favourable environments it can fix at rates similar to those of all other grain legumes (Giller 1990). Investment in breeding varieties for adaptation to adverse environmental conditions (notably soil acidity and P limitations) will give rapid success in improving $N_2$-fixation and yields. Breeding for enhanced $N_2$-fixation simply requires selection for biomass, grain and N-yields under N-limiting conditions. Experience in Malawi using such an approach took yields from about 0.2–0.5 t/ha to 1.5 t/ha within five years (CIAT, unpublished data), though this programme was not focused specifically on breeding for $N_2$-fixation. Breeding beans for drought stress has also improved rooting vigour in soils with poor fertility, with substantial increases in yields (Beebe *et al.* 2008). This gives us confidence that we can rapidly identify genotypes with superior ability to fix $N_2$, as well as materials with specific traits to feed into future breeding programmes.

## $G_R$ *in the* $(G_L \times G_R) \times E \times M$ *equation*

Legume hosts differ in the range of bacterial partners with which they form symbioses. At one end of the spectrum are legumes such as soybean and chickpea which nodulate with a restricted number of rhizobial strains or species and are considered specific in their requirement. At the other end of the spectrum is cowpea, which is considered the most promiscuous (or non-specific) of the grain legumes, nodulating with a wide range of both fast- and slow-growing rhizobia (Pueppke and Broughton 1999). Among rhizobia, similar relationships also exist. Rhizobia that were isolated from tropical soils were earlier grouped as members of the 'cowpea miscellany' because cowpea is nodulated by many strains of slow-growing rhizobia (Giller 2001). It is now clear that rhizobia in the tropics constitute a highly diverse group of both fast- and slow-growing types with a wide range of symbiotic specificities. From the farmers' perspective, the ability of many grain legumes to nodulate effectively with rhizobia found in the soil is a great benefit. Legumes that have a specific requirement for rhizobia need inoculation that can be achieved using commercially available rhizobial inoculants. However, inoculant use is very limited in smallholder farming systems of sub-Saharan Africa (SSA).

The global market for rhizobial inoculants is focused on soybean due to its strong requirement for inoculation. The specific nodulation of most commercial varieties of soybean arose because they were bred largely in North America where indigenous rhizobia compatible with soybean were absent (Giller 2001). This led to the inoculation of soybean being a standard practice. The breeding programme was highly successful in increasing the potential grain yield of soybean, originally introduced as a forage crop, by increasing the seed size and harvest index, so that varieties from North America have been used in many other countries. In essence, the breeders inadvertently turned soybean into a specifically nodulating legume. Promiscuous multi-purpose varieties of soybean were selected in Nigeria by the IITA in a research programme led by two of the authors of this chapter, Kenton Dashiell and Nteranya Sanginga (Kueneman *et al.* 1984; Sanginga *et al.* 2003). The increase in promiscuity was achieved by crossing high-yielding varieties from North America with more promiscuous materials from Asia. In Southern Africa, highly promiscuous varieties were identified from early introductions of soybean that originated from China (Mpepereki *et al.* 2000). In areas where inoculants are not readily available, the promiscuous varieties provide a particular niche for smallholder farmers, although there are indications that even the promiscuous varieties sometimes respond to inoculation (Osunde *et al.* 2003). A key question for N2Africa is whether the promiscuous soybean varieties would respond frequently and strongly to inoculation across different regions and countries in Africa.

The common bean is relatively promiscuous in its rhizobial requirement and nodulates with many different species of fast-growing rhizobia (Giller 2001). Earlier studies in Kenya found that inoculation increased nodulation but did not significantly increase yields (Ssali 1988). Large increases in nodulation and growth in acid soils were found in pot trials when common bean was inoculated, together with liming and P fertilization (Ssali 1981). Multi-locational trials in northern Tanzania showed small though consistent responses to inoculation with a mean yield increase of around 8 per cent across 30 trials (Amijee and Giller 1998; Giller *et al.* 1998). Ndakidemi *et al.* (2006), using the same inoculant strain of *Rhizobium tropici*, CIAT 899, observed stronger responses to inoculation across ten locations in northern Tanzania.

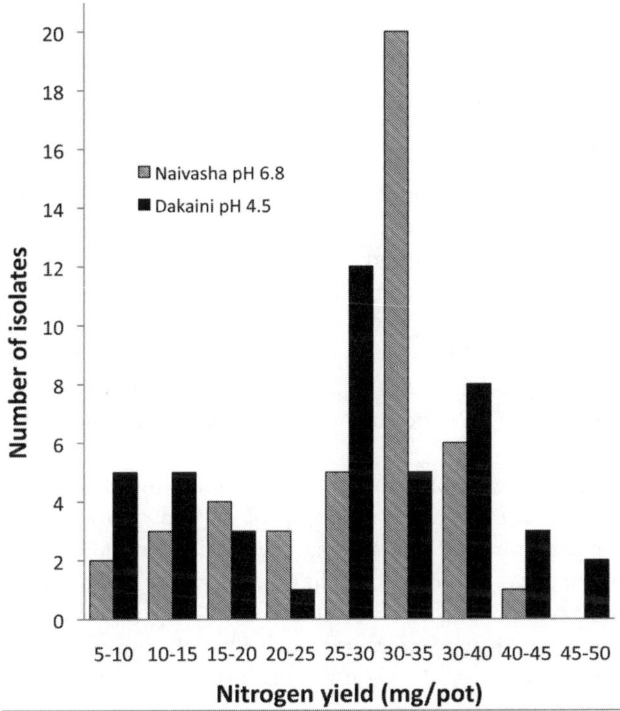


*Figure 12.1* Distribution of rhizobia in two Kenyan soils (Naivasha and Dakaini) based on their $N_2$-fixing effectiveness with beans (*Phaseolus vulgaris*). Uninoculated plants yielded 9 mg N per pot and plants inoculated with CIAT 899, the standard commercial inoculant strain, yielded 32 mg N per pot. Several of the isolates gave significantly more $N_2$-fixation than the commercial strain (source: redrawn from Anyango *et al.* 1995).

Detailed studies showed that the crop was nodulated by a wide range of rhizobia in Kenyan soils, and many of the strains forming nodules were ineffective in $N_2$-fixation, though several were more effective than CIAT 899 (Figure 12.1; Anyango *et al.* 1995). Thus past research indicates that investment in research is worthwhile to develop inoculants of improved quality and more effective strains for common beans. Other tropical grain legumes (cowpea, groundnut and pigeonpea) are promiscuous and unlikely to respond to inoculation unless inoculated with selected strains in high numbers. For these crops, increasing $N_2$-fixation inputs in smallholders' fields is more readily achieved by the selection of better varieties and agronomic approaches. The potential for the selection of better strains that may result in improvements in the yields of the promiscuous legumes through inoculation will be explored through research.

### *E in the $(G_L \times G_R) \times E \times M$ equation*

Legumes need to be targeted to the environments or agro-ecologies in which they can grow well. The major factors that determine the suitability of climates are largely the amount of rainfall, the distribution of rainfall (which in the tropics determines the length of the growing season) and the temperature, which is largely determined by the altitude. There is a huge range of varieties within all the major grain legume crops with a growth duration ranging from less than 60 days to more than 270 days for some varieties. The shorter-duration varieties are generally better suited to warmer, drier climates, although it is not easy to generalize. An interesting example is climbing beans, which originate in cool climates in the Andes and tend to grow best at altitudes of 2,000 masl or more in Central Africa. Breeding of climbing beans adapted to lower, warmer altitudes seeks to expand the range of the crop. Evaluations of these 'mid-altitude climbers' or MACs by N2Africa has identified many promising lines in western Kenya.

In terms of the soil environment, groundnut and cowpea tend to grow better on coarse sands than soybean or common bean, and cowpea is more tolerant of acid soils (and aluminium toxicity) than common beans (Giller 2001). Legume yields and $N_2$-fixation are commonly limited by nutrient deficiencies (particularly of phosphorus, but also of other nutrients). Phosphorus availability is often limited – due either to the inherently small P stocks in light-textured (sandy) soils, or to the fixation of P into unavailable forms in heavy-textured (clay) soils. Phosphorus deficiency can also be overcome by using germplasm that is efficient at mobilizing and using phosphorus (Ae *et al.* 1990; Lynch 2007; Ramaekers *et al.* 2010). Crop legumes are all (with the exception of lupin) dependent on arbuscular mycorrhiza for the efficient uptake of P from the soil. Plant genetic variation in phosphorus uptake and use efficiency is manifested through a variety of physiological adaptations including mycorrhizal dependency, root exudation and functional root architecture (Nwoko and Sanginga 1999; Lynch 2007). Genotypes of common beans differ in their ability to scavenge and accumulate molybdenum (Brodrick and Giller 1991), iron and zinc (Blair *et al.* 2009). In general, if the soil nutrient supply is strongly limiting, deficiencies in grain legumes need to be addressed through management.

### *M in the $(G_L \times G_R) \times E \times M$ equation*

Ensuring that the genetic potential of the legume–rhizobium symbiosis in the field is expressed depends on good management. Wide plant spacing or late planting reduces efficiency of the capture of light for photosynthesis and limits the potential yield. In terms of P deficiency, animal manures can supply P, but only if applied in large quantities, and plant residues contain too little P to make much of a contribution. Many deposits of rock phosphate (RP) are found in SSA, but few are sufficiently soluble to be applied directly without pre-treatment (Buresh *et al.* 1997). Although considerable attention was given to 'soil fertility recapitalization' in the 1990s, recommending one-off large applications of RP, it is clear that frequent small additions (typically 15–30 kg P per hectare) of soluble forms of P give a much more efficient and affordable use of scarce nutrient resources (Buresh *et al.* 1997). While all crops need an adequate supply of P, legumes are particularly responsive and large inputs from nitrogen-fixation are possible only when P deficiencies are corrected (Giller and Cadisch 1995). Typical P-recovery efficiencies are 20–25 per cent in a first crop and 2–3 per cent in subsequent crops (Janssen *et al.* 1987). If P is added regularly, the residual effect builds up. The addition of P with organic manures can enhance the availability of P by reducing the fixation of phosphate ions onto clay surfaces.

Soil nutrient constraints may be overcome by the use of manure and/or mineral fertilizers. In sandy soils, an over-supply of P can exacerbate incipient deficiencies of zinc, but this problem can be avoided if animal manures are used (Zingore *et al.* 2008a). In soils that have been cropped repeatedly, other nutrients are often needed in addition to P, to correct deficiencies of potassium, calcium and magnesium, or micronutrients, thus stressing the need for balanced fertilization (Smithson *et al.* 1993; Zingore *et al.* 2008a, 2008b). Attention to nutrient management for crop sequences and intercrops (using the principles of ISFM) would ensure the most efficient use of nutrient resources and maximize inputs from $N_2$-fixation.

## Key results from the first two years of N2Africa

### *N2Africa as a 'development to research' project*

As opposed to the dominant 'agricultural research for development' (AR4D) paradigm, N2Africa focuses on the delivery and dissemination (D&D) of the best available $N_2$-fixing legume technologies at the core of project activities (Figure 12.2). Monitoring and evaluation (M&E) seek to understand why certain technologies work best for particular farmers, and feedback loops through adaptive research seek to refine and improve the technologies through addressing those problems that emerge. Thus, the emphasis is on improving $N_2$-fixing legume technologies, solving problems encountered in the field, understanding how to tailor technologies to different farms and farming systems and using this understanding to refine D&D.

In practice, this is not as easy as it sounds. Particularly in countries with bi-modal rainfall and two rainy seasons each year, it is hard to collate and analyse data from field experiments in time to use the outcomes of one season for the planning of the next. The collation of data on M&E also requires significant coordination and takes time. This often means that decisions on the adaptation of country plans are often taken, based on feedback from farmers, on the first impressions of crop performance, and before deeper learning has taken place.

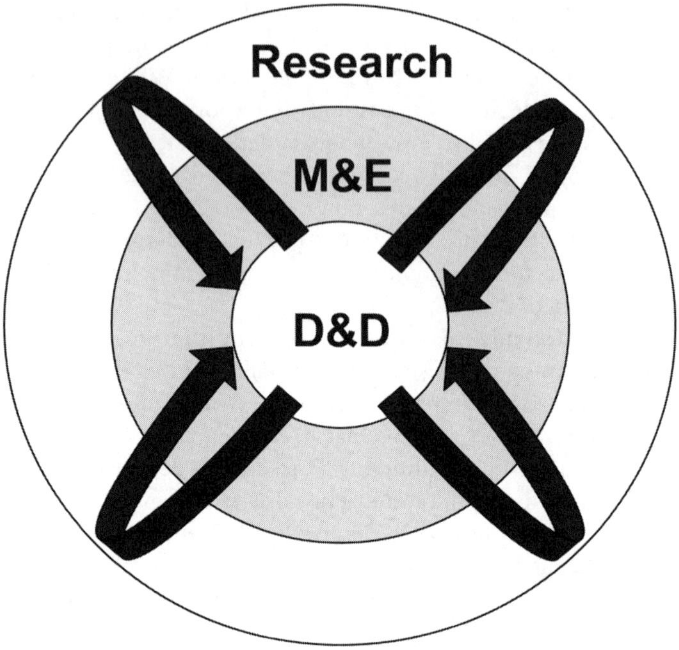


*Figure 12.2* N2Africa is a 'development to research project' in which D&D are core activities that take $N_2$-fixing technologies to thousands of farmers; M&E provide the learning of what works where, why and for whom; and research feedback loops analyse and feed back to improve the technologies and their targeting in D&D.

### *Targeting guided by farming systems analysis*

N2Africa analyses diversity in farms and farming systems using the NUANCES framework (Giller *et al.* 2011b). A detailed baseline survey of 3,400 farms in the N2Africa target areas reveals the wide range of farm sizes and resource endowments in each of the countries. In three countries in East and Central Africa (DRC, Kenya and Rwanda), farm sizes are very small with median values below 1 ha in all countries and as small as 0.2 ha in Rwanda (Figure 12.3). Farm typologies are developed and detailed characterizations are made of representative farms to assist in understanding appropriate opportunities for intensification through legumes. The opportunities differ widely for farmers with varied resource endowment and aspirations (Franke *et al.* submitted).

The diversity of farms within each location highlights the need to revisit the accepted definition of a farming system (Dixon *et al.* 2001):

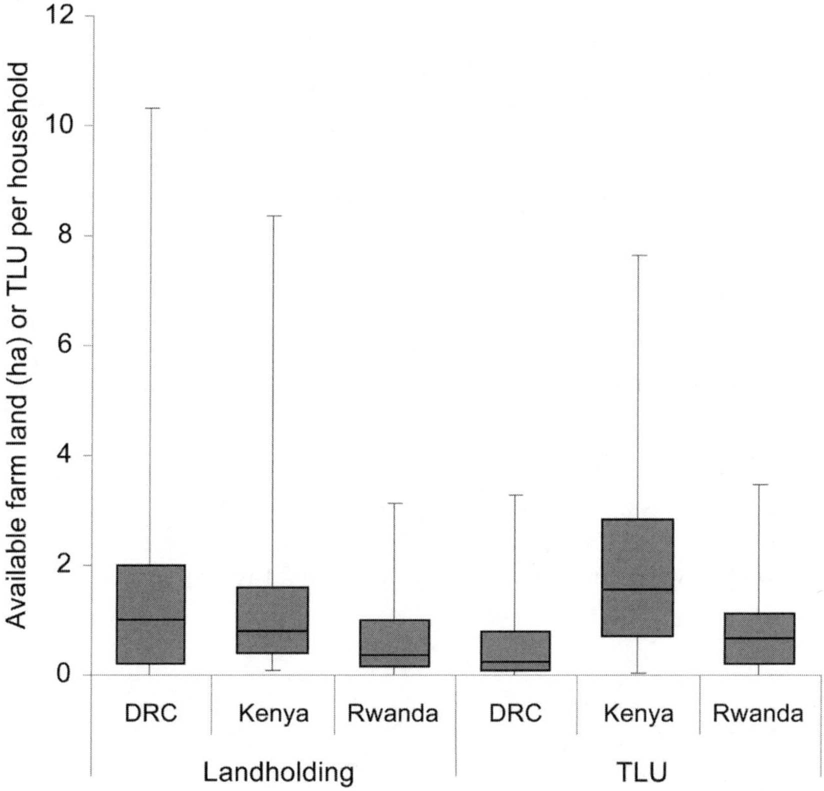


*Figure 12.3* Farm size (ha) and livestock ownership (tropical livestock units – TLU) in the N2Africa target areas in DRC, Kenya and Rwanda from the baseline survey conducted in 2010.

> A farming system . . . is defined as a population of individual farm systems that have broadly similar resource bases, enterprise patterns, household livelihoods and constraints, and for which similar development strategies and interventions would be appropriate.

It is clear that the same development strategy is not appropriate for every farmer. While some have sufficient land to invest in growing grain legumes for the market through a 'value chain' approach, poorer farmers are not able to do this. The versatility of legumes means they can still offer something for the households with limited land areas, for instance, when legumes can be intercropped with other staple crops enabling an intensification of land use, or when legumes such as climbing bean grown as a sole crop can compete with non-legume staple crops in terms of productivity.

***Multi-locational combined demonstration trials***

A major activity of N2Africa is to explore the performance and potential impact of management approaches to increase yields and the area of grain legumes planted across villages and regions. Simple, multi-locational trials are conducted on farmers' fields spanning the target areas. An example from Sud Kivu, DRC, shows the wide variability in yields found in farmers' fields (Figure 12.4a). In control plots sown at the same time soybean yields varied 0.2–2.5 t/ha. The addition of phosphorus alone gave variable increases in yield, whereas inoculation gave larger and more consistent increases. The addition of phosphorus together with inoculation obtained the largest yields, with many plots yielding as much as 3.0–3.5 t/ha. The strong positive interaction results from the increased demand for P when inoculated plants grow much more strongly. In a nearby village, Bughore, where the demonstrations were sown just one week later (Figure 12.4b), there was a drought immediately after sowing which led to poor emergence and the yield was poor in all the plots. This shows the highly risky nature of farming in relation to unreliable rainfall, and emphasizes the need for timeliness of sowing in relation to the rains.

Since the introduction of well-adapted varieties of climbing beans in the 1980s, the crop has swept through northern Rwanda and largely replaced bush beans

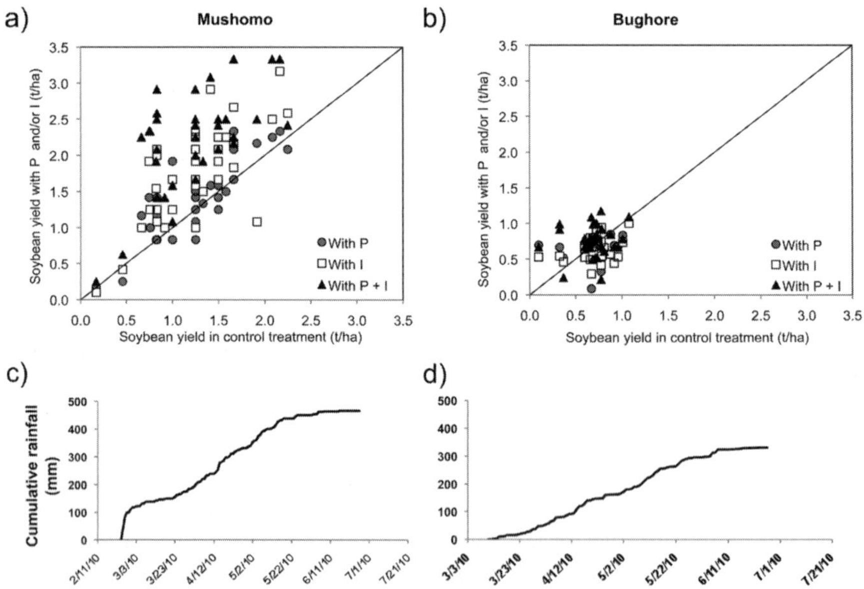


*Figure 12.4* Responses to phosphorus (P) and inoculation with rhizobia (I), and I + P in demonstration trials (a, b) and cumulative rainfall (c, d) in two villages, Mushomo (a, c) and Bughore (b, d) in Sud Kivu, DRC, in 2010.

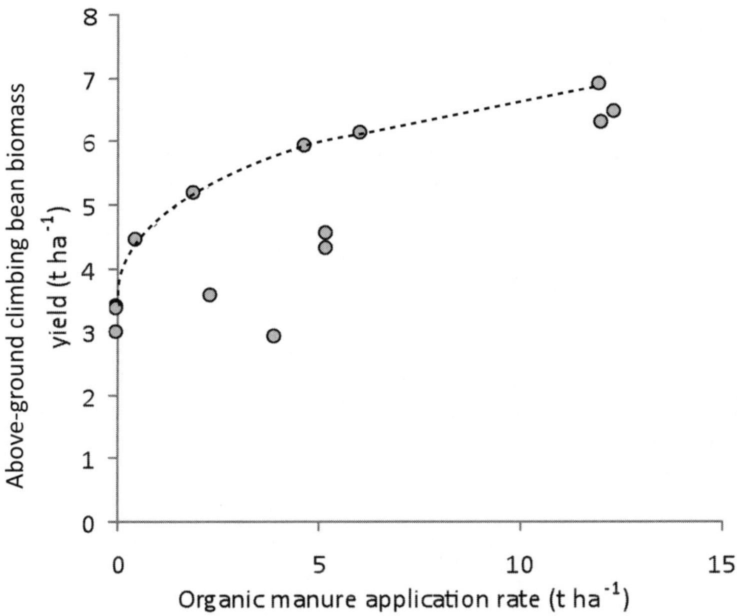


*Figure 12.5* Relation between organic manure application and above-ground biomass yield in farmers' fields in Rwanda (Pearson's $r = 0.82$). Maximum yields, connected by the dashed line in the graph, strongly increase with small applications of manure, while at higher application rates the response to manure diminishes. Yields below the dashed lines were probably constrained by other factors (source: drawn from data in Reckling 2011).

(Sperling and Muyaneza 1995). The ability to intensify production by growing a crop vertically results in a much greater yield potential, and our work indicates that yields of climbing beans are determined by the length, number and quality of the stakes used to support them (Klapwijk 2011; Reckling 2011). It has long been known that climbing beans nodulate more prolifically and have much greater $N_2$-fixation potential than bush beans (Graham 1981). Our field experience confirms the advice of farmer Marie Thérêse Nyiransekuye: 'To grow beans you have to own a cow.' A small amount of manure in the planting hole is needed to provide the nutrients and a favourable growing environment for climbing beans to yield well. On-farm studies in Rwanda indicated that the productivity of climbing beans increases with increasing rates of organic manure (Figure 12.5). So what can farmers who do not have cows do to grow climbing beans? This is obviously a problem in DRC, where there are so few cattle (Figure 12.3). Yet even when the livestock are as small as rabbits and guinea pigs (see Chapter 11), manure targeted to the planting hole can still give a significant benefit in the yield of climbing beans.

## Consequences for agro-ecological intensification

Symbiotic $N_2$-fixation is a key component of agro-ecological intensification – delivering N from the air directly into the farming systems. $N_2$-fixation through grain legumes and legume forages offers 'protein from the air' for people and livestock, and contributes to the yields of other crops through residual effects on soil fertility. N2Africa has rapidly demonstrated a 'proof of concept' in terms of the importance of integrated approaches to $(G_L \times G_R) \times E \times M$ in farmers' fields. As project activities gradually expand, we learn more about the constraints and opportunities across different farming systems and countries in SSA. Zooming in within each country, we find that crop responses to inputs – both of the legumes and other crops in rotation or intercrops – are highly variable across small distances within farms and farming systems. This emphasizes the need to understand variability, in order to tailor legume-based technologies to suit the socio-ecological niches and satisfy the needs of a diverse group of farmers. Common problems that impede the intensification of legume production relate to the consolidation of smallholder production into sufficiently large quantities, to the timely availability of inputs and to human capacity. The greatest successes have been observed where robust legume technologies are supported by well-developed institutions.

## Acknowledgements

We thank the Bill & Melinda Gates Foundation, in particular Dr Prem Warrior, and the Howard G. Buffet Foundation for partnering in N2Africa. Our work relies on a huge range of partners from NGOs, national research and extension organizations, farmers' organizations and the farmers themselves. We are grateful to all our partners for their enthusiastic collaboration.

## Notes

1 Marie Therese Nyiransekuye, a farmer in Gakenke district, Kivuruga sector, Cyintare cell/village, Rwanda; see www.n2africa.org/N2media – Climbing beans.
2 Earlier discussions had taken place through the Tropical Legumes II project on an initiative to address the need for inoculation in soybean, including a workshop held in Arusha, Tanzania, in March 2008.
3 The Bill & Melinda Gates Foundation do not accept unsolicited proposals for funding.

## References

Ae, N., Arihara, J., Okada, K., Yoshimara, T. and Johansen, C. (1990) 'Phosphorus uptake by pigeon pea and its role in cropping systems of the Indian subcontinent', *Science,* vol. 248, pp. 477–480.

Amijee, F. and Giller, K. E. (1998) 'Environmental constraints to nodulation and nitrogen fixation of *Phaseolus vulgaris* L. in Tanzania. I. A survey of soil fertility and root nodulation', *African Journal of Crop Science,* vol. 6, pp. 159–169.

Anyango, B., Wilson, K. J., Beynon, J. L. and Giller, K. E. (1995) 'Diversity of rhizobia nodulating *Phaseolus vulgaris* L. in two Kenyan soils of contrasting pHs', *Applied and Environmental Microbiology*, vol. 61, pp. 4016–4021.

Baijukya, F. P. (2006) 'Adapting to change in banana-based farming systems of northwest Tanzania: the potential role of herbaceous legumes', PhD thesis, Wageningen University, Wageningen.

Beebe, S. E., Rao, I. M., Cajiao, C. and Grajales, M. (2008) 'Selection for drought resistance in Common Bean also improves yield in phosphorus limited and favorable environments', *Crop Science*, vol. 48, pp. 582–592.

Blair, M. W., Astudillo, C., Grusak, M. A., Graham, R. and Beebe, S. E. (2009) 'Inheritance of seed iron and zinc concentrations in common bean (*Phaseolus vulgaris* L.)', *Molecular Breeding*, vol. 23, pp. 197–207.

Brodrick, S. J. and Giller, K. E. (1991) 'Genotypic difference in molybdenum accumulation affects $N_2$-fixation in tropical *Phaseolus vulgaris* (L)', *Journal of Experimental Botany*, vol. 42, pp. 1339–1343.

Buresh, R. J., Smithson, P. C. and Hellums, D. T. (1997) 'Building soil phosphorus capital in Africa', in R. J. Buresh, P. A. Sanchez and F. Calhoun (eds), *Replenishing Soil Fertility in Africa*, ASA, CSSA, SSSA, Madison, WI, pp. 111–149.

Chikowo, R., Mapfumo, P., Nyamugafata, P. and Giller, K. E. (2004) 'Maize productivity and mineral N dynamics following different soil fertility management practices on a depleted sandy soil in Zimbabwe', *Agriculture, Ecosystems and Environment*, vol. 102, pp. 119–131.

de Wolf, J. J. (2010) 'Innovative farmers, non-adapting institutions: a case study of the organization of agroforestry research in Malawi', in L. A. German, J. J. Ramisch and R. Verma (eds), *Beyond the Biophysical: Knowledge, Culture, and Politics in Agriculture and Natural Resource Management*, Springer, Düsseldorf, pp. 217–239.

Dixon, J., Gulliver, A. and Gibbon, D. (2001) *Farming Systems and Poverty: Improving Farmers' Livelihoods in a Changing World*, FAO, Rome.

Ebanyat, P. (2009) 'A road to food: efficiency of nutrient management options targeted to heterogeneous soilscapes in the Teso farming system, Uganda', PhD thesis, Wageningen University, Wageningen.

Franke, A. C., van den Brand, G. J. and Giller, K. E. (submitted) 'Which farmers benefit most from grain legumes in Malawi? An *ex ante* impact assessment based on farm characterisations and model explorations', *European Journal of Agronomy*.

Franzel, S., Wambugu, C., Tuwei, P. and Karanja, G. (2003) 'The adoption and scaling up of the use of fodder shrubs in central Kenya', *Tropical Grasslands*, vol. 37, pp. 239–250.

Giller, K. E. (1990) 'Assessment and improvement of nitrogen fixation in tropical *Phaseolus vulgaris* L.', *Soil Use and Management*, vol. 6, pp. 82–84.

Giller, K. E. (2001) *Nitrogen Fixation in Tropical Cropping Systems*, CAB International, Wallingford.

Giller, K. E. and Cadisch, G. (1995) 'Future benefits from biological nitrogen fixation: an ecological approach to agriculture', *Plant and Soil*, vol. 174, pp. 255–277.

Giller, K. E., Amijee, F., Brodrick, S. J. and Edje, O. T. (1998) 'Environmental constraints to nodulation and nitrogen fixation of *Phaseolus vulgaris* L. in Tanzania. II. Response to N and P fertilizers and inoculation', *African Journal of Crop Science*, vol. 6, pp. 171–178.

Giller, K. E., Rowe, E., de Ridder, N. and van Keulen, H. (2006) 'Resource use dynamics and interactions in the tropics: scaling up in space and time', *Agricultural Systems*, vol. 88, pp. 8–27.

Giller, K. E., Murwira, M. S., Dhliwayo, D. K. C., Mafongoya, P. L. and Mpepereki, S. (2011a) 'Soyabeans and sustainable agriculture in southern Africa', *International Journal of Agricultural Sustainability*, vol. 9, pp. 50–58.

Giller, K. E., Tittonell, P., Rufino, M. C., van Wijk, M. T., Zingore, S., Mapfumo, P., Adjei-Nsiah, S., Herrero, M., Chikowo, R., Corbeels, M., Rowe, E. C., Baijukya, F., Mwijage, A., Smith, J., Yeboah, E., van der Burg, W. J., Sanogo, O. M., Misiko, M., de Ridder, N., Karanja, S., Kaizzi, C., K'ungu, J., Mwale, M., Nwaga, D., Pacini, C. and Vanlauwe, B. (2011b) 'Communicating complexity: integrated assessment of trade-offs concerning soil fertility management within African farming systems to support innovation and development', *Agricultural Systems*, vol. 104, pp. 191–203.

Graham, P. H. (1981) 'Some problems of nodulation and symbiotic nitrogen fixation in *Phaseolus vulgaris* L.: a review', *Field Crops Research*, vol. 4, pp. 93–112.

Janssen, B. H., Lathwell, D. J. and Wolf, J. (1987) 'Modeling long-term crop response to fertilizer phosphorus. II. Comparison with field results', *Agronomy Journal*, vol. 79, pp. 452–458.

Kaschuk, G., Kuyper, T. W., Leffelaar, P. A., Hungria, M. and Giller, K. E. (2009) 'Are the rates of photosynthesis stimulated by the carbon sink strength of rhizobial and arbuscular mycorrhizal symbioses?', *Soil Biology and Biochemistry*, vol. 41, pp. 1233–1244.

Kaschuk, G., Leffelaar, P., Giller, K. E., Alberton, O., Hungria, M. and Kuyper, T. W. (2010) 'Responses of legumes to rhizobia and arbuscular mycorrhizal fungi: a meta-analysis of potential photosynthate limitation of symbioses', *Soil Biology and Biochemistry*, vol. 42, pp. 125–127.

Kerr, R. B., Snapp, S., Chirwa, M., Shumba, L. and Msachi, R. (2007) 'Participatory research on legume diversification with Malawian smallholder farmers for improved human nutrition and soil fertility', *Experimental Agriculture*, vol. 43, pp. 437–453.

Klapwijk, C. J. (2011) 'A comparison of the use of bean stakes in northern Rwanda', MSc thesis, Wageningen University, Wageningen.

Kristjanson, P., Okike, I., Tarawali, S., Singh, B. B. and Manyong, V. M. (2005) 'Farmers' perceptions of benefits and factors affecting the adoption of improved dual-purpose cowpea in the dry savannas of Nigeria', *Agricultural Economics*, vol. 32, pp. 195–210.

Kueneman, E., Root, W., Dashiell, K. and Hohenberg, J. (1984) 'Breeding soybeans for the tropics capable of nodulating effectively with indigenous *Rhizobium* spp', *Plant and Soil*, vol. 82, pp. 387–396.

Lynch, J. P. (2007) 'Roots of the Second Green Revolution', *Australian Journal of Botany*, vol. 55, pp. 493.

Mapfumo, P., Mtambanengwe, F., Giller, K. E. and Mpepereki, S. (2005) 'Tapping indigenous herbaceous legumes for soil fertility management by resource-poor farmers in Zimbabwe', *Agriculture, Ecosystems and Environment,* vol. 109, pp. 221–233.

Mpepereki, S., Javaheri, F., Davis, P. and Giller, K. E. (2000) 'Soyabeans and sustainable agriculture: "promiscuous" soyabeans in southern Africa', *Field Crops Research*, vol. 65, pp. 137–149.

Ndakidemi, P. A., Dakora, F. D., Nkonya, E. M., Ringo, D. and Mansoor, H. (2006) 'Yield and economic benefits of common bean (*Phaseolus vulgaris*) and soybean (*Glycine max*) inoculation in northern Tanzania', *Australian Journal of Experimental Agriculture,* vol. 46, pp. 571.

Nwoko, H. and Sanginga, N. (1999) 'Dependence of promiscuous soybean and herbaceous legumes on arbuscular mycorrhizal fungi and their response to bradyrhizobial inoculation in low P soils', *Applied Soil Ecology*, vol. 13, pp. 251–258.

Ojiem, J. O., de Ridder, N., Vanlauwe, B. and Giller, K. E. (2006) 'Socio-ecological niche: a conceptual framework for integration of legumes in smallholder farming systems', *International Journal of Agricultural Sustainability*, vol. 4, pp. 79–93.

Ojiem, J. O., Vanlauwe, B., de Ridder, N. and Giller, K. E. (2007) 'Niche-based assessment of contributions of legumes to the nitrogen economy of Western Kenya smallholder farms', *Plant and Soil*, vol. 292, pp. 119–135.

Osunde, A. O., Gwam, S., Bala, A., Sanginga, N. and Okogun, J. A. (2003) 'Responses to rhizobial inoculation by two promiscuous soybean cultivars in soils of the Southern Guinea savanna zone of Nigeria', *Biology and Fertility of Soils*, vol. 37, pp. 274–279.

Pueppke, S. G. and Broughton, W. J. (1999) '*Rhizobium* sp. strain NGR234 and *R. fredii* USDA257 share exceptionally broad, nested host ranges', *Molecular Plant-Microbe Interactions*, vol. 12, pp. 293–318.

Ramaekers, L., Remans, R., Rao, I. M., Blair, M. W. and Vanderleyden, J. (2010) 'Strategies for improving phosphorus acquisition efficiency of crop plants', *Field Crops Research*, vol. 117, pp. 169–176.

Reckling, M. (2011) 'Towards increased adoption of grain legumes among Malawian farmers: exploring opportunities and constraints through detailed farm characterization', MSc thesis, Wageningen University, Wageningen.

Sanginga, N., Dashiell, K., Diels, J., Vanlauwe, B., Lyasse, O., Carsky, R., Tarawali, S., Asafo-Adjei, B., Menkir, A., Schulz, S., Singh, B., Chikoye, D., Keatinge, D. and Ortiz, R. (2003) 'Sustainable resource management coupled to resilient germplasm to provide new intensive cereal–grain–legume–livestock systems in the dry savanna', *Agriculture, Ecosystems and Environment*, vol. 100, pp. 305–314.

Sanginga, P. C., Adesina, A. A., Manyong, V. M., Otite, O. and Dashiell, K. (1999) *Social Impact of Soybean in Nigeria's Southern Guinea Savanna*, IITA, Ibadan, Nigeria.

Singh, B. B., Ajeigbe, H. A., Tarawali, S. A., Fernandez-Rivera, S. and Abubakar, M. (2003) 'Improving the production and utilization of cowpea as food and fodder', *Field Crops Research*, vol. 84, pp. 169–177.

Smithson, J. B., Edje, O. T. and Giller, K. E. (1993) 'Diagnosis and correction of soil nutrient problems of common bean (*Phaseolus vulgaris*) in the Usambara Mountains of Tanzania', *Journal of Agricultural Science, Cambridge*, vol. 120, pp. 233–240.

Sperling, L. and Muyaneza, S (1995) 'Intensifying production among smallholder farmers: the impact of improved climbing beans in Rwanda', *African Crop Science Journal*, vol. 3, pp. 117–125.

Ssali, H. (1981) The effect of $CaCO_3$, inoculation and lime pelleting on the nodulation and growth of beans in five acid soils', *Plant and Soil*, vol. 62, pp. 53–63.

Ssali, H. (1988) '*Rhizobium phaseoli* inoculation trials on farmers' fields in Kenya', *East African Agriculture and Forestry Journal*, vol. 53, pp. 151–157.

Sumberg, J. E. (2002) 'The logic of fodder legumes in Africa', *Food Policy,* vol. 27, pp. 285–300.

Tarawali, S. A., Peters, M. and Schultze-Kraft, R. (1999) *Forage Legumes for Sustainable Agriculture and Livestock Production in Subhumid West Africa*, ILRI, Nairobi.

Zingore, S., Delve, R. J., Nyamangara, J. and Giller, K. E. (2008a) 'Multiple benefits of manure: the key to maintenance of soil fertility and restoration of depleted sandy soils on African smallholder farms', *Nutrient Cycling in Agroecosystems*, vol. 80, pp. 267–282.

Zingore, S., Murwira, H. K., Delve, R. J. and Giller, K. E. (2008b) 'Variable grain legume yields, responses to phosphorus and rotational effects on maize across soil fertility gradients on smallholder farms in Zimbabwe', *Nutrient Cycling in Agroecosystems*, vol. 80, pp. 1–18.

# 13 Integrating climate change adaptation and mitigation in East African coffee ecosystems

*Henk van Rikxoort,*[1,2] *Laurence Jassogne,*[3] *Peter Läderach*[4] *and Piet van Asten*[3]

[1] *International Center for Tropical Agriculture (CIAT), Colombia*
[2] *Wageningen University and Research Centre (WUR), the Netherlands*
[3] *International Institute of Tropical Agriculture (IITA), Uganda*
[4] *International Center for Tropical Agriculture (CIAT), Nicaragua*

## Introduction

Coffee is of major importance for East African countries. On a macro-level in all these countries, the crop is one of the three main export products and represents a large export share; in 2009 this was: Burundi 76 per cent, Uganda 35 per cent, Ethiopia 31 per cent and Rwanda 29 per cent (AEO 2013). As exports contribute strongly to the GDP of East African countries (in 2010 this was: Kenya 26 per cent, Uganda 24 per cent and Tanzania 24 per cent), coffee plays an important role in the national economies in the region (World Bank 2013). At the micro-level, coffee has an important place in the livelihood strategy for many of the rural poor in East Africa, as smallholder farmers are primarily responsible for producing coffee in the region (Salami *et al.* 2010).

Anthropogenic climate change represents a serious threat to this role for East Africa's national economies, supply chains and rural livelihoods. A recent study that assessed the impact of climate change on Kenya's coffee production zones predicts a temperature increase of 1.0°C in 2020 and 2.3°C in 2050, combined with a less seasonal climate in terms of both temperature and precipitation. The suitability of coffee-growing zones in Kenya is expected to decrease by approximately 30 per cent in 2020 and drastically decrease in 2050 by over 50 per cent (Läderach *et al.* 2010).

Besides suffering from the impacts of climate change, coffee production itself can also contribute significantly to climate change. Coffee roaster Tchibo assessed the carbon footprint along one complete supply chain from growing to consumption. The study revealed that the carbon footprint of coffee in this supply chain is 8.4 kg of $CO_2$-e per kilogram of coffee produced, processed and consumed. Coffee farming was identified in the study as one of the hot spots for greenhouse gas (GHG) emissions, primarily due to the use of synthetic fertilizers and pesticides (Tchibo 2008).

The impacts of climate change call first for an adaptation of East African coffee ecosystems to the external climate-related stresses. Adaptation is essential in securing the economic significance of the sector on a macro-level and contributing to improving the prospects of many of the rural poor in the region who primarily depend on the revenues made with this cash crop. Second, carbon footprint policies put pressure on international coffee buyers to reduce GHG emissions in their supply chains. Therefore, coffee systems that are also able to mitigate climate change will have a comparative advantage in future markets. In this chapter we seek to explore the potential for combined adaptation and mitigation in coffee ecosystems. We do so by first proposing a classification for different East African coffee production systems. Afterwards we assess the adaptive capacity of those systems and their contribution to mitigation by adopting a resilience framework. We conclude with outlining pathways for coffee systems to adapt to climate change that contain strong synergies with mitigation.

## Methods

This study draws on three different data sets: (1) data compiled at 40 different smallholder coffee farms and estates in Kenya as part of the German International Cooperation's (GIZ) Sangana Public–Private Partnership (PPP) project (Van Rikxoort 2010); (2) a survey of 257 coffee farms in Uganda conducted by IITA in the framework of an ongoing USAID Livelihoods and Enterprises for Agricultural Development (LEAD) project (Jassogne 2011); and (3) data collected at 117 coffee farms in five Latin American countries as part of a study executed by CIAT (Van Rikxoort *et al.* 2013). On-farm carbon monitoring was part of the three studies whereby allometric equations were used to account for carbon sequestration and the Cool Farm Tool was used to quantify GHG emissions arising from, for example, fertilizer use and post-harvesting processes (Hillier *et al.* 2011).

Next to data analysis, an extensive literature review was conducted to characterize the main coffee production systems in East Africa and management practices with coinciding adaptation and mitigation effects.

## Results and discussion

### *Classification for East African coffee*

We distinguish seven coffee-growing systems in East Africa, divided into two major groups: estate and smallholder farming (Figure 13.1). Under the estate group, the unshaded monoculture is a system grown in full sun without any form of tree cover. This plantation system requires high amounts of external inputs, such as synthetic fertilizers and pesticides. Especially in Kenya, the unshaded estates are evident and are estimated to represent 45 per cent of

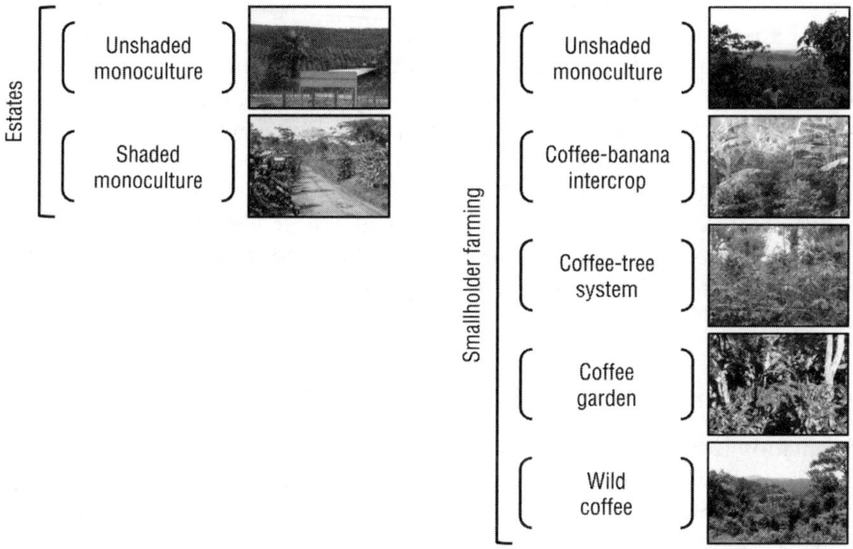


*Figure 13.1* Classification for East African coffee, showing the differences in shade, canopy and vegetational diversity.

the national coffee production (KCTA 2008). Estate coffee is also produced in shaded monocultures. The main difference from the unshaded monocultures is that these systems adopt a managed shade cover usually consisting of one single species of shade trees. Such systems can be observed in Robusta systems in Uganda, where often *Acacia* spp. are grown between the coffee plants.

In the smallholder farming group we also distinguish an unshaded monoculture. Apart from being grown in full sun, this system is entirely different from its estate counterpart. This coffee system is part of a mixed cropping system at the farm level, for example containing food crops and small numbers of livestock. Most of Kenya's smallholder coffee is grown using this system. The coffee–banana intercrop is a system where coffee is grown with banana on the same plot. The system is part of a mixed smallholder farming system where coffee is treated as a cash crop and banana is used partly as a cash crop and partly for subsistence, depending on the availability of other food crops and market dynamics. Coffee–banana intercropping systems are common in the more densely populated areas of the East African highlands (Van Asten *et al.* 2011). In some areas, coffee can be found to be intercropped with other food crops such as beans. A coffee–tree system consists of coffee grown between a diversity of shade trees. Often species are found that can be used either commercially or for subsistence. A high level of management characterizes the coffee–tree system whereby the growth of certain shade tree species is favoured and others are eliminated. The coffee garden is a multi-layered agroforestry system with a high amount of vegetational diversity. In the lower canopies, food crops can also be found, such as taro and beans. This system is maintained with minimal management and a low use of external inputs due to higher shade levels and a reduced amount of coffee plants per hectare. Examples include the Chagga homegardens on the slopes of Mt. Kilimanjaro in Tanzania (Hemp 2005). Lastly, in wild coffee systems, coffee is grown in primary forests and is harvested by smallholders that do not possess any ownership of the crop. Wild coffee is consequently unmanaged and no external inputs are applied. Wild Arabica coffee is typical in the Afromontane rainforests of Ethiopia (Schmitt and Grote 2006). In Uganda wild Robusta is found in the Kibale forest (Lilieholm and Weatherly 2010).

## *Adaptive capacity*

Adaptation is treated in several ways in the literature on environmental change. In this study we adopt a resilience framework, which is systems oriented and considers adaptive capacity as the central element of resilient agro-ecological systems (Nelson *et al.* 2007). First, several possible measures to adapt to climate variability and change in coffee systems are discussed. Afterwards, East Africa's different coffee production systems characterized earlier are evaluated on the basis of the availability of those sources of adaptation.

It is widely recognized that shade in coffee systems helps adaptation to extreme weather events by providing a more stable and moderated microclimate, reducing evapotranspiration and conserving water (Lin 2007). Furthermore, intercropping with shade trees and food crops is an adaptive practice that can buffer the effects of turbulent markets and decrease input costs by providing other crops and products for coffee farmers to sell (Beer *et al.* 1998). Shading can also improve the revenues made by coffee farmers as it positively affects bean size, composition, as well as beverage quality (Vaast *et al.* 2006). Mulching and using soil cover in the form of litter, weeds and food crops protect the soil from sun, rain and erosion, while at the same time adding nutrients, preventing unwanted weeds and retaining moisture (De Rouw *et al.* 2010). We recognize that factors such as social and human capital and governance structures are also important in influencing adaptive capacity (Brooks *et al.* 2005) but have left these out of the assessment as the physical coffee production systems are the unit of analysis in this chapter.

Apart from the estate and smallholder monocultures, all the described coffee production systems adopt some form of shade, although densities and numbers of shade tree species differ substantially among and also within the systems. Intercropping is inherent only in the coffee–banana intercrop, coffee–tree system and the coffee garden. The data from east and south-central Ugandan coffee systems show the effect of intercropping on mulch depth: unshaded monoculture (0.7 cm) vs. coffee–banana intercrop (2.8 cm). The level of different intercropping species differs among systems; the coffee–banana intercrop uses

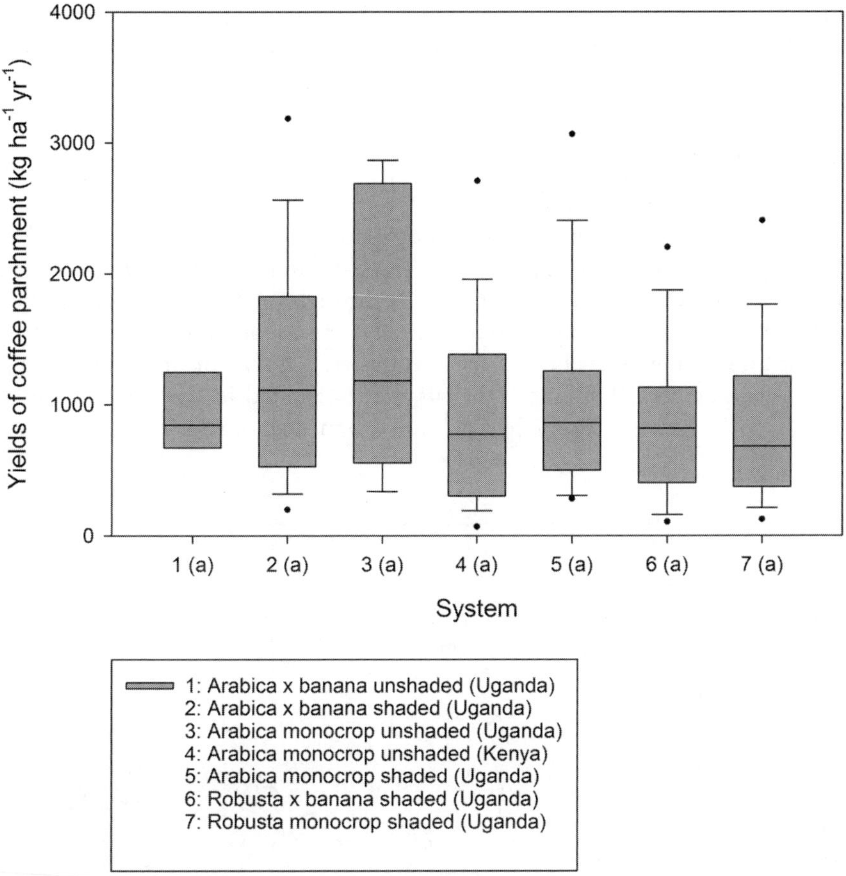


*Figure 13.2* Boxplot of yields of intercrop and monocrop coffee production systems. The figure is based on a sample drawn from 257 smallholder coffee farms in Uganda and Kenya. Systems followed by the same letter are not significantly different according to the ANOVA test ($P < 0.05$).

predominantly banana whereas in a coffee garden a variety of other food crops can be found. Increasing income and food security from crops such as banana, taro or beans largely compensate for yield losses in coffee due to competition for nutrients in these intercropping systems (Van Asten *et al.* 2011). At the existing production levels there are no significant differences in yield ($P < 0.05$) noted between coffee monocultures and intercropping systems (Figure 13.2). This means that in smallholder systems intercropping provides an attractive adaptation strategy without sacrificing coffee yields significantly. Data show that the availability of mulching material such as leaf litter and pruning residues coincides with the occurrence of shade trees and levels of intercropping.

Summarizing, we argue that the more densely shaded and intercropped production systems provide multiple sources of adaptive capacity that contribute to the resilience of these systems, whereas unshaded monocultures lack these sources.

### *Contribution to mitigation*

Coffee production systems that incorporate trees are known for their potential for carbon sequestration and are constantly reported as sequestering more carbon than systems that grow coffee in full sun (Soto-Pinto *et al.* 2010). Data collected in different production systems in Uganda through the USAID-LEAD project confirm that trees play an important role in carbon sequestration as shade trees account for 86 per cent of the carbon sequestered, coffee plants 8 per cent and bananas 7 per cent (Jassogne 2011). Residues applied in the form of mulch or soil cover increase the levels of soil carbon and organic matter, thereby

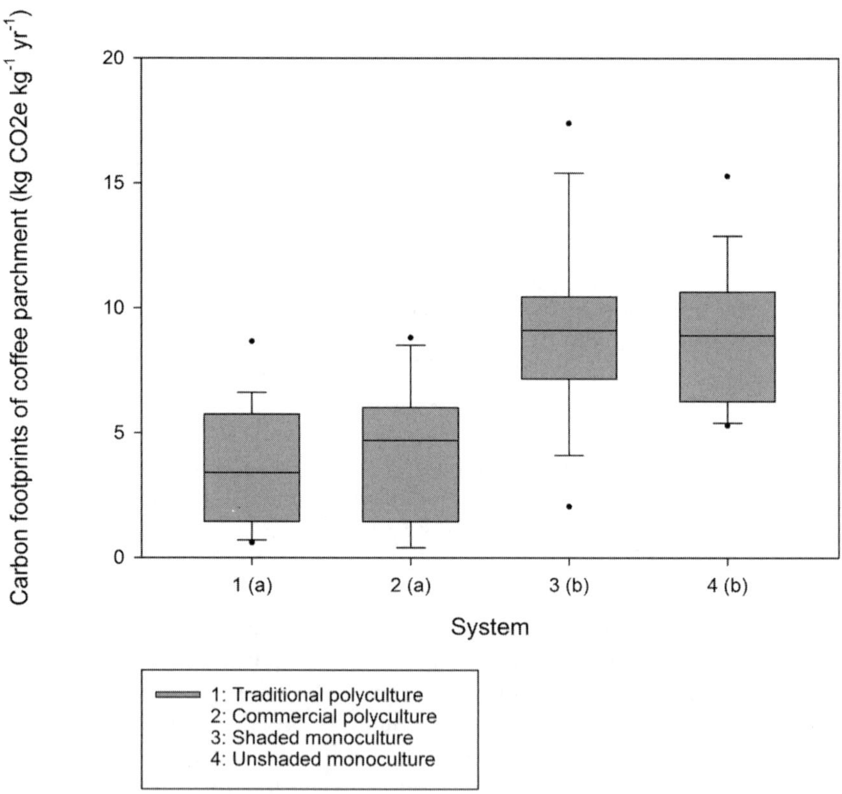


*Figure 13.3* Boxplot of carbon footprints of two polycultural and two monocultural coffee production systems. The figure is based on a sample drawn from 117 coffee farms in Latin America. Systems followed by the same letter are not significantly different according to the ANOVA test ($P < 0.05$).

increasing soil fertility and the productivity of the crop and decreasing the amount of chemical inputs needed (Sarno *et al.* 2004). Besides increasing carbon stocks in biomass and soils, coffee production systems have a different profile for GHG emissions depending on the levels of carbon sequestration, fertilizer applications and the post-harvesting processes applied. To evaluate the GHG emissions occurring in the outlined production systems we compared them with similar systems that were distinguished in Mesoamerica by Moguel and Toledo (1999). Figure 13.3 shows that polycultures emit less GHG emissions and consequently have a lower carbon footprint than monocultures (Van Rikxoort *et al.* 2013). Furthermore, the standing carbon stocks are higher for polycultures (70.9 t carbon per hectare) compared with monocultures (17.8 t carbon per hectare).

In short, this review of literature and the analysis of data from Uganda, Kenya and five Latin American countries show that polycultural coffee production systems – especially the coffee–tree systems and coffee gardens – contain higher carbon stocks and emit fewer GHG emissions than systems that do not contain any shade trees.

## Conclusions

Coffee can be found in seven different systems in East Africa, divided into an estate group with unshaded and shaded monocultures, and a smallholder farming group with unshaded monocultures, coffee–banana intercrops, coffee–tree systems, coffee gardens and wild coffee. An assessment of the availability of sources of adaptation shows that ecologically complex and diversified systems have an increased adaptive capacity, which contributes to the resilience of these production systems. The same systems present higher-standing carbon stocks and a reduction in GHG emissions that arise, for example, from synthetic fertilizer use and post-harvesting processes. For this reason we argue that shaded and intercropped production systems are sustainable pathways of intensifying coffee ecosystems in East Africa and are more adapted to withstand the various impacts that are predicted to come with climate change. We propose incremental system adjustments to coffee production systems that already contain one element of shade or intercropping. System transformations are needed in unshaded monocultures to secure and extend the role of coffee for East Africa's national economies and the livelihoods of smallholder farmers in a changing climate.

## Acknowledgements

This chapter was realized as a joint collaboration between IITA and CIAT. We thank the Consultative Group on International Agricultural Research (CGIAR) Research Program on Climate Change, Agriculture and Food Security (CCAFS) for funding the study. The chapter benefited from the valuable comments and suggestions made by two anonymous reviewers.

## References

AEO (2013) *African Exports 2009: Three Main Exports with Their Share in Total Exports*. www.africaneconomicoutlook.org/en/data-statistics/table-7-exports-2010 (accessed 8 June 2013).

Beer, J., Muschler, R., Kass, D. and Somarriba, E. (1998) 'Shade management in coffee and cacao plantations', *Agroforestry Systems*, vol. 38, no. 1, pp. 139–164.

Brooks, N., Adger, N. W. and Kelly, M. P. (2005) 'The determinants of vulnerability and adaptive capacity at the national level and the implications for adaptation', *Global Environmental Change*, vol. 15, no. 2, pp. 151–163.

De Rouw, A., Huon, S., Soulileuth, B., Jouquet, P., Pierret, A., Ribolzi, O., Valentin, C., Bourdon, E. and Chantharath, B. (2010) 'Possibilities of carbon and nitrogen sequestration under conventional tillage and no-till cover crop farming', *Agriculture, Ecosystems & Environment*, vol. 136, no. 1, pp. 148–161.

Hemp, C. (2005) 'The Chagga Home Gardens: Relict areas for endemic Saltatoria Species (Insecta: Orthoptera) on Mount Kilimanjaro', *Biological Conservation*, vol. 125, no. 2, pp. 203–209.

Hillier, J., Walter, C., Malin, D., Garcia-Suarez, T., Mila-i-Canals, L. and Smith, P. (2011) 'A farm-focused calculator for emissions from crop and livestock production', *Environmental Modelling & Software*, vol. 26, no. 9, pp. 1070–1078.

Jassogne, L. (2011) *Climate Change Impact and Opportunities in Coffee–Banana Systems in Uganda*, International Institute of Tropical Agriculture (IITA), Kampala, Uganda.

KCTA (2008) *Kenya Coffee Directory*, Kenya Coffee Traders Association (K.C.T.A.) and Camerapix Magazines Ltd, Nairobi, Kenya.

Läderach, P., Eitzinger, A., Ovalle, O., Ramírez, J. and Jarvis, A. (2010) *Climate Change Adaptation and Mitigation in the Kenyan Coffee Sector*, International Center for Tropical Agriculture (CIAT), Cali, Colombia.

Lilieholm, R. J. and Weatherly, W. P. (2010) 'Kibale forest wild coffee: challenges to market-based conservation in Africa', *Conservation Biology*, vol. 24, no. 4, pp. 924–930.

Lin, B. B. (2007) 'Agroforestry management as an adaptive strategy against potential microclimate extremes in coffee agriculture', *Agricultural and Forest Meteorology*, vol. 144, no. 1, pp. 85–94.

Moguel, P. and Toledo, V. M. (1999) 'Biodiversity conservation in traditional coffee systems of Mexico', *Conservation Biology*, vol. 13, no. 1, pp. 11–21.

Nelson, D. R., Adger, N. W. and Brown, K. (2007) 'Adaptation to environmental change: contributions of a resilience framework', *Annual Review of Environment and Resources*, vol. 32, pp. 395–419.

Salami, A., Kamara, A. B. and Brixiova, Z. (2010) 'Smallholder agriculture in East Africa: trends, constraints and opportunities', *Working Papers Series no. 105*, African Development Bank, Tunis, Tunisia.

Sarno, M. Iijima, Lumbanraja, J., Sunyoto, Yuliadi E., Izumi, Y. and Watanabe, A. (2004) 'Soil chemical properties of an Indonesian red acid soil as affected by land use and crop management', *Soil and Tillage Research*, vol. 76, no. 2, pp. 115–124.

Schmitt, C. and Grote, U. (2006) *Wild Coffee Production in Ethiopia: The Role of Coffee Certification for Forest Conservation*, Institute of Forest and Environmental Policy, Freiburg, Germany.

Soto-Pinto, L., Anzueto, M., Mendoza, J., Ferrer, G. J. and Jong, B. (2010) 'Carbon sequestration through agroforestry in indigenous communities of Chiapas, Mexico', *Agroforestry Systems*, vol. 78, no. 1, pp. 39–51.

Tchibo (2008) *Case Study Tchibo Privat Kaffee Rarity Machare*, Öko-Institut e.V, Berlin, Germany.

Vaast, P., Bertrand, B., Perriot, J., Guyot, B. and Génard, M. (2006) 'Fruit thinning and shade improve bean characteristics and beverage quality of coffee (*Coffea arabica* L.) under optimal conditions', *Journal of the Science of Food and Agriculture*, vol. 86, no. 1, pp. 197–204.

Van Asten, P. J. A., Wairegi, L. W. I., Mukasa, D. and Uringi, N. O. (2011) 'Agronomic and economic benefits of coffee–banana intercropping in Uganda's smallholder farming systems', *Agricultural Systems*, vol. 104, no. 4, pp. 326–334.

Van Rikxoort, H. (2010) *Carbon Footprinting as a Part of the Add-on Climate Module*, German International Cooperation (GIZ), Nairobi, Kenya.

Van Rikxoort, H., Läderach, P. and van Hal, J. (2013) 'The Potential of Latin American Coffee Production Systems to Mitigate Climate Change', in W. Leal Filho (ed.), *Climate Change and Disaster Risk Management*, Berlin, Germany: Springer. pp. 655–679.

World Bank (2013) *Export of Goods and Services (% of GDP)*. http://data.worldbank.org/indicator/NE.EXP.GNFS.ZS (accessed 8 June 2013).

# Part III

# Drivers for adoption

# 14 Agricultural technology diffusion and adoption in banana- and legume-based systems of Central Africa

*Emily Ouma,*[1] *Ibrahim Macharia,*[2] *Eliud Birachi,*[3] *Hildegard Garming,*[4] *Martha Nyagaya,*[3] *Pieter Pypers,*[3] *Justus Ochieng,*[1,5] *Guy Blomme,*[4] *Piet van Asten*[1] *and Bernard Vanlauwe*[3]

[1] *International Institute of Tropical Agriculture, Burundi and Uganda*
[2] *Kenyatta University, Kenya*
[3] *International Centre for Tropical Agriculture, Kenya and Rwanda*
[4] *Bioversity International, Uganda and Costa Rica*
[5] *University of Kassel, Germany*

## Introduction

The important role of technological change and innovation in increasing agricultural productivity, economic growth and poverty reduction in sub-Saharan Africa (SSA) has been widely acknowledged and documented (World Bank 2007). The majority of the populace in SSA live in rural areas and rely on agriculture for their livelihood. Agricultural research, the main source of technological innovation, is therefore critical in the improvement of productivity with high potentials for poverty reduction and for meeting food security needs without irreversible degradation of the natural resource base. Many interventions to improve the productivity of agricultural systems have been promoted in the Great Lakes region of Africa through technological change.

The Consortium for Improving Agriculture-based Livelihoods in Central Africa (CIALCA)[1] aims at improving the livelihoods of rural households through the identification, evaluation and promotion of technological options with the objective of enhancing the productivity of banana and legume-based systems and creating an enabling environment for their adoption. The potential of the technologies developed and disseminated by the project for a marked increase in the productivity of the production systems has been demonstrated based on field trials. However, until now, not much has been known about the level of awareness of the technologies and the adoption rates. Assessing the technology adoption rates and the factors influencing them are important in priority setting, providing feedback to the research programmes, guiding policy-makers and those involved in technology transfer to have a better understanding of the modalities of the assimilation and diffusion of new technologies.

This chapter examines the adoption rates of CIALCA technologies and the factors influencing their uptake from the perspective of modern evaluation theory by employing the treatment effects approach (Heckman 1996; Wooldridge 2002). The approach is essential because the commonly used estimators for adoption rates suffer from either non-exposure or selection bias. Much of the adoption literature focuses on technology-related determinants or farmers' characteristics to estimate adoption rates (Feder *et al.* 1985; Doss 2006). Such models are based on the assumption that, once introduced, the knowledge about the new technology spreads or somehow 'diffuses' within the farming communities. The non-exposure bias arises from the fact that farmers who have not been exposed to a technology cannot adopt it, even though they might have done so if they had known about it. This leads to the underestimation of the population adoption rate (Diagne and Demont 2007). Due to the non-exposure bias, the normally computed sample adoption rate (the proportion of sampled farmers who have adopted) does not consistently estimate the true population adoption rate, even with a random sample.[2] This is because farmers self-select into exposure, and researchers and extension agents tend to target progressive farmers first (Diagne 2006). Similarly, the effects of the determinants of adoption cannot be consistently estimated using simple probit, logit or Tobit adoption models that cannot control for exposure. To account for selection bias, some authors have employed a latent variable correction procedure (e.g. Dimara and Skuras 2003). However, this approach has been criticized by Diagne and Demont (2007), who argued that the parametric latent variable formulation is not efficient since the adoption outcome variable is binary, rendering the resulting estimates 'messy'.

The true population adoption rate corresponds to what is defined in modern evaluation literature as the average treatment effect (ATE). The ATE parameter measures the effect of a 'treatment' on a person randomly selected in the population (Wooldridge 2002). In our adoption context, 'treatment' corresponds to the exposure to a technology. The consistent estimation of ATE requires controlling for exposure status and the use of a set of covariates which, in the adoption context, correspond to the determinants of adoption status commonly used in probit or logit models of adoption.

## Empirical framework

The analysis in this chapter follows the modern treatment effect estimation literature using a counterfactual outcome framework proposed by Diagne and Demont (2007) to control for exposure bias in the estimation of technology adoption rates. The adoption of CIALCA technologies is assumed to be a dichotomous choice, where the technology is adopted by farmers when the perceived net benefit from adoption is greater than the result of not adopting the technology. The difference between the farmers' perceived net benefit from the adoption of the CIALCA technologies and from non-adoption may be denoted as $I^\star$, such that $I^\star > 0$ indicates that the net benefit from adoption exceeds that of non-adoption. $I^\star$ is unobservable, but can be expressed as a function of observable elements in the latent variable model:

$$I^\star_i = \beta X_i + \mu_i, \qquad I_i = 1\ [I^\star_i > 0] \tag{1}$$

where $I_i$ is a binary indicator variable that equals 1 for farmer $i$ in case of adoption and 0 otherwise, $\beta$ is a vector of parameters to be estimated, $X_i$ is a vector of explanatory variables, and $\mu_i$ is an error term assumed to be normally distributed. The probability of adopting CIALCA technologies can be represented as:

$$\Pr(I_i = 1) = \Pr(I^\star > 0) = \Pr(\mu_i > -\beta X_i) = 1 - F(-\beta X_i) \tag{2}$$

where $F$ is the cumulative distribution function for $\mu_i$. Different models, such as logit or probit, normally result from the assumptions that are made on the functional form of the cumulative distribution function $F$ (Maddala 1983). However, these models yield biased and inconsistent estimates even when based on a randomly selected sample. This is due to 'non-exposure' bias or 'selection bias'. Farmers may not adopt a technology because they were not exposed to it, but might have adopted it had they been exposed to it. Non-separation of exposure and adoption decisions leads to an underestimation of population adoption rates. Selection bias arises because technologies are not randomly assigned to farmers. This leads to farmers' self-selection into exposure.

The true population adoption rate corresponds to what is defined in the modern treatment effect literature as the ATE. The ATE parameter measures the effect or impact of a 'treatment' on a person randomly selected in the population. With the treatment effect framework, every farmer in the population has two potential outcomes: with and without exposure to a technology. Let $I_1$ be the potential adoption outcome of a farmer when exposed to the CIALCA technology and $I_0$ be adoption outcome when not so exposed.[3] The treatment effect for farmer $i$ is measured by the difference $I_{i1} - I_{i0}$. Hence, the expected population adoption impact of exposure to CIALCA technology is given by the mean value $E(I_1 - I_0)$, which is, by definition, the ATE.[4] However, $I_1$ is observed only for farmers exposed to the CIALCA technology. It is impossible to observe both the adoption outcome and its counterfactual, making it impossible to measure $I_1 - I_0$ for any given farmer.

Since exposure to the CIALCA technology is a necessary condition for its adoption, we have $I_0 = 0$ for any farmer, whether exposed to the technology or not. Hence, the adoption impact of a farmer $i$ is given by $I_{i1}$ and the average adoption impact of exposure is given by $ATE = EI_1$. Unfortunately, $I_1$ is observed only for farmers exposed to the CIALCA technology, therefore $EI_1$ cannot be estimated by the sample average of a randomly drawn sample. If we let the binary variable $w$ be an indicator for exposure to CIALCA technology, where $w = 1$ denotes exposure and $w = 0$ otherwise, the average adoption impact on the exposed sub-population is given by the conditional expected value $E(I_1 | w = 1)$, which is by definition the ATE on the treated (ATE1). Since we do observe $I_1$ for all the exposed farmers, the sample average of $I_1$ from the sub-sample of exposed farmers will consistently estimate ATE, provided the sample is random. The ATE can be decomposed as a weighted sum of the average treatment effect on the treated (ATE1) $E(I_1 | w = 0)$, the expected adoption impact in the non-exposed sub-population (ATE0):

$$ATE = EI_1 = P(w = 1)\ X\ ATE1 + (1 - P(w = 1))ATE0 \tag{3}$$

where $P(w = 1)$ is the probability of exposure.[5] From (3) the expected non-exposure bias, the expected bias from using the sample average adoption rate among the exposed and the expected adoption impact in the non-exposed sub-population, can be estimated. We can also obtain the observed adoption outcome $I$ as a function of the potential outcomes $I_1$ and $I_0$ and the treatment status (exposure) variable $w$ as:

$$I = wI_1 + (1 - w)I_0 = wI_1 \tag{4}$$

For consistent estimation of population adoption parameters, we identify ATE based on the conditional independence (CI) assumption involving potential outcomes (Wooldridge 2002; Imbens and Wooldridge 2009). The CI assumption postulates that a set of observed covariates determining exposure, when controlled for, renders the treatment status $w$ independent of the potential outcomes $I_1$ and $I_0$. Based on the CI assumption, ATE parameters can be estimated either with parametric or with non-parametric regression methods. We estimate ATE, ATE1 and ATE0 with parametric procedures by specifying a model for the conditional expectation of the observed variables $w$, $x$ and $I$ (Diagne and Demont 2007):

$$E(I | X, w = 1) = g(X, \beta) \tag{5}$$

where $g$ is a known function of the vector of covariates $X$, determining the adoption of CIALCA technologies and $\beta$ is the unknown parameter vector which can be estimated by maximum likelihood procedures using observations $(I, X)$ from the exposed sub-sample with $I$ as the dependent variable. With the estimated parameters $\beta$, the predicted values are computed for all observations in the sample, including the non-exposed. The average of these predicted values, $(X, \beta)$, is used to compute ATE for the full sample, and ATE1 and ATE0 for the exposed and non-exposed sub-samples.

## Contextual background

CIALCA has been operating since 2006 in ten mandate areas in DRC, Rwanda and Burundi.[6] These areas have some of the largest population densities in Africa, with average values ranging between 238 and 514 people/km$^2$. CIALCA has promoted the technology options of integrated soil fertility management (ISFM) and integrated pest management (IPM) as a framework for the enhanced productivity of banana and legume system components. These technology options comprise improved germplasm, the promotion of efficient fertilizer use, optimized organic matter management and local adaptation. The production–consumption continuum has been applied to link the actors in the system value chain, from the inputs required for production to delivery to the consumers. In order to catalyse the adoption of technologies to enhance productivity, interventions have been promoted that improve the security of income and nutrition. Interventions intended for nutrition security include the promotion of foods enriched with soybean and dietary diversification. Income enhancement has been promoted by applying market linkage approaches through collective efforts among smallholder farmers.

## The data

The data used in our analysis were derived from a cross-sectional household survey in 2011 covering 913 farmers in Burundi, DRC and Rwanda. The sample design followed a multi-stage procedure to select mandate area, village and households. The survey covered seven mandate areas, intentionally selected because of the intensity of CIALCA interventions. Five villages were then randomly selected in each of the mandate areas stratified into three categories. The first, also known as 'action sites', consisted of villages that hosted demonstrations of the technologies. Action sites are selected sites in each mandate area in which field activities related to technology identification, evaluation and adaptation are implemented. The second, also known as 'satellite sites', involved neighbouring villages where development partners were involved in scaling out CIALCA technologies. The third consisted of control villages which had agro-ecological conditions similar to those of the action and satellite sites but without any CIALCA interventions. They were within a 10–15 km radius of the action and satellite sites. Households were then randomly sampled, proportional to size, to yield a sample of approximately 130 households/mandate area.

Descriptive statistics of adoption status for several variables are presented in Table 14.1. The full sample consisted of 913 households, of which 32 per cent were adopters of CIALCA technologies. Adoption was defined as the use of any of the ISFM and/or IPM options promoted by CIALCA during the last season. The variables in Table 14.1 were used in the estimation of the econometric models.

*Table 14.1* Descriptive statistics: mean of variables by status of adoption

| *Variables* | *Full sample* $n = 913$ | *Adopters* $n = 290$ | *Non-adopters* $n = 623$ | *t-values* |
|---|---|---|---|---|
| Gender (1 = male) | 0.82 | 0.85* | 0.80 | 1.82 |
| Years of agricultural experience | 23.92 | 22.11*** | 24.75 | –4.61 |
| Education0 (1 = no formal education of household head) | 0.25 | 0.17*** | 0.29 | –3.99 |
| Education1 (1 = primary level of household head) | 0.49 | 0.51 | 0.48 | 0.88 |
| Education2 (1 = secondary level of household head) | 0.23 | 0.29*** | 0.19 | 3.37 |
| Education3 (1 = post-secondary level of household head) | 0.01 | 0.01 | 0.00 | 0.76 |
| Farmers' group membership (1 = yes) | 0.41 | 0.61*** | 0.31 | 7.99 |
| Cultivated land area in acres | 276.62 | 334.51 | 248.4 | 0.49 |
| Number of plots | 2.54 | 2.71*** | 2.46 | 2.82 |
| Household size | 5.84 | 6.16*** | 5.69 | 2.69 |
| Dependency ratio (%) | 41.9 | 42.09 | 41.88 | 0.12 |
| Log (value of assets owned in USD) | 6.73 | 7.29*** | 6.47 | 3.95 |
| Off-farm income access (1 = yes) | 0.24 | 0.24 | 0.23 | 0.12 |
| Agricultural extension (1 = yes) | 0.74 | 0.83*** | 0.69 | 4.61 |
| Occupation of household head (1 = agriculture) | 0.91 | 0.89 | 0.92 | –1.04 |
| Radio ownership (1 = yes) | 0.62 | 0.68*** | 0.59 | 2.48 |
| CIALCA training participation (1 = yes) | 0.32 | 0.45*** | 0.24 | 5.82 |
| Bas-Congo mandate area | 0.15 | 0.11** | 0.16 | –2.12 |
| North Kivu mandate area | 0.14 | 0.13 | 0.14 | –0.27 |
| South Kivu mandate area | 0.13 | 0.20*** | 0.09 | 4.34 |
| Umutara mandate area | 0.14 | 0.13 | 0.14 | –0.48 |
| Gitega mandate area | 0.15 | 0.11** | 0.16 | –2.00 |
| Kigali-Kibungo mandate area | 0.16 | 0.23 | 0.12 | 4.26 |
| Rusizi mandate area | 0.15 | 0.09*** | 0.17 | –3.55 |

*Notes*: ***, ** and * means that mean values for adopters are significantly different from those of non-adopters at the 1, 5 and 10 per cent level.

The difference between adopters and non-adopters of CIALCA technologies seemed to be related to institutional factors rather than to financial, natural or physical endowments. Mandate areas, specifically Kigali-Kibungo and South Kivu, had a higher proportion of adopters than the rest. This may be due to linkages with development partners working in these areas that scale out CIALCA technologies. In Kigali-Kibungo, for instance, the government of Rwanda is implementing projects that promote improved crop productivity through improved access to seeds. In South Kivu, several non-government organizations (NGOs) are working on agricultural development projects.

## Results

The analysis followed two stages which were estimated simultaneously. In the first, probit models were used to analyse the determinants of CIALCA technology awareness. In the second, probit models that control for awareness exposure using the ATE framework were used to estimate unbiased adoption parameters. Table 14.2 presents the results from the probit estimation of the determinants of CIALCA technologies awareness. The log likelihood function of −192.8 and the highly significant likelihood ratio statistic show good model fitness.

The positive and significant coefficient on group membership points to the important role played by social networks in disseminating technology information. The importance in agricultural technology dissemination of social networks, such as interactions with neighbours, farmers' groups, churches and input suppliers, has been widely documented (Bandiera and Rasul 2006; Matuschke and Qaim 2009). Similarly, the positive coefficient on radio ownership shows the importance of this communication equipment in raising awareness of CIALCA technology. The probability of awareness of CIALCA technologies was 8.3 percentage points higher for farmers with a radio compared to those without, *ceteris paribus*. This is expected, since information on some of the CIALCA technologies was largely transmitted through the mass media by CIALCA or its partners.

The location-specific variables revealed a difference in the exposure of CIALCA technology over space.[7] For instance, the probability of awareness of CIALCA technologies was 40 percentage points lower for farmers in Bas-Congo and 13–33 percentage points lower in Gitega, Rusizi and Umutara compared with farmers located in the Kivus, *ceteris paribus*. The marginal effect on Kigali-Kibungo was not statistically significant. In North and South Kivu, CIALCA has used local radio programmes to inform the population on some of its technologies. This result may be a reflection of the effectiveness of radio programmes in reaching out to farmers in rural areas.

Table 14.3 presents the results of CIALCA technology adoption models with two alternative specifications: the ATE corrected model for exposure and the classical adoption probit that does not account for exposure bias. Similarities in terms of the coefficient signs, significant levels of the coefficients and

*Table 14.2* Determinants of exposure to CIALCA technologies

| *Variables* | *Marginal effects*† | *z-value* |
|---|---|---|
| Gender | 0.031 | 0.84 |
| Age of the household head | –0.000 | –0.20 |
| Farmers' group membership | 0.074** | 2.52 |
| Agricultural extension | 0.047 | 1.34 |
| Radio ownership | 0.083** | 2.39 |
| CIALCA trainings | 0.132*** | 5.09 |
| Log assets | 0.004 | 0.86 |
| Bas-Congo | –0.400** | –2.03 |
| Umutara | –0.133* | –1.45 |
| Gitega | –0.331*** | –3.20 |
| Kigali-Kibungo | –0.132 | –1.45 |
| Rusizi | –0.191* | –1.82 |
| *Summary statistics* | | |
| Pseudo R-squared | 0.161 | |
| LR Chi-square | 73.98*** | |
| Number of observations | 519 | |
| Log likelihood function | –192.786 | |

*Notes*: ***, ** and * denote statistical significance at the 1, 5, and 10 per cent level. † Marginal effects evaluated at sample means.

marginal effects are observed for the two models. Access to agricultural extension and membership in farmers' groups positively influenced the likelihood of adoption of CIALCA technologies. This implies that the two pathways were the main avenues for access to CIALCA technologies.

Although some of the coefficient signs and levels of significance were similar in the two models, the magnitudes of the coefficients and marginal effects tended to be rather different. In the ATE treatment, the effects are smaller because the variable on exposure has already been accounted for. For example, group membership increases the probability of exposure to the technologies, and those among the exposed in the group are more likely to adopt. The coefficient on off-farm income was negative and significant in the ATE-corrected model though insignificant in the classical model. This implies that access to off-farm income had a negative effect on adoption. The probability of adopting CIALCA technologies was 14 percentage points lower among farmers with access to off-farm income compared with those without it. While off-farm income may provide the financial liquidity needed for technology adoption, it may also be an indication of a specialization away from agriculture which could result in a lower level of interest in new technologies.

Farm size and the ownership of productive assets did not influence adoption significantly, implying scale-neutrality of the technologies. This has also been

*Table 14.3* Determinants of CIALCA technologies adoption

| *Variables* | *ATE corrected model for exposure* | | | *Classical adoption model* | | |
|---|---|---|---|---|---|---|
| | *Estimated coefficients* | *S.E.* | *Marginal effects* | *Estimated coefficients* | *S.E.* | *Marginal effects* |
| Gender | 0.041 | 0.199 | 0.025 | 0.144 | 0.183 | 0.053 |
| Age | 0.001 | 0.035 | 0.004 | 0.010 | 0.033 | 0.004 |
| Age-squared | –0.000 | 0.252 | 0.000 | –0.000 | 0.237 | –0.000 |
| Education0 | –0.204 | 0.176 | –0.079 | –0.353 | 0.164 | –0.126 |
| Education1 | 0.124 | 0.242 | 0.049 | 0.099 | 0.235 | 0.037 |
| Education2 | 0.537 | 0.164 | 0.212 | 0.562 | 0.152 | 0.218 |
| Occupation of household head | –0.339 | 0.531 | –0.135 | –0.371 | 0.499 | –0.144 |
| Household size | 0.027 | 0.289 | 0.011 | 0.029 | 0.277 | 0.011 |
| Dependency ratio | –0.001 | 0.144 | –0.000 | –0.002 | 0.137 | –0.000 |
| Farmers' group membership | 0.376*** | 0.335 | 0.149*** | 0.523*** | 0.321 | 0.196*** |
| Agricultural extension | 0.359** | 0.308 | 0.081 | 0.378** | 0.365 | 0.135*** |
| Radio | 0.081 | 0.275 | 0.032 | 0.127 | 0.260 | 0.047 |
| Log of asset value | 0.032 | 0.250 | 0.013 | 0.037 | 0.233 | 0.014 |
| Cultivated land area | 0.001 | 0.057 | 0.000 | 0.000 | 0.053 | 0.000 |
| No. of plots | 0.001 | 0.026 | 0.000 | 0.015 | 0.025 | 0.005 |
| Off-farm income | –0.361* | 0.168 | –0.142** | –0.254* | 0.235 | –0.093* |
| Bas-Congo | –0.458 | 0.000 | –0.145 | –0.477 | 0.000 | 0.094 |
| Gitega | –0.563* | 0.033 | –0.027 | –0.792** | 0.031 | –0.036 |
| Kigali-Kibungo | 0.043 | 0.003 | 0.265 | 0.084 | 0.003 | 0.238 |
| Umutara | –0.396 | 0.364 | 0.066 | –0.499* | 0.343 | 0.077 |
| Rusizi | 0.593* | 0.292 | –0.225** | –0.720** | 0.271 | –0.234*** |
| Constant | –1.576 | 0.965 | – | –1.740** | 0.921 | – |
| *Summary statistics* | | | | | | |
| Pseudo $R$-squared | 0.135 | | | 0.167 | | |
| LR Chi-square | 73.808*** | | | 104.465*** | | |
| $N$ | 398 | | | 472 | | |
| Log likelihood function | –237.15 | | | –261.52 | | |

*Notes*: ***, ** and * denote statistical significance at the 1, 5, and 10 per cent level. Marginal effects evaluated at sample means.

found in many other studies related to the adoption of agricultural technologies, such as Edmeades and Smale (2006) and Kabunga *et al.* (2011). Besides, most of the farmers in the region were small-scale farmers with small land holdings.

Some of the location-specific variables had significant coefficients. Gitega, Umutara and Rusizi mandate areas had negative and statistically significant coefficients in the model, implying that the likelihood of adoption was lower in these areas than in the Kivu area. The marginal effects were, however, significant only for the Rusizi mandate area. The coefficient on Kigali-Kibungo was not statistically significant compared to the Kivus. In the Kivu area, membership in farmers' groups was relatively high compared to other mandate areas and the high likelihood of adoption of CIALCA technologies from the model results may have been due to information exchanges among farmers.

Table 14.4 presents the predicted adoption rates with and without ATE correction for exposure bias. The observed adoption rate for the entire sample (Na/N) was 37 per cent. The joint exposure and adoption rate (JEA) was 38 per cent in the ATE-corrected model. The similarity between the observed adoption rate and the JEA is expected, as indicated by Diagne and Demont (2007), although they are not good indicators of the potential population adoption rate due to non-exposure bias.[8]

The predicted adoption rate for the full population (ATE) corrected for awareness exposure is 46 per cent. Thus, if all farmers were aware of the CIALCA technologies, the adoption rate would be 8 per cent higher than that

*Table 14.4* Predicted adoption rates

| *Variables* | *Estimate* | *S.E.* | *z-value* |
|---|---|---|---|
| *ATE-corrected population estimates* | | | |
| Predicted adoption rate in the full population (ATE) | 0.456*** | 0.022 | 20.94 |
| Predicted adoption rate in exposed sub-population (ATE1) | 0.489*** | 0.021 | 22.23 |
| Predicted adoption rate in unexposed sub-population (ATE0) | 0.336*** | 0.028 | 12.20 |
| Joint exposure and adoption rate (JEA) | 0.378*** | 0.017 | 22.23 |
| Population adoption gap (PAG) | –0.078*** | 0.006 | –12.48 |
| Population selection bias (PSB) | 0.033*** | 0.004 | 7.17 |
| *Observed sample estimates* | | | |
| Exposure rate (Ne/N) | 0.773*** | 0.018 | 41.96 |
| Adoption rate (Na/N) | 0.374*** | 0.021 | 17.58 |
| Adoption rate among exposed sub-sample (Na/Ne) | 0.483*** | 0.027 | 17.58 |

*Note*: ***, ** and * denote statistical significance at the 1, 5 and 10 per cent level. Robust standard errors are reported.

observed in the sample, i.e. the population adoption gap, JEA minus ATE, is 8 per cent. The predicted adoption rate among the presently exposed sub-population (ATE1) was estimated at 49 per cent, being slightly higher than that of the full population (ATE), indicating a positive population selection bias (PSB) in the magnitude of 3.3 per cent. This is not surprising, as most innovative farmers self-select into exposure. The predicted adoption rate in the non-exposed sub-population was calculated as the average treatment effect on the untreated (ATE0), and was 33 per cent. This is the adoption rate among the non-exposed population after becoming exposed to the technologies.

### *Adoption constraints*

The exposure rate to CIALCA technologies was relatively high, with 77 per cent of the sampled households being aware of them. However, awareness did not necessarily translate into adoption. Figure 14.1 shows the reasons for non-adoption by those who were aware of the technologies.

Poor access to technologies was one of the major reasons for non-adoption. For instance, the supply of the improved germplasm could not meet the demand. The technology package for IPM was not adopted by farmers in areas where banana pests and diseases were not perceived as a major challenge. However, farmers had no access to some of the components of IPM, such as those for sterilizing the farm equipment.

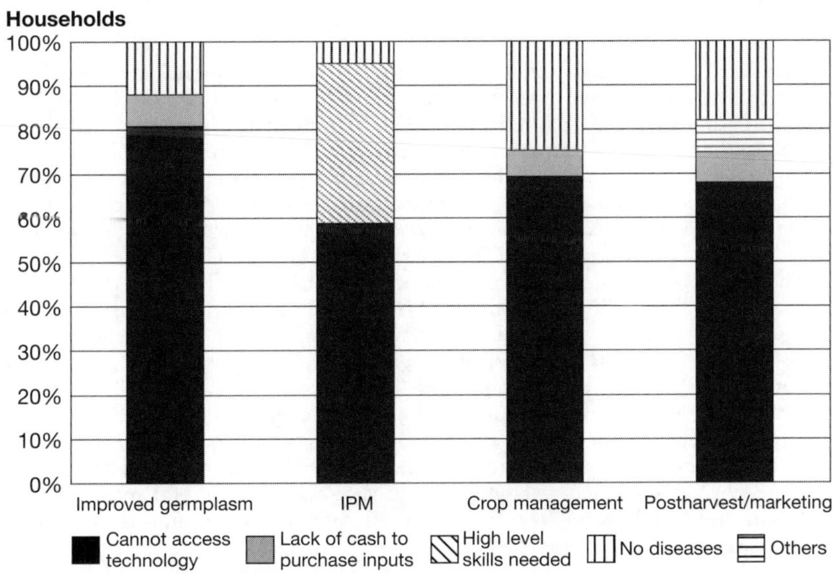


*Figure 14.1* Constraints to adoption.

## Summary and conclusion

We have analysed the adoption rates and factors influencing the adoption of CIALCA technologies in banana- and legume-based farming systems of Central Africa using data from a survey of farming households. The ATE framework has been applied to control for selection bias that may arise from a lack of awareness or a partial exposure to the technologies. At the population level, the results show that with exposure, the adoption rates could be slightly higher at 46 per cent. Generally, exposure rates to CIALCA technologies are high, with about 77 per cent of the sampled households being exposed. However, the corresponding adoption rates of 38 per cent are relatively low. The major constraint to adoption still remains poor access to the technologies. The marginal effects from the model estimations show the important role of farmers' groups in exposure to and adoption of technologies. Extension services from governments and NGOs are significant in the adoption models. These are the main routes for farmers to gain access to improved technology. This implies that the implementation of sustainable technological change in smallholder agriculture at scale needs to consider more efficient models of technology delivery. Such delivery models should build on existing social structures and networks at community levels to complement the traditional extension services. To reduce outreach costs, greater targeting of communication tools, such as radios, needs to be emphasized to increase exposure. For agricultural intensification to be sustainable, through efforts such as ISFM and IPM, the farmers' access to these technologies needs to be improved as some of the necessary inputs required may be unavailable.

## Notes

1 CIALCA is a consortium of the International Institute of Tropical Agriculture (IITA), Bioversity International and the International Centre for Tropical Agriculture (CIAT) and their national research and development partners, supported by the Belgian Directorate General for Development Cooperation.

2 The standard sample adoption rate confounds information on diffusion and adoption of a technology, as it estimates the proportion of households who are exposed to the technology and have adopted it. By confounding the two types of information, the sample adoption rate provides unclear policy or research messages.

3 The adoption outcome in this case is the adoption status, a dichotomous 0–1 variable.

4 $I_1$ and $I_0$ are considered random variables, representing the potential outcome of any farmer randomly selected from the underlying population of farmers.

5 Used strictly to mean awareness of the existence of the new technology and does not necessarily imply any learning of its characteristics.

6 Mandate areas are defined as areas with similar agro-ecological conditions and poverty profiles that have nonetheless relatively good access to large urban markets. Mandate areas are different in surface area between the two target countries. The number of people living in each mandate area can vary between 300,000 and 1,200,000.

7 North and South Kivu have been left out of the model and are used as the comparison base.

8 The observed sample adoption rate and the joint exposure and adoption rate are similar since a random sample should yield consistent estimates of the population counterpart.

## References


Bandiera, O. and Rasul, I. (2006) 'Social networks and technology adoption in northern Mozambique', *The Economic Journal*, vol. 116, pp. 869–902.

Diagne, A. (2006) 'The diffusion and adoption of NERICA rice varieties in Côte d'Ivoire', *Developing Economies*, vol. 44, pp. 208–231.

Diagne, A. and Demont, M. (2007) 'Taking a new look at empirical models of adoption: average treatment effect estimation of adoption rates and their determinants', *Agricultural Economics*, vol. 37, nos 2–3, pp. 201–210.

Dimara, E. and Skuras, D. (2003) 'Adoption of agricultural innovations as a two stage partial observability process', *Agricultural Economics*, vol. 28, pp. 187–196.

Doss, C. (2006) 'Analyzing technology adoption using microstudies: limitations, challenges, and opportunities for improvement', *Agricultural Economics*, vol. 34, pp. 207–219.

Edmeades, S. and Smale, M. (2006) 'A trait-based model of the potential demand for a genetically engineered food crop in a developing economy', *Agricultural Economics*, vol. 35, no. 3, pp. 351–362.

Feder, G., Just, R. E. and Zilberman, D. (1985) 'Adoption of agricultural innovations in developing countries: a survey', *Economic Development and Cultural Change*, vol. 33, no. 2, pp. 255–298.

Heckman, J. (1996) 'Randomization and social program evaluation as an instrumental variable', *Review of Economics and Statistics*, vol. 77, no. 2, pp. 336–341.

Imbens, G. and Wooldridge, J. (2009) 'Recent developments in the econometrics of program evaluation', *Journal of Economic Literature*, vol. 47, no. 1, pp. 5–86.

Kabunga, N., Dubois, T. and Qaim, M. (2011) *Information Asymmetries and Technology Adoption: The Case of Tissue Culture Bananas in Kenya*, Discussion Paper No. 74, Courant Research Centre, Georg-August-Universität Göttingen.

Maddala, G. S. (1983) *Limited-Dependent and Qualitative Variables in Econometrics*, Cambridge University Press, Cambridge.

Matuschke, I. and Qaim, M. (2009) 'The impact of social networks on hybrid seed adoption in India', *Agricultural Economics*, vol. 40, no. 5, pp. 493–505.

Wooldridge, J. (2002) *Econometric Analysis of Cross Section and Panel Data*, MIT Press, Cambridge, MA.

World Bank (2007) *Agriculture for Development: World Development Report 2007*, World Bank, Washington, DC.

# 15 Supply and demand drivers of the sustainable intensification of farming systems through grain legumes in Central, Eastern and Southern Africa

*J. Rusike,[1] S. Boahen,[2] K. Dashiell,[3] S. Kantengwa,[4] J. Ongoma,[5] D. M. Mongane[6] and G. Kasongo[7]*

[1] *International Institute of Tropical Agriculture, Tanzania*
[2] *International Institute of Tropical Agriculture, Mozambique*
[3] *International Institute of Tropical Agriculture, Nigeria*
[4] *N2 Africa, Rwanda*
[5] *Kleen Homes and Gardens, Kenya*
[6] *International Center for Tropical Agriculture, DRC*
[7] *International Institute of Tropical Agriculture, Malawi*

## Introduction

There is compelling evidence that sustainable agro-ecological intensification is necessary in the farming systems in sub-Saharan Africa (SSA) for an increase in agricultural productivity and profitability, food availability, farm incomes and nutrition, and a reduction in poverty (Pretty *et al.* 2011). Grain legumes are potentially valuable crops for agricultural intensification and the diversification of farming systems. The middle- and high-altitude areas in Central, Eastern and Southern Africa are endowed with favourable agro-ecological conditions for their production. Grain legumes have the potential to permit households to exploit the comparative advantage of their areas in order to sustainably intensify their production systems and diversify into livestock, using increased crop residues as feed.

There is some debate about the relative importance of the factors driving changes in agricultural production systems at different locations and how to design, evaluate and promote agricultural research-for-development interventions for grain legumes that respond to the most important factors and have a positive impact on outcomes of interest at a large scale (Hazell and Wood 2011; Tripp 2011; Akibode and Maredia 2011). This chapter contributes to this literature by using value chain analysis (VCA) and time-series econometric modelling to analyse the key drivers in agricultural production systems in the mid- and high-altitude areas in Central, Eastern and Southern Africa and to draw implications for the targeting of agricultural research-for-development investments to shift the farming systems towards intensification. The study focuses on South Kivu, Democratic Republic of Congo, Kenya, Malawi, Mozambique, Rwanda, Tanzania, Uganda and Zambia. These countries are chosen because they have a wide range of low-, mid- and high-altitude farming systems and data are available.

The objectives are to (1) identify the key demand and supply factors driving changes in the grain legume supply systems in the study area and assess their effects on market prices and production outcomes; (2) identify end-market opportunities for and constraints on smallholders in exploiting the opportunities; and (3) identify nodes of leverage and entry points for targeting research-for-development investments to resolve the constraints and permit large numbers of smallholders to exploit the opportunities, capture benefits and shift to the sustainable intensification of farming systems.

To achieve these objectives the following hypotheses are tested: (1) increasing liberalization, urbanization and globalization of agricultural markets are inducing private sector agribusiness firms to expand investments in the procurement of raw commodities at the farm gate, bulking, transportation, storage, processing, packaging, export, wholesaling and retailing, thereby driving the integration of markets and the transmission of international and urban market prices to farm-gate prices through arbitrage; (2) grain legume development projects focusing on the delivery to farmers of high-quality seeds of improved varieties, improved crop and post-harvest management technologies and the provision of information have a positive impact on the supply response of smallholder farmers.

## Materials and methods

Hazell and Wood (2011) developed a theoretical framework for analysing the drivers of change in global agriculture and land use. The factors are categorized into global, country and local-scale drivers. Local-scale factors are specific to geographical areas and different types of agricultural production systems. They include population pressure and demographic structure, poverty, health, technology design, property rights, natural resources, infrastructure, market access and non-farm opportunities to facilitate part-time farming and an exit from agriculture. Country-scale drivers include per capita income growth, urbanization, the commercialization of market chains, agricultural policy, insecurity and water scarcities. Global-scale drivers include international trade and the globalization of markets, world prices for agricultural products, energy prices and the agricultural policies of countries in the Organisation for Economic Co-operation and Development (OECD) and the World Trade Organization (WTO).

Applying this framework to the context of farming systems in Central, Eastern and Southern Africa generates a large list of factors that can be hypothesized to drive changes in agriculture. Because we are focusing on the leverage nodes and entry points for targeting agricultural research investments to shift the systems towards intensification, we concentrate on the liberalization of agricultural markets, urbanization, globalization and the integration of value chains and agricultural policies relating to grain legume development projects.

The first hypothesis is first tested by identifying the most important factors on the demand side that are driving changes in grain legume production systems in the study area by tallying responses obtained through qualitative interviews with the representatives of organizations engaged in the grain legume value chains. The hypothesis is further tested by estimating if prices in representative markets in the most important production areas and centres of consumption in the study countries and in international markets are co-integrated, using the Engle-Granger procedure.

The second hypothesis is first tested by identifying the most important factors driving supply responses to economic, biophysical and institutional factors through qualitative interviews with participants in the grain legume value chains. The hypothesis is further tested by estimating the impact of past projects for grain legume development on acreage and the yield response of farmers, using time-series intervention analysis.

Primary data were collected through a rapid appraisal VCA. This was implemented using questionnaires and interviews with key players along the value chains for common bean, cowpea, groundnut and soybean in the focus countries. The qualitative interviews were supplemented with secondary time-series data on price and production which were collected from ministries of agriculture and national statistical offices, FAOSTAT, the International Monetary Fund and the US Department of Labor, Bureau of Labor Statistics.

## Results and discussion

### *Co-integration analysis*

The rapid appraisal value chain surveys revealed that the production and consumption of grain legumes are concentrated geographically. The most important areas of production of common bean are in the high-altitude and high-rainfall zones. The major production areas for soybean are in the mid-altitude areas with moderate annual rainfall and a long drying season. The main cropping zones for cowpea and groundnut are in the low-altitude areas with low rainfall.

The main centres of consumption are other rural deficit areas, domestic and regional urban centres and the international export markets. Because of rapid urbanization and the income growth of households in urban areas, urban markets are emerging as the major drivers of change in the farming systems throughout Africa.

Grains are transferred from centres of production to centres of consumption through entrepôts for moving produce from food sheds through five channels

(Figure 15.1). Respondents explained that there was an increasing trend towards the integration of the domestic value chains into the global value chains, especially for industrial crops such as soybean and export crops such as groundnut and cowpea. This is being driven by private sector firms which expand their activities to procure raw commodities at the farm gate and arrange local and central assembly, transportation, storage, processing, packaging, export, wholesaling and retailing. Key informants reported that markets were becoming more integrated and globalized and international and urban market prices were being transmitted to farm-gate prices through arbitrage. This was providing a market pull to induce farmers to expand investments in the adoption of new technologies, increase productivity and profitability and earn higher incomes in a sustainable way without depleting the natural resource base, instead of the continuous cropping of cereals.

But the qualitative interview responses may have been biased. This is because respondents have limited knowledge, preconceptions and a stake in the outcome. Therefore we tested the perceptions of respondents by estimating an error correction model. Following Enders (2010) we tested for a co-integrating relationship for the domestic markets of each of the four grain legumes in each of the study countries using the Engle-Granger procedure by estimating the following equations:

$$\Delta y_t = \alpha_1 + \alpha_y \hat{e}_{t-1} + \sum \alpha_{11}(i)\Delta y_{t-1} + \sum \alpha_{12}(i)\Delta z_{t-1} + \epsilon_{yt}$$
$$\Delta z_t = \alpha_2 + \alpha_z \hat{e}_{t-1} + \sum \alpha_{21}(i)\Delta y_{t-1} + \sum \alpha_{22}(i)\Delta z_{t-1} + \epsilon_{zt}$$
$$\Delta w_t = \alpha_3 + \alpha_w \hat{e}_{t-1} + \sum \alpha_{31}(i)\Delta y_{t-1} + \sum \alpha_{32}(i)\Delta z_{t-1} + \epsilon_{wt}$$

where $y_t$, $z_t$, and $w_t$ are the natural logarithms of the prices of the grain legumes in representative markets in the food shed, entrepôt and consumption area; $\Delta$ is the first difference; $\varepsilon_{yt}$, $\varepsilon_{zt}$, $\varepsilon_{wt}$ are white-noise disturbances; and $\alpha_1$, $\alpha_2$, $\alpha_3$, $\alpha_y$, $\alpha_z$, $\alpha_w$, $\alpha_{11}(i)$, $\alpha_{12}(i)$, $\alpha_{11}(i)$, $\alpha_{21}(i)$, $\alpha_{31}(i)$ and $\alpha_{32}(i)$ are parameters.

We first pre-tested the prices to determine whether or not they contained unit roots using the Dickey-Fuller and augmented Dickey-Fuller tests with different lag lengths. We found that the absolute values of the $t$-statistics are below –2.89, which is the critical value for the Dickey-Fuller test with 100 observations. Therefore we cannot reject the null hypothesis of a unit root in the price series. Because the price series are non-stationary, they may be co-integrated. We next estimated the long-run regression $y_t$, $z_t$, or $w_t$ as the left-hand-side variables. We carried out the Engle-Granger test on residuals from the equation to determine whether these are stationary. Based on these data we concluded that the prices are co-integrated. We estimated an error-correcting model using two lagged changes of each variable. Table 15.1 presents the error-correction terms using the residuals from the estimate of the co-integrating relationships using the Engle-Granger procedure. The signs of the speed-of-adjustment coefficients showed that prices converge towards a

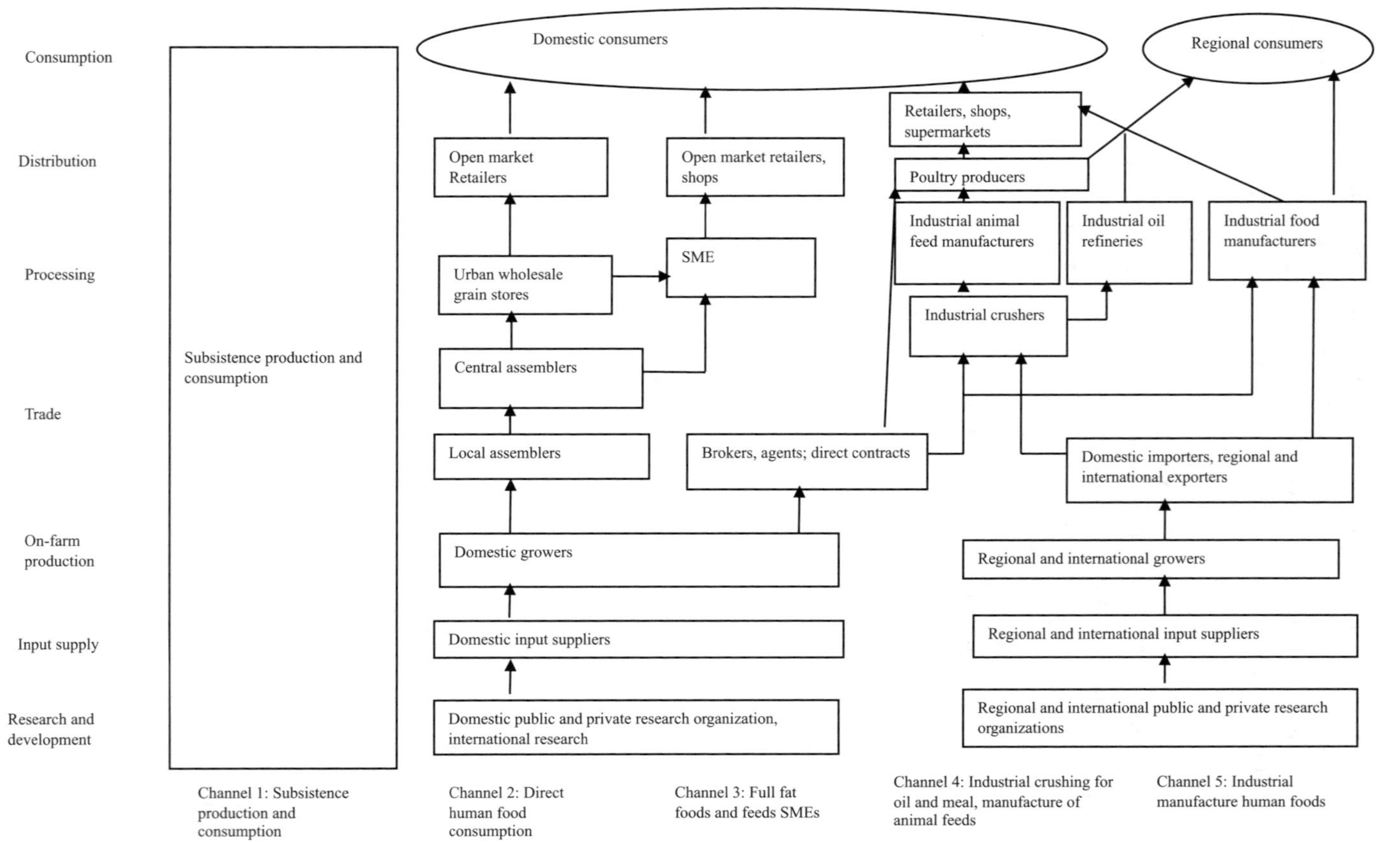


*Figure 15.1* Grain legume value chains in Central, Eastern and Southern African countries (source: author's representation based on interviews with value chain participants).

*Table 15.1* Estimates of error-correction terms

| *Country* | *Error corrections* | *Common bean* | *Cowpea* | *Groundnut* | *Soybean* |
|---|---|---|---|---|---|
| South Kivu-Bukavu | ΔBukavu | –0.052<br>(–0.54) | | | |
| | ΔUvira | 0.187<br>(1.65) | | | |
| Kenya | ΔNairobi | –0.258**<br>(–3.05) | –0.398**<br>(–2.91) | –0.241<br>(0.97) | |
| | ΔKisumu | 0.306196*<br>(2.45) | 0.172<br>(1.21) | 0.460*<br>(1.79) | |
| | ΔMombasa | 0.317**<br>(2.84) | –0.280<br>(1.57) | –0.520*<br>(2.27) | |
| Malawi | ΔLilongwe | –0.201*<br>(–1.91) | –0.414***<br>(–3.78) | –0.508***<br>(–4.54) | –0.0742803<br>(–1.22) |
| | ΔBlantyre | 0.164*<br>(1.98) | 0.004<br>(–0.04) | –0.031<br>(–0.30) | 0.0089288<br>(0.12) |
| | ΔLizulu | 0 .027<br>(0.27) | –0.025<br>(–0.24) | –0.166<br>(–1.55) | 0.1620885<br>(1.35) |
| Mozambique | ΔMaputo | –0.164***<br>(–4.19) | –0.075<br>(–1.52) | –0.069*<br>(–1.84) | |
| | ΔChimoio | 0.043<br>(0.80) | 0.275**<br>(3.15) | 0.201*<br>(2.37) | |
| | ΔQuelimane | 0.205*<br>(2.41) | 0.131*<br>(2.15) | 0.306***<br>(4.18) | |
| Rwanda | ΔKigali | –0.437<br>(–0.98) | –0.027<br>(–0.20) | –0.517*<br>(–2.24) | |
| | ΔUrban east | –0.650<br>(–0.81) | 0.841**<br>(3.18) | 0.020<br>(0.06) | |
| | ΔRural east | –0.183386<br>(–0.24) | 1.067**<br>(2.75) | –0.349<br>(–0.66) | |
| Tanzania | ΔDar es Salaam | –0.335***<br>(–4.11) | –0.104<br>(–1.49) | –0.083<br>(–1.17) | |
| | ΔArusha/ Dodoma | 0.007<br>(0.07) | 0.061<br>(0.67) | 0.265***<br>(4.03) | |
| | ΔMbeya/ Singida | 0.236<br>(1.61) | 0.185*<br>(1.88) | 0.146*<br>(2.01) | |
| Uganda | ΔKampala | 0.390*<br>(–1.92) | –0.291*<br>(–1.82) | 0.295*<br>(–2.42) | |
| | ΔBusia | 0.530***<br>(4.23) | | 0.292*<br>(2.10) | 0.073<br>(0.70) |
| | ΔArua | 0.290<br>(1.29) | | 0.137<br>(0.83) | 0.174*<br>(1.72) |
| Zambia | ΔLusaka | –0.044<br>(–0.65) | | –0.043<br>(–0.65) | |
| | ΔKasama | 0.155<br>(1.48) | | 0.223*<br>(2.44) | |
| | ΔSolwezi/ Chipata | 0.344**<br>(2.88) | | 0.281***<br>(3.50) | |

*Notes*: *t*-statistics for the error-correction term in parentheses. Statistical significance at * $p < 0.05$; ** $p < 0.01$; *** $p < 0.001$.

long-run equilibrium. We concluded that the markets are integrated. The error corrections were not statistically significant for some markets. This showed that some markets are weakly exogenous and price takers.

We estimated the long-run co-integrating relationship between domestic and international markets with the following equation:

$$\log(p\_cy_t) = \alpha + \log(p\_\mathrm{int}_t) + \log(ex\_cy_t) \qquad (1)$$

where $p\_cy_t$ is the retail price of the grain legume in the representative market in the consumption centre in the country, $p\_\mathrm{int}_t$ is the international price and $ex\_cy_t$ is the bilateral exchange rate with the United States. The logarithms of the exchange rate, domestic prices and international prices appear to be unit root processes. We first estimated the long-run equilibrium relationship between the variables. We used the residuals to carry out the Engle-Granger test for co-integration using three lagged changes. Table 15.2 reports the estimates of the coefficients for the first lag of the residuals from the estimate of the long-run relationships. The results show that there is long-run parity. Based on these data, we concluded that domestic and international markets are integrated.

### *Intervention analysis of past grain legume development projects*

Respondents interviewed for this study reported their belief that projects for grain legume development implemented to expand production in areas with the highest potential for production have had the most impact on acreage and the farmers' yield response. Starting in the 1980s and 1990s, government and non-governmental organizations (NGOs) in most of the focus countries implemented agricultural development projects to expand the supply of grain legumes. The projects supported production through the development, evaluation and promotion of higher-yielding, disease-resistant varieties; appropriate crop and post-harvest management technologies; improved access to input markets (seeds, fertilizer and credit) and output markets; capacity development for farmers to participate in competitive markets; and an enabling environment for business. Some respondents explained that public sector investments were necessary because private sector firms work with limited resources and are uncertain about capturing the benefits from investing in improving on-farm production. Respondents rated that agricultural research investments should be prioritized on the development, testing and promotion of new varieties; crop and post-harvest management practices; input supply systems for seeds, inoculants, fertilizers and agro-chemicals; output marketing systems; the provision of information; the development of micro-finance markets; farmers' organizations; and the creation of an enabling business environment.

Because the perceptions of respondents may be limited, we tested their explanations by estimating the impact of the projects for grain legume development, implemented from the mid-1990s, using intervention analysis. Following

Table 15.2 Estimates of the coefficients for the first lag of the residuals from the estimate of the long-run relationships

| *Grain legume* | | | |
|---|---|---|---|
| *Country* | *Common bean* | *Groundnut* | *Soybean* |
| South Kivu: Bukavu | –0.139** (–2.84) | | |
| Kenya: Nairobi | –0.139** (–3.31) | –0.028 (–1.08) | |
| Malawi: Lilongwe | –0.419 *** (–5.10) | –0.160** (–3.55) | 0.153* (–1.96) |
| Mozambique: Maputo | –0.156*** (–4.34) | –0.189*** (–5.15) | |
| Rwanda: Kigali | –0.408*** (–5.18) | –0.076** (–2.73) | –0.466* (–2.26) |
| Tanzania: Dar es Salaam | –0.306** (–2.91) | –0.183 (–1.65) | |
| Uganda: Kampala | –0.129 (–0.89) | –0.084 (–0.84) | –0.117 * (–1.95) |
| Zambia: Lusaka | –0.102* (–2.37) | –0.097* (–2.46) | |

*Notes*: *t*-statistics for the error-correction term in parentheses. Statistical significance at * $p < 0.05$; ** $p < 0.01$; *** $p < 0.001$.

Enders (2010) we implemented the test by estimating the following first-order autoregressive equation:

$$y_t = a_0 + a_1 y_{t-1} + c_0 z_t + \varepsilon_t \tag{2}$$

where $y_t$ is the national aggregate area planted or the yield of the grain legume for the country; $z_t$ is the intervention (or dummy) variable that takes on the value of zero prior to the implementation of the development project and unity beginning with the first year of implementation; and $\varepsilon_t$ is the white-noise disturbance error term.

The initial or impact effect of the project's interventions is the magnitude of $c_0$. We tested the statistical significance of $c_0$ using a standard *t*-test. The long-run effect of the intervention is estimated by $c_0/(1{-}a_1)$. The results are presented in Table 15.3. Based on the data we concluded that the interventions increased the area planted and the yield response to market and biophysical factors because $c_0$ is positive and statistically different from zero in most cases. The negative coefficient on yields is because the adoption of improved varieties has not been accompanied by the adoption of improved soil fertility management practices and this has resulted in declining yields due to nutrient depletion. This implies that future interventions in grain legume development projects need to place

*Table 15.3* Estimates of the impact of past grain legume development project interventions on area planted and yield

| *Country* | *Impact effect* | *Common bean* | | *Cowpea* | | *Groundnut* | | *Soybean* | |
|---|---|---|---|---|---|---|---|---|---|
| | | *Area* | *Yield* | *Area* | *Yield* | *Area* | *Yield* | *Area* | *Yield* |
| Kenya | Initial | | | 2.464★<br>(2.45) | 0.857<br>(1.75) | 0.964★★★<br>(5.74) | 0.090<br>(0.38) | 1.390<br>(1.32) | –0.396<br>(–1.17) |
| | Long-run | | | 2.618 | 1.024 | 1.544 | 0.230 | 2.121 | –0.787 |
| Malawi | Initial | 0.090<br>(0.81) | 0.129<br>(1.22) | | | 0.308★<br>(2.68) | 0.705★★<br>(3.40) | 1.680★<br>(1.76) | –0.091<br>(–1.26) |
| | Long-run | 0.550 | 0.379 | | | 0.972 | 0.821 | 2.207 | –0.154 |
| Rwanda | Initial | 0.194<br>(1.40) | 0.025<br>(0.56) | | | | | 0.444★<br>(2.78) | –0.134<br>(–0.82) |
| | Long-run | 0.313 | 0.082 | | | | | 1.597 | –0.135 |
| Tanzania | Initial | 0.130<br>(1.42) | 0.061<br>(0.89) | | | 0.236★<br>(2.28) | 0.156★<br>(1.88) | | |
| | Long-run | 0.623 | 0.107 | | | 1.442 | 0.197 | | |
| Uganda | Initial | 0.0556<br>(0.66) | –0.132<br>(–0.78) | | | 0.065<br>(0.53) | –0.122<br>(–1.08) | 0.267★★★<br>(6.39) | –0.008<br>(–0.39) |
| | Long-run | 0.200 | –0.212 | | | 0.188 | –0.171 | 0.652 | –0.015 |
| Zambia | Initial | 0.259★<br>(2.00) | 0.364★<br>(2.64) | | | 0.059<br>(0.36) | –0.908<br>(1.128) | 0.106<br>(0.86) | 0.113<br>(–1.56) |
| | Long-run | 0.874 | 0.460 | | | 0.176 | –0.235 | 0.718 | –0.190 |

*Notes*: *t*-statistics for the error-correction term in parentheses. Statistical significance at ★ $p < 0.05$; ★★ $p < 0.01$; ★★★ $p < 0.001$.

an emphasis on integrated variety and crop and natural resource management practices in order to have a positive impact on yields.

## Conclusion

There is growing evidence that sustainable agro-ecological intensification in the farming systems in SSA is required to increase agricultural productivity, food availability, farm incomes and nutrition and to reduce poverty. Grain legumes have a significant potential to permit households to sustainably intensify and diversify their agricultural production systems in the mid-altitude and high-altitude areas in Central, Eastern and Southern Africa because these areas are endowed with the factors of production that are favourable for growing these crops.

The liberalization of agricultural markets, urbanization, globalization and the integration of value chains and agricultural policies are the major factors that are driving changes in global agriculture and land use and also in the farming systems in the study areas. Smallholders' supply responses to these global, biophysical and institutional factors are being driven by the agricultural development projects supporting the supply of grain legumes. To permit households to exploit the end-market opportunities, research investments need to be aligned with the global market drivers of change in agriculture and location-specific projects for grain legume development. They should target the development, testing and promotion of new varieties; crop and post-harvest management practices; input supply systems for seeds, inoculants, fertilizers and agro-chemicals; output marketing systems; the provision of information; the development of micro-finance markets; farmers' organizations; and the creation of an enabling environment for business.

## Acknowledgement

This study was carried out as a component of baseline studies under the N2Africa project.

## References

Akibode, S. and Maredia, M. (2011) 'Global and regional trends in production, trade and consumption of food legumes', http://impact.cgiar.org/sites/default/files/images/Legumetrendsv2.pdf (accessed 28 March 2011).

Enders, W. (2010) *Applied Econometric Time Series*, John Wiley and Sons, New York.

Hazell, P. and Wood, S. (2011) 'Drivers of change in global agriculture', *Philosophical Transactions of the Royal Society B*, vol. 363, pp. 495–515.

Pretty, J., Toulmin, C. and Williams, S. (2011) 'Sustainable intensification in African agriculture', *International Journal of Agricultural Sustainability*, vol. 9, no. 1, pp. 5–24.

Tripp, R. (2011) 'The impacts of food legume research in the CGIAR: a scoping study', http://impact.cgiar.org/sites/default/files/images/LegumeScoping2011.pdf (accessed 28 March 2011).

# 16 Assessing and improving the nutritional diversity of cropping systems

*Roseline Remans, Kyle DeRosa, Phelire Nkhoma and Cheryl Palm*

*The Earth Institute, Columbia University*

## Introduction

Life cannot exist without continuous supplies in adequate amounts of all essential nutrients. If even one nutrient is limiting or missing from the nutrient medium or diet of an organism, the organism will suffer and ultimately die. The magnitude of the effects of malnutrition is increasingly being recognized. Estimates suggest that nutrition-related factors are responsible for about 35 per cent of child deaths and 11 per cent of the total global disease burden (Black *et al.* 2008). Almost 870 million people are currently under-nourished (FAO 2011a), over two billion people are afflicted by a deficiency in one or more micronutrients (WHO 2011) and over one billion are over-weight (WHO 2011).

A human diet requires at least 51 nutrients in adequate amounts (Graham *et al.* 2007). In addition to the production of sufficient calories, a major, often overlooked challenge in agricultural intensification and food systems is the need to provide an adequate diversity of the nutrients necessary for a healthy life. It is hypothesized that changes in agricultural production systems from diversified cropping systems towards high yielding but often ecologically more simple cereal-based systems contribute to poor diet diversity, micronutrient deficiencies and the resulting malnutrition in the developed as well as the developing world (Welch and Graham 1999; Frison *et al.* 2006; Graham *et al.* 2007; Negin *et al.* 2009). Adverse interactions among health, agriculture and environment can result in vicious 'under-nutrition cycles', contributing to poverty traps (Figure 16.1). The vicious cycles that cause loss of nutrients in agricultural systems, soil erosion and decreasing biodiversity lead to the environmental degradation which results in decreased food production. Lack of food is associated with malnutrition and illness and especially with declining labour productivity, which exacerbates poor agricultural management (Deckelbaum *et al.* 2006).

For agriculture, environment and nutrition programmes to be successful and lift communities out of the malnutrition cycle, it is critical to understand

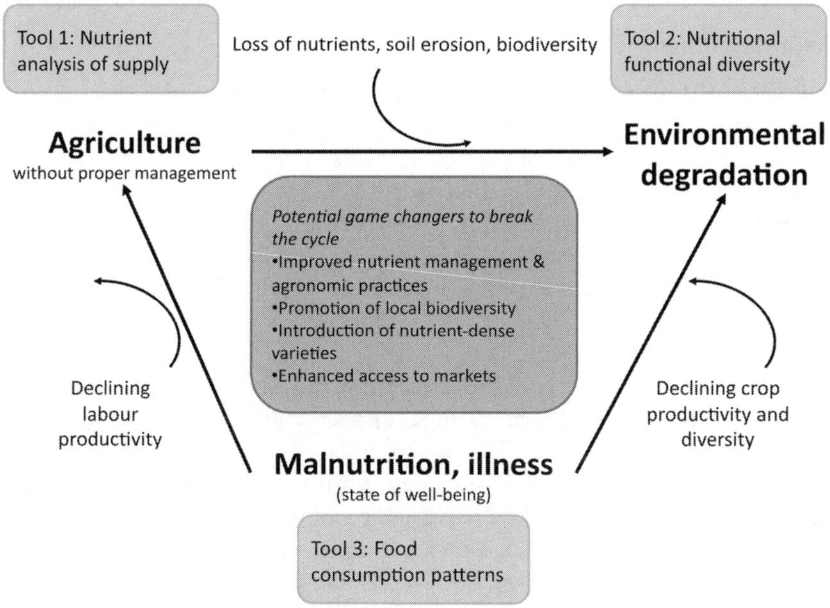


*Figure 16.1* The cycles of agricultural mismanagement, agricultural degradation and malnutrition and three complementary tools to study nutritional diversity along these cycles (source: adapted from Deckelbaum *et al.* 2006).

and target the interactions linking the three sectors. The various pathways between these sectors (World Bank 2007; IFPRI 2011), the complexity of human nutrition and its multiple determinants (UNICEF 1990; Bhutta *et al.* 2008) make this a challenging task. Furthermore, the success of agricultural systems has historically been evaluated primarily on the metrics of crop yield, economic output and cost–benefit ratios (IAASTD 2009). However, these metrics do not reflect the diversity of nutrients provided by the system that is critical for human health. Consequently, evidence linking agricultural and environmental processes to nutrition outcomes is still scarce (Penafiel *et al.* 2011; Masset *et al.* 2011; Barrett *et al.* 2011).

In this study we explore three complementary tools (Figure 16.1) that aim to help to evaluate how certain interventions can play a role in breaking the under-nutrition cycle. We illustrate the application of these tools using data from the Millennium Villages Project (MVP), an ongoing multi-country, multi-sectoral development initiative to demonstrate progress towards the Millennium Development Goals (MDGs) over the course of ten years (Sanchez *et al.* 2007). Agricultural intensification is part of the MVP's general approach to poverty reduction.

## Materials and methods

### *Research settings*

Sites for MVP were chosen to represent the major agro-ecological zones and farming systems in sub-Saharan Africa (SSA), presenting a range of challenges to income generation, food security, disease ecology, infrastructure and health system development (Sanchez *et al.* 2007). They were drawn from hunger 'hotspots' with a regional baseline prevalence of child under-nutrition of at least 20 per cent (Sanchez *et al.* 2007). This study involves ten project sites: Ethiopia – Koraro, Ghana – Bonsaaso, Kenya – Dertu, Kenya – Sauri, Malawi – Mwandama, Mali – Tiby, Nigeria – Pampaida, Nigeria – Ikaram, Tanzania – Mbola and Uganda – Ruhiira.

Most of the results are based on data from the Mwandama site in Malawi. This site is in the Southern Zomba district and covers a population of approximately 35,000 people. The region, once characterized by native *Miombo* woodlands, is now intensively cultivated, with little tree cover. Smallholders grow mainly maize, pigeon pea, cassava and groundnut, while commercial estates produce tobacco and maize. Livestock management, restricted to chickens and goats in this setting, is practised by a very small percentage of the households (~5 per cent).

### *Sampling and data collection*[1]

A set of surveys was administered in 2006 to a sample of 300 households that were randomly selected from strata defined by sub-village, wealth terciles and gender of the household head (MVP 2008). A household questionnaire, including sections about agricultural production, food security, consumption and expenditures, was administered to the household head. A food frequency questionnaire (FFQ) assessed diet diversity based on the reported frequency of consumption (number of times per day, week, month or year) of 120–150 locally available food items. Anthropometric measurements, including height, weight and middle-upper arm circumference, were taken using standard best practices (Cogill 2001) among children under the age of five years in sampled households. Stunting is defined as height-for-age z-score below –2. Wasting is defined as weight-for-age z-score below –2.

In addition to the surveys, all plots cultivated by the households, including home gardens, were sampled from 60 random households to document the occurrence of crop, plant and tree species, with different species and varieties according to local definitions (Remans *et al.* 2011a). In addition, the adoption of agricultural diversification activities was evaluated using a specific survey on adoption with a sub-sample of 100 farmers.

### *Data analysis*

*Nutrient analysis of agricultural production.* Amounts of crops and animal-based products produced per household were generated from the agricultural production table of the household. The total amounts of nutrients (in grams) provided by the crops and animal-based products were calculated using available food composition tables (Lukmanji *et al.* 2008; FAO 2011b) and qualitative information was collected on the most common ways of food preparation. Food composition data were standardized by converting values to percentage of the Dietary Reference Intake (DRI) (NAS 2009).

*Nutritional functional diversity.* Species richness was defined by the number of identified and previously described edible species per farm. Petchey and Gaston's functional diversity (FD) metric (Petchey and Gaston 2002) was used as a measure of nutritional FD, with 17 nutrients from 77 crops as described previously (Remans *et al.* 2011a). Analysis script and databases are available upon request. We did not include livestock in the calculations for the FD metric because (1) the percentage of households owning livestock was small (~5 per cent) in Mwandama, and (2) available information was limited on nutritional composition data provided by whole living animals.

*Food consumption patterns.* Food consumption patterns were calculated from individual food frequency surveys and household food consumption tables from surveys (Remans *et al.* 2011b).

## Results

### *Site characteristics: under-nutrition and agro-ecological zone*

Two measures of under-nutrition – child stunting, an indicator for chronic under-nutrition, and wasting, an indicator for acute malnutrition – across diverse MV sites allow the potential links between agro-ecological characteristics and levels and types of child under-nutrition to be identified (Figure 16.2). The prevalence of wasting is highest in pastoralist and agro-pastoralist communities, while the prevalence of stunting is highest in the maize- and plantain-based systems (Malawi – Mwandama, Tanzania – Mbola, Kenya – Sauri, Uganda – Ruhiira). The public health significance of these levels of stunting in maize- and banana-based systems is considered very high according to the WHO categorization, indicating potential examples of under-nutrition traps (Figure 16.1). To illustrate the nutrition-assessment tools, we focus on Mwandama – Malawi as the example.

### *Tool 1: nutrient analysis of agricultural production*

This tool provides insights on the amount of nutrients produced by local agriculture at the community level (35,000 people in Mwandama), compared with the amount required based on demographic composition and

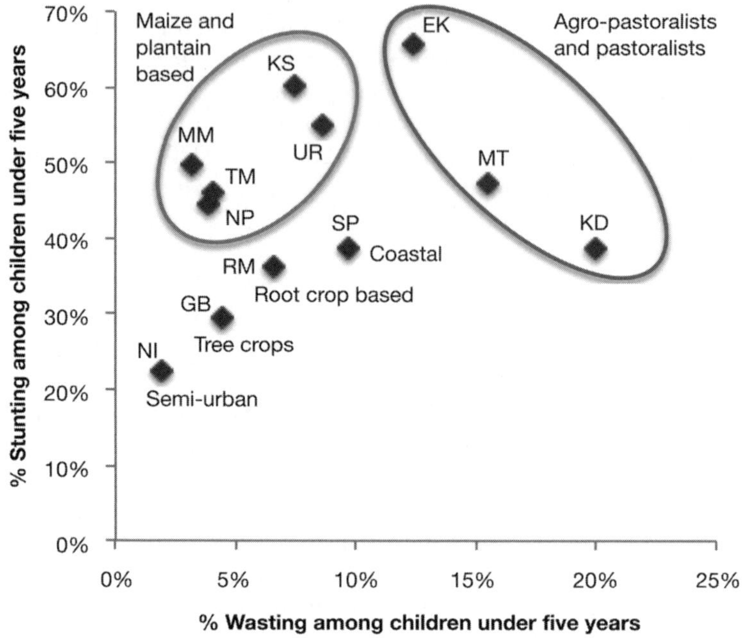


*Figure 16.2* Prevalence of stunting (height-for-age z-score less than –2 SD, an indicator for chronic under-nutrition) and wasting (weight-for-age z-score less than –2 SD, an indicator for acute malnutrition) for children under five years in age across MVP sites at baseline of the project (2005–2006).

Notes: Agro-ecological characteristics are indicated. EK: Ethiopia – Koraro; GB: Ghana – Bonsaaso; KD: Kenya – Dertu; KS: Kenya – Sauri; MM: Malawi – Mwandama; MT: Mali – Tiby; NI: Nigeria – Ikaram; NP: Nigeria – Pampaida; RM: Rwanda – Mayange; TM: Tanzania – Mbola; UR: Uganda – Ruhiira.

DRI (Figure 16.3a). In Mwandama, baseline local agricultural production produced sufficient proteins, carbohydrates, fibre, vitamin B1 (thiamine), B3 (niacin), iron and zinc to meet the demand of the local population. Nutrient gaps from the local agricultural supply were identified for fat, vitamin A, vitamin B2, vitamin C, calcium and potassium.

The data for year 3 (Figure 16.3a) show an increase in the production of calories, carbohydrates and fibre compared to the baseline. This reflects the emphasis of the MVP to increase staple crop – i.e. maize – production in the initial years of the project. Other nutrient gaps still remain at year 3, including those for vitamin A, vitamin C and calcium, and a different approach from increasing local staple crop production is required to meet local needs.

The villages are not closed food systems. In Mwandama, approximately 70 per cent of the food consumed is produced locally, and 30 per cent is produced

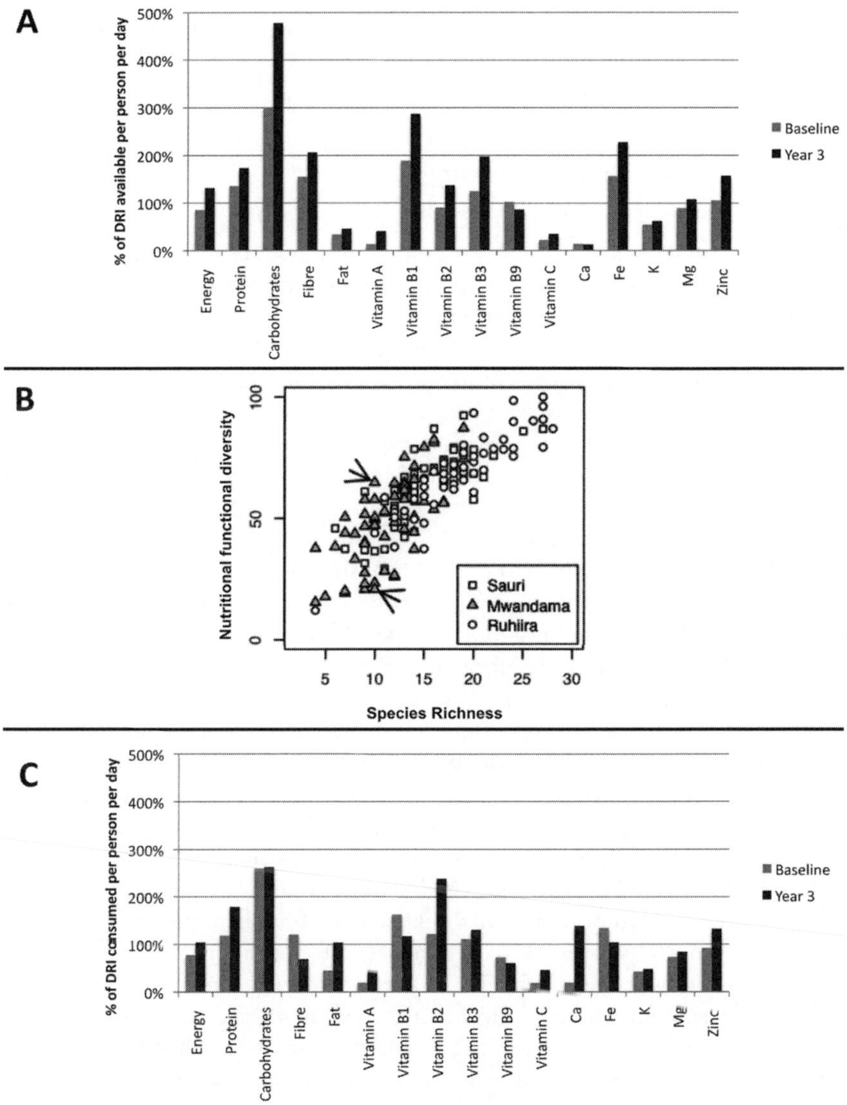


*Figure 16.3* Example of application of tools using data from Mwandama – Malawi. (a) Nutrient availability per capita per day at MVP baseline and year 3, from local agricultural production and compared to DRI values (= 100 per cent). (b) Nutritional functional diversity plotted against species richness on-farm in Mwandama, Malawi as compared to Ruhiira, Uganda and Sauri, Kenya. The arrows indicate two farms that have the same number of species but differ in nutritional functional diversity. (c) Nutrient consumption per capita per day at MVP baseline and year 3, from household food consumption tables, and as compared to DRI values (= 100 per cent).

or processed elsewhere. This raises the question of how nutrient gaps from local production systems can be addressed most efficiently: by importing food items/nutrients from elsewhere or by adapting local production. The other tools help answer this question.

### *Tool 2: nutritional functional diversity*

A second tool calculates nutritional FD (Remans *et al.* 2011a). This metric refers to the diversity of edible nutrients provided by a system, whether the farming or the larger agro-ecological system. The nutritional FD metric is based on the species composition in a farm or ecosystem and the nutritional composition of these species. The metric quantifies functional diversity and allows the identification of variability in nutritional diversity across farms and agro-ecological systems (e.g. food deserts vs. diversity hotspots). The metric provides insights on which species/varieties can add nutritional diversity to the farming system and what happens if certain species would be lost (providing insights on the resilience of the system to produce a diversity of nutrients).

An example of the application and interpretation of this metric is made for Mwandama compared with two other MVP sites. Data on edible plant species diversity and their nutritional composition were collected for 170 farms in Mwandama, Malawi; Ruhiira, Uganda and Sauri, Kenya, to calculate the nutritional FD for each of the farms. Regression of FD against species richness reveals several patterns (Figure 16.3b). (1) There is a strong positive correlation ($p < 0.001$; $r^2 = 0.68$) between FD and species richness, independent of the site. Thus, as the number of edible species increases, the diversity of nutritional functions that the farm provides also increases. (2) At around 25 species per farm, the relationship between FD and species richness levels off, meaning that additional species do little to increase nutritional diversity. (3) Although their species richness and $FD_{total}$ are correlated, farms with the same number of species can have very different nutritional FD scores. For example, two farms in Mwandama (indicated by arrows on Figure 16.3b), both with ten species, show an FD of 23 and 64, respectively. The difference in FD is linked to a few species with different nutritional traits. Both farms grow maize, cassava, beans, banana, papaya, pigeon pea and mango, but the farm with the higher FD score also grows pumpkin, mulberry and groundnut, while the farm with the lower FD score has avocado, peaches and black jack (*Bidens pilosa*). Pumpkin (including the leaves, fruits and seeds, which are all eaten) adds nutritional diversity to the system by its relatively high contents of vitamin A, Zn and S-containing amino acids (methionine and cysteine); mulberry adds vitamin B complexes (thiamin, riboflavin) and groundnut adds fat, Mn and S. Black jack, avocado and peaches found in the farm with the lower FD do not contain the vitamin B or S complexes, and thus are less complementary to the other plants on that farm.

This example illustrates that the nutritional FD metric identifies differences in nutritional diversity as well as the species that are critical for providing certain nutrients (e.g. mulberry for vitamin B complexes). The results also emphasize that the nutritional composition of the species available in the system determines whether the introduction or removal of certain species will contribute to the nutritional diversity of the farm or ecosystem.

### *Tool 3: food consumption patterns*

A third tool provides insights in food consumption patterns using food frequency surveys and household food consumption tables. As for production, nutrient gaps in consumption at the baseline were identified for vitamin A, vitamin C, calcium and potassium (Figure 16.3c). From the baseline to year 3, targets for nutrient intake are reached for calories, fat, calcium and zinc. The initial focus of the project on the production of major staple crops which was captured in the increased production of calories did not result in an increase in dietary caloric intake (Figure 16.3c). The data suggest that households have sold their surplus carbohydrates and bought more diversified and nutritious foods – for example, sources rich in calcium such as milk. This is in line with the analysis of food frequency surveys that show an increase in diet diversity from the baseline to year 3 (Remans *et al.* 2011b).

## Discussion

This chapter describes the assessment of nutrient gaps and nutritional diversity in a changing agricultural landscape using a suite of tools. The strength of the study is that it provides a methodology to integrate nutritional outputs and diversity in the evaluation and planning of agricultural intensification. Assessing the multiple nutrients in production, biodiversity and consumption also sheds light on the interconnectedness among agricultural management, environmental degradation and human nutrition (Figure 16.1). More concretely, the tools can contribute to the planning and targeting of interventions that aim to meet nutrient needs. Specific soil management practices can be identified to increase the amount of nutrients in the production system, e.g. of zinc through the use of zinc fertilizer (Graham *et al.* 2007). In addition, an adequate composition of species and varieties will be needed to take up the nutrients from the soil efficiently and make them available for human consumption. Promoting local agro-diversity, e.g. green leafy vegetables (Shackleton *et al.* 2009), improved varieties high in micronutrients (e.g., varieties developed by the programme HarvestPlus: www.harvestplus.org) or small animals and livestock, provides promising potential to increase the quality and diversity of nutrients from the production system. Furthermore, improved post-harvest storage and processing significantly avoid the loss of nutrients and increase access to food and nutrient bio-availability before consumption. Finally, creating the awareness of nutrient gaps in local diets can guide the improvement of consumption patterns.

Several limitations of the study and the tools need to be noted. First, the tools are very data-intensive and the quality of the results and insights depends on the quality of the data. Secondary food composition data are available at the species level, and sometimes the variety level, in international food composition databases (e.g. FAO-INFOODS). However, the nutritional composition of crops is dependent on the environmental and agronomic conditions under which the crop is grown and the specific variety that was used.

Second, although animal-based products were included in the nutrient gap assessment of production and consumption, livestock were not included in the nutritional FD metric (because of the small number of farmers with livestock in Mwandama and the limited available information or guidelines on the nutritional composition provided by whole, living animals). For future studies, it will be important to take animal-based products into account in the nutritional FD calculations to gain a comprehensive picture of nutritional diversity.

Third, several other analyses can further enhance our understanding of the nutrient gaps and dynamics. For example, the identification of 'hunger' months, i.e. months in which over 20 per cent of households report not having sufficient food to meet family needs (FANTA 2007; Remans *et al.* 2011b), sheds light on seasonality. Insights on economic access to a nutritious diet that meets household needs can be obtained by applying a novel approach developed by Save the Children UK and called the 'cost of the diet tool' (Perry 2008). The estimated minimum cost of the diet can subsequently be compared with the household's income and food expenditures to assess food access (Curran *et al.*, unpublished, for Mwandama, Malawi).

Studies on adoption of agricultural activities that aim to enhance nutrition (e.g. Shackleton *et al.* 2009; Verzivolli 2010; Masset *et al.* 2011; Penafiel *et al.* 2011; Impens 2011) provide important lessons on the drivers of and barriers to nutrient changes in the system. In Mwandama, Malawi, for example, the percentage of farmers growing and consuming orange-fleshed sweet potato (OFSP) increased from 3 to 33 in three years (Verzivolli 2010). The larger availability of OFSP was reflected in the production and consumption tables (consumption of roots and leaves), and related to increased vitamin A intake at MVP year 3 (Figure 16.3c). Some of the key drivers for OFSP adoption included the multiple benefits provided by the crop: high nutritional value of roots and leaves, drought tolerance, good market value, no extra labour required (Verzovolli 2010). In addition, training of farmers and demonstration sessions by other farmers were facilitated by project staff and resulted in a relatively large number of adopters in a three-year period.

In the past, agro-ecological intensification has been mainly evaluated based on metrics of yield, economic output and sometimes environmental outcomes. These metrics do not adequately address the diversity of nutrients that is critical for human health. A major lesson learned from this study is that assessing nutrient gaps in patterns of food production and consumption can and should be a critical component of planning and evaluating agricultural intensification. The inter-connectedness between agricultural management, environmental degradation and malnutrition (Figure 16.1) underlines nutritional diversity for human and environmental health.

## Note

1 Ethical statement: the study was approved by institutional review boards at Columbia University (AAAA-8202) and in all partner countries. Community consent was obtained prior to each survey round in each village. Informed consent was obtained from individuals or parents for all survey procedures.

## References

Barrett, C. B., Travis, A. J. and Dasgupta, P. (2011) 'On biodiversity conservation and poverty traps', *Proceedings of the National Academy of Science of the USA*, vol. 108, pp. 13907–13912.

Bhutta, Z. A., Ahmed, T., Black, R. E., *et al.* (2008) 'What works? Interventions for maternal and child under-nutrition and survival', *Lancet*, vol. 371, pp. 417–440.

Black, R. E., Allen, L. H., Bhutta, Z. A., *et al.* (2008) 'Maternal and child under-nutrition: global and regional exposures and health consequences', *Lancet*, vol. 371, pp. 243–260.

Cogill, B. (2001) *Anthropometric Indicators Measurement Guide*, Academy for Educational Development, Washington, DC.

Deckelbaum, R. J., Palm, C., Mutuo, P. and DeClerck, F. (2006) 'Econutrition: implementation models from the Millennium Villages Project in Africa', *Food and Nutrition Bulletin*, vol. 27, no. 4, pp. 335–342.

FANTA (2007) *Guidelines for Measuring Household and Individual Dietary Diversity*, FAO, Rome.

FAO (2011a) *The State of Food Insecurity 2011*, FAO, Rome.

FAO (2011b) 'The International Network of Food Data Systems (INFOODS)'. www.fao.org/infoods/index_en.stm (accessed 5 December 2011).

Frison, E., Smith, I. F., Johns, T., Cherfas, J. and Eyzaguirre, P. B. (2006) 'Agricultural biodiversity, nutrition, and health: making a difference to hunger and nutrition in the developing world', *Food and Nutrition Bulletin*, vol. 25, pp. 143–155.

Graham, R. D., Welch, R. M., Saunders, D. A., *et al.* (2007) 'Nutritious subsistence food systems', *Advances in Agronomy*, vol. 92, pp. 1–72.

IAASTD (2009) *International Assessment of Agricultural Knowledge, Science and Technology for Development*, Island Press, Washington, DC.

IFPRI (2011) *Leveraging Agriculture for Improved Health and Nutrition*, International Food Policy Research Institute Vision 2020 Conference, New Delhi, India.

Impens, R. (2011) 'Agricultural diversification: a case-study in the Bonsaaso Millennium Village', Master's thesis, Bioscience Engineering, Katholieke Universiteit Leuven, Belgium.

Lukmanji, Z., Hertzmark, E., Mlingi, N., Assey, V., Ndossi, G. and Fawzi, W. (2008) *Tanzania Food Composition Tables*, MUHAS-TFNC, HSPH, Dar es Salaam Tanzania. http://www.hsph.harvard.edu/nutritionsource/files/2012/10/tanzania-food-composition-tables.pdf.

Masset, E., Haddad, L., Cornelius, A. and Isaza-Castro, J. (2011) *A Systematic Review of Agricultural Interventions that Aim to Improve Nutritional Status of Children*, EPPI-Centre, Social Science Research Unit, Institute of Education, University of London, London.

MVP (2008) *The Millennium Villages Project Evaluation Protocol: Integrating the Delivery of Health and Development Interventions and Assessing the Impact on Child Survival in Sub-Saharan Africa.* www.thelancet.com/protocol-reviews/09PRT-8648 (accessed 11 June 2013).

NAS (2009) *Dietary Reference Intakes: Recommended Intakes for Individuals,* National Academy of Sciences, Washington, DC.

Negin, J., Remans, R., Karuti, S. and Fanzo, J. C. (2009) 'Integrating a broader notion of food security and gender empowerment into the African Green Revolution'. *Food Security,* vol. 1, no. 3, pp. 351–360.

Penafiel, D., Lachat, C., Espinel, R., Van Damme, P. and Kolsteren, P. (2011) 'A systematic review on the contribution of edible plant and animal biodiversity to human diets', *EcoHealth,* doi:10.1007/s10393-011-0700-3.

Perry, A. (2008) *Cost of the Diet: Novel Approach to Estimate Affordability of a Nutritious Diet,* Save the Children UK.

Petchey, O. L. and Gaston, K. J. (2002) 'Extinction and the loss of functional diversity', *Proceedings of the Royal Society of Biological Sciences,* vol. 269, pp. 1721–1727.

Remans, R., Flynn, D., DeClerck, F., *et al.* (2011a) 'Assessing nutritional diversity of cropping systems in African villages', *PLoS ONE,* vol. 6, no. 6: e21235. doi:10.1371/journal.pone.0021235.

Remans, R., Pronyk, P. M., Fanzo, J. C., *et al.* (2011b) 'Multisector intervention to accelerate reductions in child stunting: an observational study from 9 sub-Saharan African countries', *American Journal of Clinical Nutrition,* vol. 94, pp. 1632–1642.

Sanchez, P. A., Palm, C., Sachs, J. D., Denning, G. and Flor, R. (2007) 'The African Millennium Villages', *Proceedings of the National Academy of Sciences of the USA,* vol. 104, pp. 16775–16780.

Shackleton, C. M., Pasquini, M. W. and Drescher, A. W. (eds) (2009) *African Indigenous Vegetables in Urban Agriculture,* Earthscan, London.

UNICEF (1990) *Strategy for Improved Nutrition of Women and Children in Developing Countries: A UNICEF Policy Review,* UNICEF, New York.

Verzovolli, I. (2010) 'Livelihood diversification strategies: on-farm diversification strategies in the Millennium Villages Project in Malawi', Master in Development Studies, University of Geneva, Switzerland.

Welch, R. D. and Graham, R. D. (1999) 'A new paradigm for world agriculture: meeting human needs. Productive, sustainable, nutritious', *Field Crops Research,* vol. 60, pp. 1–10.

WHO (2011) *World Health Statistics Report 2011,* World Health Organization, Geneva.

World Bank (2007) *From Agriculture to Nutrition: Pathways, Synergies, and Outcomes,* World Bank, Washington, DC.

# 17 Disseminating agroforestry innovations in Cameroon

## Are relay organizations effective?

*Ann Degrande,*[1] *Yannick Yeptiep Siohdjie,*[2] *Steven Franzel,*[3] *Ebenezer Asaah,*[1] *Bertin Takoutsing,*[1] *Alain Tsobeng*[1] *and Zac Tchoundjeu*[1]

[1] *World Agroforestry Centre – West and Central Africa, Cameroon*
[2] *Dschang University, Cameroon*
[3] *World Agroforestry Centre HQ, Kenya*

## Introduction

Because 75 per cent of the poor in developing countries live in rural areas, strengthening the agricultural sector and supporting agro-ecological intensification not only improves access to nutritious food; it does more – at least twice as much – to reduce rural poverty than investment in any other sector (FAO 2011). The role of extension in this battle is clear; there is a great need for information, ideas and organization to develop an agriculture that will meet complex demand patterns, reduce poverty and preserve or enhance ecological resources.

After many years of under-investment in agriculture and particularly in agricultural extension, the tide has fortunately changed and more funding is becoming available for extension. With this renewed interest, there is also growing awareness that farmers get information from many sources and that public extension is one source, but not necessarily the most efficient (Feder *et al.* 2010). Therefore, over the last decade, there have been many reforms to extension and advisory services to make them more pluralistic, demand-driven, cost-effective, efficient and sustainable. However, there is limited or conflicting evidence as to their effect on productivity and poverty, as well as on their financial sustainability.

While ineffective dissemination methods have contributed to the low adoption of agricultural innovations in general, this is particularly true for innovations in agroforestry, which are known to be complex and knowledge intensive, involving several components (crops, livestock and trees), requiring the learning of new skills, such as nursery establishment, and often providing benefits only after a long period (Franzel *et al.* 2001). To face the challenges of inappropriate extension methods the World Agroforestry Centre (ICRAF) in West and Central Africa has been experimenting for the last five years with relay organizations (ROs) and rural resource centres for the dissemination of innovations, particularly participatory tree domestication (Simons and Leakey 2004; Tchoundjeu *et al.* 2006; Asaah *et al.* 2011).

This chapter assesses the performance of ROs in disseminating innovations and enhancing their adoption in Cameroon, identifies the factors that affect their performances, and formulates implications for scaling up the approach.

## The concept of ROs and rural resource centres: ICRAF's experience in Cameroon

After a decade of research by ICRAF in Cameroon, innovations ready for dissemination include vegetative propagation techniques (marcotting, rooting of cuttings and grafting), integration of trees through the development of multistrata agroforests, soil fertility management techniques and improved marketing strategies for the commercialization of tree products, mainly through the organization of group sales. To accelerate the uptake of these new techniques in a context where public agricultural extension is hardly operational, ICRAF established collaboration with local organizations that were already involved in extension in different areas of Cameroon, but not necessarily linked to a research body or familiar with new developments in the area of agroforestry.

These ROs are boundary-spanning actors that link research organizations, such as ICRAF, and farmers' communities. They join with researchers in conducting participatory technology development, implying a two-way interaction of capacity building and institutional support, on the one hand, and feedback on the technology development, on the other. The ROs disseminate innovations to farmers using demonstrations, training and technical assistance, after which farmers provide feedback and, by so doing, help to develop the innovations further. At the same time, some ROs use the rural resource centre concept in their extension approach. These are places where agricultural and agroforestry techniques are practised and where farmers can come for information, experimentation and training. A typical rural resource centre in the context of this study consists of a tree nursery, demonstration plots, a small library, a training hall and eventually accommodation facilities. Depending on which innovations are relevant to the area, the centre may also host a unit for the processing of products and seed multiplication plots. Rural resource centres are managed by community-based organizations, which can be non-governmental organizations (NGOs) or farmers' groups, and provide a platform for bringing together different actors for exchange and shared learning.

## Methodology

The methodology used to assess the performance of ROs in the dissemination of agroforestry innovations in Cameroon was inspired by the framework for designing and analysing agricultural advisory services, developed by Birner *et al.* (2006), where the performance of the agricultural advisory services is explained as a function of (1) characteristics of the advisory services and the linkages with research and education; (2) frame conditions and the 'fit' of the service with those frame conditions; and (3) the ability of clients to speak up and hold the service providers accountable. In practice, the following indicators were used: number of groups and number of farmers technically supported by ROs, the diffusion and adoption rate of innovations disseminated, knowledge and the level of mastery among the farmers trained, and their level of satisfaction.

Furthermore, the degree to which ROs are able to achieve their objectives and satisfy their clients was hypothesized to depend on both internal and external factors. At the internal level, it was assumed that their capacity to disseminate innovations depends on the human, financial and material resources at their disposal, as well as the technical capacities of their staff, their experience with agroforestry, and the type and duration of their collaboration with agroforestry research. On the other hand, external factors that are likely to affect the outcome of their work included the need for agroforestry in the area or the existence of problems that it can address, road infrastructure, market access and farmers' experience in collective action. Therefore, the 18 ROs from three agro-ecological zones of Cameroon (western Hauts Plateaux, forest zone with mono-modal rainfall, and forest zone with bi-modal rainfall), were grouped in four categories, based on whether internal and external factors were favourable or unfavourable[1] (Table 17.1), and a sample of two ROs in each of the categories was taken for in-depth investigation. Information was collected in July–August 2010 at five different levels: ICRAF staff as resource persons; leaders of the eight ROs selected for this study; leaders of 27 farmers' groups supported by the ROs selected randomly from the list of farmers groups; 76 group members (35 of them women) selected at random; and seven farmers (all male) who had stopped collaborating with the ROs.

## Results and discussion

### *Presentation of ROs and their functioning*

All ROs had more or less the same objectives, i.e. diffusing technologies developed in collaboration with ICRAF (100 per cent) and improving local people's livelihoods in general (62 per cent). The following activities were mentioned by all ROs: tree domestication, tree planting, establishing demonstration plots and organizing study and exchange visits for farmers. Seventy-five per cent of the ROs were involved in soil fertility management and half of them accompanied their groups in collective action for the marketing of agroforestry products. For the achievement of their objectives, ROs developed partnerships with international, national and local organizations. However, less than half of those ROs studied collaborated with governmental institutions or programmes.

*Table 17.1* Categorization of ROs according to qualitative internal and external factors likely to affect performance

| | | *External factors* | |
|---|---|---|---|
| | | *Favourable* | *Unfavourable* |
| *Internal factors* | *Favourable* | *Category I*<br>PROAGRO (2000)<br>RIBA (2002)*<br>APADER (2004)*<br>MIFACIG (1998) | *Category II*<br>FONJAK (2000)*<br>ADEAC (2003)*<br>CAMECO (2005) |
| | *Unfavourable* | *Category III*<br>RAGAF (2008)<br>PIPAD (2008)*<br>AJPCEDES (2008)<br>FOEPSUD (2005)*<br>CIMAR (2009)<br>FEPROFCAO (2009) | *Category IV*<br>GICAL (2000)*<br>APED (2008)<br>SAGED (2008)*<br>CAFT (2004)<br>CANADEL (2010) |

*Notes*: Brackets show the year of start of collaboration with ICRAF. PROAGRO, RIBA, APADER, MIFACIG and CIMAR have a rural resource centre. * RO selected for the study.

All ROs in this study got part of their financial resources from ICRAF's support, revenues from the nursery and through service provision (e.g. training). Contributions from members constituted a source of income for one-third of the ROs. The sales of livestock (one RO) and of agricultural products (one other RO) were used to finance day-to-day operations. In terms of qualified staff, there were big differences among the ROs. Logically, categories I and II (favourable internal factors) were composed of the most experienced staff, while categories III and IV were less well endowed. Likewise, categories I and II were better equipped with office furniture and transport facilities than categories III and IV.

### *Role of ROs in the development and dissemination of agroforestry innovations*

All ROs recognized that their feedback on the technologies disseminated was valued by ICRAF and taken into account in the further development of the innovations. However, only 25 per cent identified their active involvement in participatory research as a distinctive role, in addition to the dissemination of innovations. Participation of the ROs in the development of the innovations, however, seemed to be facilitated by the presence of the rural resource centres, where focus is put on interactive learning and farmers' experimentation. All eight ROs studied used a combination of approaches to disseminate agroforestry, including theoretical and practical training of farmers, open-door events to publicize and demonstrate new technologies to a wide public, and the establishment of demonstration plots showing the benefits of the innovations. In addition, two of the ROs (APADER and RIBA, belonging to category I) operated a rural resource centre, which is equipped with a nursery, experimental and demonstration plots, a training hall and accommodation amenities.

## *Performance of relay organizations*

### *Number of groups supported and farmers reached*

We noticed that ROs of category I provided assistance to the highest number of groups, although ROs of categories II and III reached more farmers (Figure 17.1). This can be explained by the larger size of the farmers' groups that the latter category supported. In terms of gender, when all ROs are taken together, 46 per cent of the farmers reached were women. This proportion is high, compared with many development interventions, and is a combined result of deliberate efforts to bring women to training sessions and to target women's groups for particular technologies. Such technologies include soil fertility improvement and the commercialization of tree products that traditionally belong to the women's domain (e.g. *Irvingia gabonensis* and *Ricinodendron heudelotii* kernels for spices and *Gnetum africanum* leaves for consumption as a vegetable). The proportion of female farmers trained was particularly high for RIBA (67 per cent), which mainly disseminates soil fertility improving

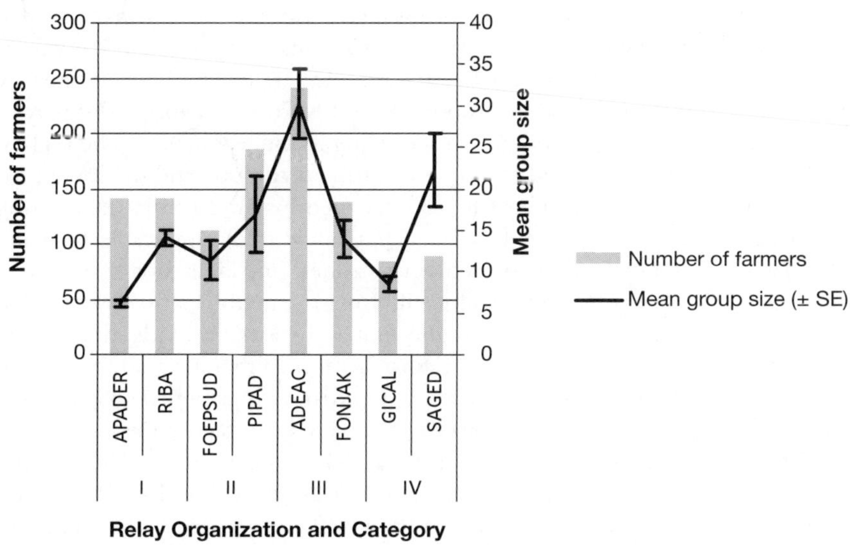


*Figure 17.1* Number of groups and farmers supported by ROs and mean group size.

techniques, and ADEAC (67 per cent), which predominantly promotes post-harvest and the group marketing of *R. heudelotii* (*njansang*) kernels. On the other hand, the proportion of women in groups supported by GICAL was the lowest (18 per cent), which can be explained by the strong focus on vegetative propagation. Altogether, 36 per cent of the farmers trained by the ROs studied were younger than 35 years old. There was no clear relationship between the number of young people reached and the category to which a RO belongs.

### *Knowledge and mastery of agroforestry techniques*

Altogether, 44 per cent of the respondents had basic knowledge on all technologies, 14 per cent said to have mastery and 6 per cent could also teach other farmers on the topic. Knowledge domains that were best acquired by farmers, however, included the rooting of cuttings and tree spacing; topics less mastered were post-harvest technologies, group sales and conflict management; all of these required the development of marketing strategies for products. This is related to the fact that ROs had been introduced to tree propagation earlier than to marketing-related aspects and were therefore able to gain more expertise on the first issue. Looking at differences per category of ROs, we noticed that the proportion of farmers who had no knowledge of agroforestry innovations was much higher for ROs from category IV (49 per cent) than from category I (21 per cent), while the opposite was true for farmers who could teach others (Figure 17.2).

### *Diffusion and adoption of innovations*

In terms of adoption, the farmers interviewed mainly applied the following techniques: marcotting, rooting of cuttings, grafting, soil fertility management and the use of a *njansang* cracking machine (post-harvest technique). Average rates of adoption varied from 52 per cent for farmers trained by ROs of category III to 61 per cent for those backstopped by ROs from category I. However, the rate of adoption varied very much according to the techniques. The highest rate of adoption was recorded for the marcotting technique and the lowest for soil fertility improvement. This can be explained by the fact that marcotting is a divisible technique (it can be done on a single tree), applicable to many different species independent of ecological zones, and does not need much equipment. On the other hand, soil fertility management is a technology that requires land tenure security and a higher initial investment in planting a large number of trees or shrubs, and therefore is more difficult to adopt. It was interesting to note that, out of the seven respondents who stopped collaborating and left the farmers' groups, 65 per cent continued to practise marcotting, 43 per cent continued the rooting of cuttings and 43 per cent continued grafting. On the other hand, only 14 per cent were still using trees and shrubs for soil fertility management after they left the groups. The reasons for this drop-out were not examined but are likely to be similar to those mentioned above, explaining low adoption.

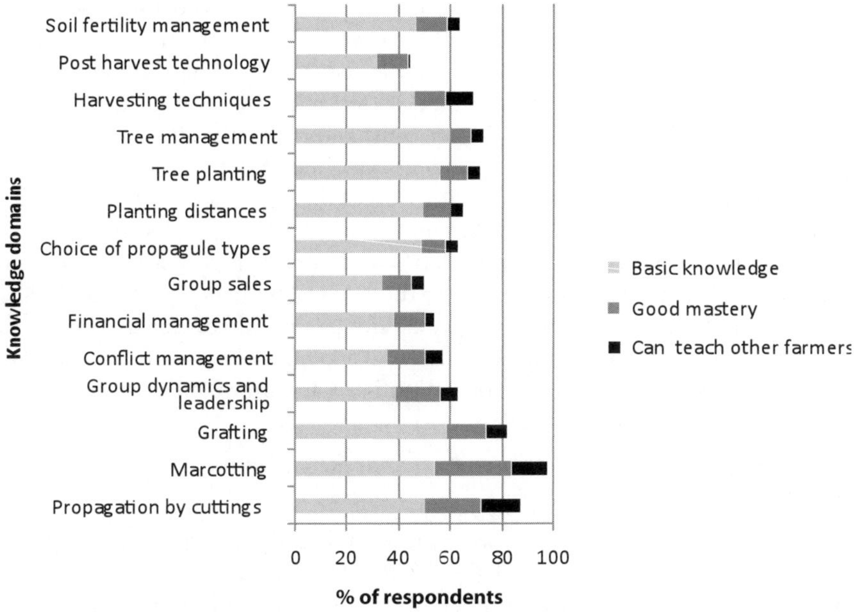


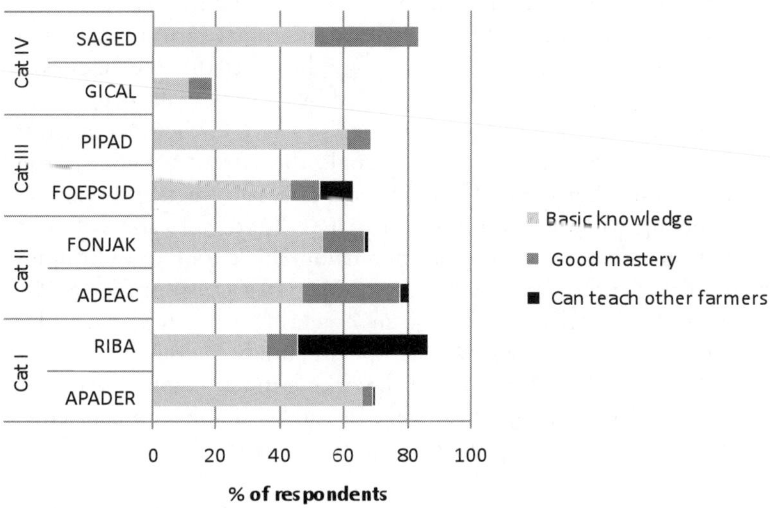


*Figure 17.2* Knowledge and mastery of agroforestry techniques, (a) by technique and (b) by RO.

*Farmers' perception of performance*

During interviews, 11 per cent of the group members said they were very satisfied with the performance of the ROs and 67 per cent were satisfied. In fact, 78 per cent of the respondents mentioned good technical support as one of the strong points of the work by ROs, followed by the regular follow-up of group activities (reported by 39 per cent) and contact with a range of other partners (26 per cent). Most respondents also felt that the language used by ROs was adapted to the target population. Moreover, the staff were said to be patient and tolerant, and the techniques disseminated were relevant to farmers' needs. Only 10 per cent of those interviewed complained that they were not satisfied at all with the performance of the ROs. The major points of dissatisfaction among these respondents were failure to find buyers for their products (86 per cent), delays in the implementation of activities (70 per cent), the absence of financial assistance (69 per cent) and missed appointments (56 per cent).

### *Analysis of the effect of internal and external factors on the performance of ROs*

On the whole, the ROs studied were successfully diffusing agroforestry innovations to farmers' groups. Nevertheless, differences have been observed between the categories of ROs for a number of performance indicators. Though not statistically significant, these results suggest that those which operate under favourable internal and external factors (category I) perform best for most of the performance indicators, but this would need confirmation in further studies involving more ROs. Also, the study suggests that external factors, such as the existing opportunities for agroforestry, strong farmers' associations and good road and communication networks (category III), might affect the effectiveness of ROs more than their internal capacity, reflected by their human, material and financial resources (category II). However, this also needs further investigation.

From the study, it was difficult to disentangle the effect of having a rural resource centre from other internal and external factors influencing effectiveness. This is not surprising because the fact that ROs are able and willing to create a rural resource centre already reflects a favourable external environment and a certain level of internal resources. As a matter of fact, from the 18 ROs working with ICRAF, the five that have such centres (PROAGRO, RIBA, APADER, MIFACIG and CIMAR) all operate under favourable external conditions. Four out of those five were also working under favourable internal conditions; only CIMAR was put in category III (unfavourable internal factors), mostly because its collaboration with ICRAF was recent.

## Conclusion

Effective extension systems are required if agriculture is to ensure sufficient good-quality food for the long term under the challenges of increased and changing demands, shrinking natural resources and climate change. Current trajectories for land use intensification clearly involve more complex and knowledge-intensive practices and hence necessitate extension approaches that support farmers' learning and adaptive management. The approach described in this chapter was found to meet some of these requirements.

In general, the ROs studied were successfully diffusing agroforestry innovations to farmers' groups. Differences in their performances could not be easily explained by either external or internal factors. However, the fact that ROs operating in areas with relatively good road and communication networks and opportunities for agroforestry, and that also have adequate internal human, material and financial capacity seemingly performed better provides us with some indications of the support that might be required to further strengthen these ROs and increase their extension capacity. In-depth studies involving more of them are necessary to increase our understanding of what factors affect the performance of organizations in disseminating agricultural innovations.

The involvement of grassroots organizations in the extension of agroforestry has increased the relevance of the techniques and the quality of services rendered to the beneficiaries. Already, farmer-led experimentation and adaptation are common in the rural resource centres. The approach has also succeeded in reaching a relatively high number of women and young people; often overlooked in 'traditional' extension systems. One challenge, however, in this approach remains the level of technical expertise which calls for continuous training, coaching and upgrading of extension staff.

Last but not least, one of the generic problems of agricultural extension, the difficulty of cost recovery, has not been addressed in the current study. It is expected that community-based extension would be more cost-efficient than other approaches. However, our current understanding of the sustainability and financial viability of the approach is not sufficient to draw any conclusions and more research is required in this domain.

## Notes

1 Factors are considered favourable when they positively affect the outcome of the ROs' work. For example, the existence of good roads and proximity to an urban market are considered favourable factors. On the other hand, factors are called unfavourable when they are expected to negatively affect performance. For instance, ROs are expected to have more difficulties in disseminating innovations in an area where farmers are not used to forming groups and working together (low experience in collective action) because more efforts are required to reach farmers individually.

## References

Asaah, E. K., Tchoundjeu, Z., Leakey, R. R. B., Takoutsing, B., Njong, J. and Edang, I. (2011) 'Trees, agroforestry and multifunctional agriculture in Cameroon', *International Journal of Agricultural Sustainability*, vol. 9, no. 1, pp. 110–119.

Birner, R., Davis, K., Pender, J., Nkonya, E., Anandajayasekeram, P., Ekboir, J., Mbabu, A., Spielman, D., Horna, D., Benin, S. and Cohen, M. (2006) 'From "best practice" to "best fit": a framework for analysing pluralistic agricultural advisory services worldwide', *International Service for National Agricultural Research Discussion Paper No. 5*, IFPRI, Washington, DC.

FAO (2011) 'Increased agricultural investment is critical to fighting hunger'. www.fao.org/investment/whyinvestinagricultureandru/en (accessed 22 September 2011).

Feder, G., Anderson, J. R., Birner, R. and Deininger, K. (2010) 'Promises and realities of community-based agricultural extension', *IFPRI Discussion Paper No. 00959*, IFPRI, Washington, DC.

Franzel, S., Coe, R., Cooper, P., Place, F. and Scherr, S. J. (2001) 'Assessing the adoption potential of agroforestry practices in sub-Saharan Africa', *Agricultural Systems*, vol. 69, nos 1–2, pp. 37–62.

Simons, A. J. and Leakey, R. R. B. (2004) 'Tree domestication in tropical agroforestry', in P. K. R. Nair, M. R. Rao and L. E. Buck (eds), *New Vistas in Agroforestry: A Compendium for the 1st World Congress of Agroforestry*, Kluwer Academic Publishers, Boston, MA, pp. 167–182.

Tchoundjeu, Z., Asaah, E. K., Anegbeh, P., Degrande, A., Mbile, P., Facheux, C., Tsobeng, A., Atangana, A. R., Ngo, M., Peck, M. L. and Simons, A. J. (2006) 'Putting participatory domestication into practice in West and Central Africa', *Forests, Trees and Livelihoods*, vol. 16, pp. 53–69.

# 18 Participatory re-introduction of *Vicia faba* beans in resource-poor farming systems

## Adoption of a farmer-led initiative

*Erik Karltun,*[1] *Tesfanesh Gichamo,*[2] *Tizazu Abebe,*[3] *Mulugeta Lemenih,*[4] *Motuma Tolera*[3] *and Linley Chiwona-Karltun*[2]

[1] *Swedish University of Agricultural Sciences (SLU), Department of Soil and Environment, Sweden;* [2] *Swedish University of Agricultural Sciences (SLU), Department of Urban and Rural Development, Sweden;* [3] *Wondo Genet College of Forestry and Natural Resources, Ethiopia;* [4] *International Water Management Institute, Ethiopia*

## Introduction

Excessive theft and conflict between communities were the reasons for a complete abandonment of faba bean (*Vicia faba* L.) cultivation in the Beseku Illala Peasant Association, Ethiopia. We studied the extent of and determinants for the adoption or non-adoption of bean cultivation following a participatory farmer-led reintroduction of the crop. The aims were to (1) assess the extent of adoption following the enforcement of collective by-laws and re-introduction of faba bean cultivation, (2) identify the determinants of farmers' households that took up the cultivation of faba beans and (3) clarify the reasons behind inter-village differences in the rate of adoption. The re-introduction resulted in an area being cultivated that was larger than an estimated area expected from the amount of seeds distributed and the recommended seed rate. The decision by a household to re-adopt bean cultivation was positively influenced by previous experience in cultivating the crop and contact with extension services. The spatial separation of the homestead from farmland due to villagization (local resettlement) was the underlying reason for non-adoption in some villages since the possibilities of guarding the fields against theft were limited. Fear of conflict as a result of reporting theft was another important reason not to cultivate beans. We conclude that traditional institutions may play an important role in interventions and up-scaling. Support is also needed from government institutions in the form of extension and the provision of high-quality inputs such as seeds.

Participation and mobilization of farmers may be a pathway for resolving local conflicts that are manifest as obstacles to agricultural intensification.

Cultivation of the faba bean has a long tradition in Ethiopia, where it is used to produce a flour *shiro*, which is used to cook a relish sauce, *shirowat*. In the typical Ethiopian diet, the faba bean is a meat extender, rich in protein and micronutrients (Fe, Mg, Mn and Zn). Due to their high content of energy and protein, an increased inclusion of beans in the diet is a vital way to reduce protein–energy malnutrition (PEM) in young children (Latham 1997). The faba bean is also one of the legume crops that have a high potential to contribute to symbiotic $N_2$-fixation; N-fixing rates in the range of 45–550 kg of N per hectare have been reported (Jarso and Kenini 2006). The net effect on the soil's N balance from N-fixation by faba beans ranges between 8 and 271 kg of N per hectare (mean 113 kg of N per hectare) (Evans *et al.* 2001). Faba beans were among the common crops grown by the farmers of the Beseku Illala Peasant Association (PA) less than two decades ago. The farmers were aware of the positive effect of the beans on soil fertility since they stated that 'Any crop grown after beans used to give a good yield' (Karltun *et al.* 2008). However, in recent years the cultivation of beans gradually declined and had been almost entirely eliminated from the cropping system because of excessive theft. In earlier times snacking on beans from your neighbours' field was a more or less acceptable habit. There was a proverb that farmers used to describe this practice:

> You cannot pass by a bean field or a beautiful woman without stopping by and enjoying for a moment.

With time, this habit became a serious problem when the frequency of pilfering increased. The rate of theft did not stop at snacking but extended to large-scale harvesting and sale on the market. The small-scale production of the crop meant that, if many people were harvesting or stealing, there was little left for the family at the end. Consequently, farmers stopped cultivating beans to avoid the social conflicts arising from the extensive theft of fresh beans directly from the fields (Chiwona-Karltun *et al.* 2009).

To curb bean theft, a consensus was reached by the people living in the area. They identified the *Iddirs* as the best institution to deal with the problem. An *Iddir* is a local informal institution for voluntary insurance and social security. Members contribute fees in exchange for the right to financial support, for example, to cover the expenses for funerals. There are several *Iddirs* in the PA. Several of the *Iddirs* came together and formulated collective (agreed upon by all) by-laws that specified a fine for theft of beans (200 Birr – approximately $12). The members of the *Iddirs* also committed themselves to stop protecting the thieves. To support the community's collective initiative, the project worked initially as a facilitator by providing improved seeds. The aims of the study presented in this chapter were to (1) assess the extent cultivation following the enforcement of the collective by-laws and the re-introduction of faba beans,

(2) identify the factors that determined whether farmers took up cultivation of faba beans or not and (3) clarify the reasons behind the inter-village differences in the rate of adoption.

## Materials and methods

The study was conducted in Beseku Ilala PA in South Central Ethiopia. The population in 2007 was 11,210, with an average household size of 5.5 members. Based on the local administrative system, Beseku is divided in 61 *gotes* (small villages), and in each *gote* there are about 30–34 households.

The first year of re-introduction was 2007 and one tonne of bean seeds was distributed to farmers free of charge. The selection of farmers to participate in the first year of re-introduction was made through a random lottery process. In 2008 another two tonnes of seeds were made available but were sold to the farmers at the market price. In 2008 all fields in the PA that were cultivated with beans during the two years were located and their area was determined by a GPS survey.

For the adoption study conducted in 2010, all farm households in the PA were categorized into adopters and non-adopters. An adopter was defined as a household that had cultivated faba beans at least once over the four years following the re-introduction; non-adopters were those who had never attempted to grow beans. A multi-stage sampling technique (Elder 2009) was used to select farm households. Out of 1,310 households in the PA, 138 were classified as adopters. A proportional allocation to size method was employed to determine the number of adopter households from each part of the PA that should be interviewed and 69 households were selected. Finally, individual adopter and non-adopter farm households from each category and part of the PA were selected using simple random sampling. Factors affecting the adoption of faba bean cultivation and the extent of cultivation among adopters were assessed using the adoption decision as the dependent variable. The adoption decision was treated as a dummy variable taking the value of 1 if the household cultivated faba beans or 0 if otherwise. Explanatory variables of importance hypothesized to have an influence on the decision to adopt or not adopt bean cultivation included household factors (age, gender and educational level of the household head, and household family size), immigration (whether the household head was native or a settler from another region). Others were economic factors (farm size, livestock holding), previous experience (knowledge of the benefits of bean cultivation and the positive effect on soil fertility) and institutional factors (contact with the government's extension service and access to credit facilities). The interview also included the ranking of factors that hampered bean cultivation, using a standard pair-wise ranking procedure. Quantitative data were analysed using a $\chi^2$-test, pair-wise comparison and descriptive statistics.

As a complement to the objectively sampled adoption study, a cross-sectional survey using qualitative methods was conducted in three selected villages: *Shibeshi*

*Gasha I* and *Shibeshi Gasha II*, where there were few adopters, and *Boye*, with a high proportion of adopters. Key informants with special knowledge about bean theft were identified for in-depth interviews. Subsequent respondents were identified using the snowball technique, which utilizes subjects who may be able to recommend other potential candidates for the study (Bernard 2000). The in-depth and key informant interviews consisted of closed and open-ended interview questions that were based on earlier findings in the area related to bean theft (Chiwona-Karltun *et al.* 2009). Focus group discussions were conducted in all three study areas (Kumar 1987). To ensure the freedom to talk openly (Nichols 1991; Dawson *et al.* 1993) discussions with women and men were held separately. Content analysis (Taylor-Powell and Renner 2003) was used to analyse the qualitative data.

## Results

### *Extent of bean cultivation*

The cultivated area for the two years (2007 and 2008) was 20.3 ha, in a total of 206 fields, making the average size of a bean field 0.1 ha. Bean cultivation

*Figure 18.1* Approximate extension of the three *gotes* used in study 2 (exact formal boundaries could not be established). The white polygons are bean fields recorded in 2007 and 2008.

*Table 18.1* Standard pair-wise ranking of constraints associated with faba bean cultivation ($n$ = 138)

| *No.* | *Constraints* | *Constraint number* | | | | | | | *Score* | *Rank* | *n* | *%* |
|---|---|---|---|---|---|---|---|---|---|---|---|---|
| | | *1* | *2* | *3* | *4* | *5* | *6* | *7* | | | | |
| 1 | Land shortage | – | 1 | 3 | 4 | 5 | 1 | 1 | 3 | 3 | 10 | 7.2 |
| 2 | Seed access | | – | 3 | 2 | 5 | 2 | 2 | 3 | 3 | 9 | 6.5 |
| 3 | Susceptibility to weather, pathogens and animals | | | – | 3 | 5 | 3 | 7 | 4 | 2 | 31 | 23 |
| 4 | Low yield | | | | – | 5 | 6 | 4 | 2 | 4 | 3 | 2.2 |
| 5 | Fresh bean theft from crop field | | | | | – | 5 | 5 | 6 | 1 | 66 | 48 |
| 6 | Labour shortage | | | | | | – | 7 | 0 | 5 | 5 | 3.6 |
| 7 | Lack of cooperation/ unity | | | | | | | – | 2 | 4 | 14 | 10 |

in the village was unevenly distributed and almost all cultivation was located close to residential settlements (Figure 18.1).

### *Factors hampering faba bean cultivation*

Among the factors that constrained or hampered the cultivation of faba beans, theft was ranked as the major problem by 48 per cent of the respondents, followed by damage by animals, pathogens or unfavourable weather conditions (23 per cent). Lack of cooperation and unity among the members of the community ranked rather low (Table 18.1).

### *Determinants of adoption*

The $\chi^2$ test indicated that the most significant differences between adopters and non-adopters were previous experience of faba bean cultivation and then exposure to agricultural extension (Table 18.2). Households with previous experience and those receiving agricultural extension were evidently more likely to adopt bean cultivation. Families that had immigrated into the area recently were also more likely to be adopters than families that had lived in the area for more than one generation. Other factors, such as the age, gender or educational level of the household head, access to credit and family or farm size, did not seem to have any influence on whether the household adopted bean cultivation or not.

*Table 18.2* Influence of different household characteristics on the decision to adopt/not adopt faba bean cultivation. The two groups are compared using the chi-square independence test ($n$ = 138). The variables are sorted according to probability level

| *Variables* | *Category/range* | *Adopters (n)* | *Non-adopters (n)* | $\chi^2$ | $p$ |
|---|---|---|---|---|---|
| Previous experience of faba bean cultivation | Yes | 58 | 41 | 10 | 0.001 |
| | No | 11 | 28 | | |
| Extension | Yes | 35 | 21 | 5.9 | 0.015 |
| | No | 34 | 48 | | |
| Farmer origin | Native | 34 | 24 | 3.6 | 0.058 |
| | Immigrant | 35 | 45 | | |
| Livestock assets | Few | 36 | 31 | 4.5 | 0.10 |
| | Intermediate | 29 | 26 | | |
| | Many | 4 | 12 | | |
| Educational level | Non-formal | 27 | 20 | 4.4 | 0.22 |
| | Grades 1–3 | 23 | 19 | | |
| | Grades 5–8 | 12 | 22 | | |
| | High school or higher | 7 | 8 | | |
| Age | <36 | 11 | 19 | 2.7 | 0.25 |
| | 36–50 | 37 | 33 | | |
| | >50 | 21 | 17 | | |
| Credit | Yes | 27 | 23 | 0.50 | 0.48 |
| | No | 42 | 46 | | |
| Gender | Man | 58 | 60 | 0.23 | 0.63 |
| | Woman | 11 | 9 | | |
| Family size | <6 persons | 12 | 13 | 0.80 | 0.67 |
| | 6–10 persons | 51 | 47 | | |
| | >10 persons | 6 | 9 | | |
| Farm size | Small (<1.5 ha) | 37 | 39 | 0.13 | 0.94 |
| | Medium (1.5–3.5 ha) | 23 | 22 | | |
| | Large (>3.5 ha) | 9 | 8 | | |

### *Inter-village variation in adoption*

For the clear geographical pattern in the adoption of bean cultivation (Figure 18.1), 76 per cent of the respondents from the villages with few adopters indicated dislocation from their farmland area due to villagization as the main reason. One farmer stated in an individual interview that 'in this area, a few farmers who have their land near to their homes grow beans but we don't. … Our farm is far away from our home; it is not possible to guard the farm all the time. Thieves may steal by day or night'. In *Boye*, the *gote* with a large proportion of adopters, farmers lived close to their farms and villagization was never implemented. Respondents in *Boye* stated that it was possible to guard their bean fields because the fields were near to their homes. Out of 34 interviewees from *Shibeshi Gasha I* and *Shibeshi Gasha II*, the two *gotes* where there was little observed adoption of bean cultivation, 76 per cent of the respondents mentioned that villagization (forced local resettlement) changed the way farming land and homesteads were positioned. The policy resulted in the physical separation of homestead and cultivated land and this was cited as one of the most important reasons for not adopting bean cultivation. Fear of conflict between families, neighbours and different tribes was taken up by 44 per cent of the respondents. To avoid conflict and retaliation, respondents stated that they did not expose the thieves even if they knew the perpetrators:

> Most people do nothing when they see a thief stealing their beans; they do not want to quarrel with the person who is stealing. This is due to fear of creating an enemy for the future. The thief may come to steal from our farm in the future in revenge. So we prefer not to say anything.
>
> (Individual interview, *Shibeshi Gasha*, February 2009)

In *Boye*, the responses indicated there was a strong sense of unity and confidence in the agreement to not tolerate bean theft;

> In this gote [*Boye*] we have unity; there are no thieves among us and everybody wants to work hard. We agreed with each other and have a common understanding. We agreed not to hide the thieves. We expose the thieves if we find them in our fields.
>
> (Women's focus group discussion, *Boye*, February 2009)

Lack of sufficient land was also mentioned by 32 per cent of respondents as a reason not to adopt bean cultivation. Some women (17 per cent of the respondents in *Shibeshi Gasha gotes*) mentioned that they did not receive any information about the re-introduction of beans.

## Discussion

It is generally accepted that the presence of crime in an area may discourage legitimate business, hence further aggravating poverty (Fafchamps and Minten 2003). The theft of crops and livestock assets during a food crisis is often associated with seasonal hunger and transitory disasters (Chiwona-Karltun *et al.* 1998; Mkumbira *et al.* 2003) but the systematic effect of thieving on crop rotation and legume cultivation, which may have direct negative effects on soil fertility and the sustainability of the cropping system, was first reported through our studies in Beseku (Karltun *et al.* 2008; Chiwona-Karltun *et al.* 2009). Although there are other factors that might affect a farmer's decision of whether to cultivate beans or not, the results of the farmers' ranking of constraints to bean cultivation and from numerous interviews and discussions show that it is the problem of theft that led to farmers abandoning bean cultivation in Beseku. By jointly identifying and understanding the underlying causes of faba bean thieving in their communities, the farmers realized that the best way to curb theft was by undertaking the following: (1) the creation of by-laws; (2) the enforcement of these by-laws through a legitimate and respected institution; and (3) the identification of a local institution, the *Iddir*, as the most appropriate non-discriminatory institution to enforce the by-laws.

The re-introduction project achieved a significant level of bean cultivation. The recommended seed rate for faba bean in Ethiopia is 175–200 kg/ha. If we assume that the beans are planted at the recommended seed rate, the seeds distributed will be enough for only 15–17 ha, compared with the 20.3 ha that were recorded. There are three possible explanations for the discrepancy: (1) the farmers might have planted at a seed rate lower than that recommended; (2) some farmers might have saved seeds from 2007 that they planted in 2008; (3) farmers that did not get any through the project might have purchased and sown seeds.

In the process of making an adoption decision, farmers usually weigh the consequences of the adoption of an innovation against its economic, social and technical feasibility (Kebede *et al.* 1990). The most significant difference in household characteristics found between the adopting and non-adopting groups was that previous experience of faba bean cultivation led to a significantly higher adoption rate than in the non-adopting households, followed by access to agricultural extension services. It is promising to find that these factors were more important than more evident indicators of poverty such as educational level, access to credit or farm size, since lack of experience and information are factors that are more easy to influence than poverty structures. It is also interesting to see that some of the government's policies, such as villagization, affected the adoption of the crop. Villagization has been defined as 'the grouping of population into centralized planned settlements' (Survival International 1988). One of the effects was the spatial separation of the crop fields and the homestead. Instead of having easy access to their fields within walking distance, with villagization it could take hours every day walking between the fields and the house. This also made it more difficult to protect crops from theft or animals (Lorgen 2000). Several respondents explicitly mentioned that the few farmers that had adopted bean cultivation in *Shibeshi Gasha gotes* were those that had land close to their homesteads. However, villagization cannot have been the only factor, since bean cultivation had been practically abandoned in the whole PA, including *gotes* that were, like Boye, not affected by villagization.

The fear of conflict related to bean thieving remains a major explanation for not confronting thieves and for not re-introducing bean growing on the farm. In many of the interviews, the fear of '*buying yourself an enemy*' is mentioned. In *Boye*, the *gote* where bean re-introduction was successful, the social cohesion in the *gote* was repeatedly identified as a decisive factor for the successful re-introduction. Conflict as a result of bean theft was also linked with other local conflicts. The complex pattern of various conflicts and tensions in the village goes beyond the scope of this study. However, in the interviews, problems related to inter-generational differences (the young vs. elders), land tenure (landless vs. farmers) and ethnicity were mentioned as some of the underlying factors that affected the stealing of beans. Avoiding conflict over theft, i.e. by choosing not to cultivate beans, had become an accepted strategy simply adopted not to aggravate conflicts.

The importance of strategically including women as recipients of agricultural information and extension services is underscored by the observation that some women's groups were not involved in the creating of by-laws and the bean seed intervention. When women are left out as recipients of new knowledge and users of new agricultural technologies this puts them at a higher risk of not withstanding shocks.

There are some conflicting results between the two studies on the re-adoption of bean cultivation. In the adopter/non-adopter comparison, farm-size distribution did not differ between the two groups. However, some farmers have indicated in the interviews in study 2 that their land-holdings were too small to allow for diversification. It is quite possible that the classification of the small farm-size in study 1 was too wide and that a category of farms considerably smaller than 1.5 ha was hidden in the larger group.

## Conclusions

Creating awareness by extension and participation will increase the adoption rate. Success or failure of an intervention aiming at improving crop production and soil fertility management can depend on local social conditions which may vary considerably in a seemingly similar environment. Traditional institutions, like the *Iddir*, may play an important role in interventions and up-scaling, but support is needed also from governmental institutions in the form of extension and the provision of inputs such as seeds. The participation and mobilization of farmers, especially women, in improving agriculture may be a pathway for resolving local conflicts that are obstacles to agricultural intensification.

## Acknowledgements

These studies would not have been possible without funding from FAO to the project 'Farmer led re-introduction of *Vicia faba* beans in Ethiopian highland farming systems' and the Swedish Ministry of Foreign Affairs as part of their special allocation towards global food security.

## References

Bernard, H. R. (2000) *Social Research Methods: Qualitative and Quantitative Approaches*, Sage Publications Inc., Thousand Oaks, CA.

Chiwona-Karltun, L., Mkumbira, J., Saka, J., Bovin, M., Mahungu, N. M. and Rosling, H. (1998) 'The importance of being bitter: a qualitative study on cassava cultivar preference in Malawi', *Ecology of Food and Nutrition*, vol. 37, no. 3, pp. 219–245.

Chiwona-Karltun, L., Lemenih, M., Tolera, M., Berisso, T. and Karltun, E. (2009) *Soil Fertility and Crop Theft: Changing Rural Dimensions and Cropping Patterns*. http://event.future-agricultures.org

Dawson, S., Manderson, L. and Tallo, V. L. (1993) *Manual for the Use of Focus Groups*, International Nutrition Foundation for Developing Countries (INFDC), Boston, MA.

Elder, S. (2009) *ILO School-to-Work Transition Survey: A Methodological Guide. Sampling Methodology*, International Labour Office, Geneva.

Evans, J., McNeill, A. M., Unkovich, M. J., Fettell, N. A. and Heenan, D. P. (2001) 'Net nitrogen balances for cool-season grain legume crops and contributions to wheat nitrogen uptake: a review', *Australian Journal of Experimental Agriculture*, vol. 41, no. 3, pp. 347–359.

Fafchamps, M. and Minten, B. (2003) 'Theft and rural poverty: results of a natural experiment', Paper presented at the International Association of Agricultural Economists 2003 Annual Meeting.

Jarso, M. and Kenini, G. (2006) '*Vicia faba* L.', in M. Brink and G. Belay (eds), *Plant Resources of Tropical Africa 1. Cereals and Pulses*, PROTA Foundation/Backhuys Publishers/CTA, Wageningen/Leiden, pp. 195–199.

Karltun, E., Röing de Nowina, K., Chiwona-Karltun, L., Lemenih, M., Tolera, M. and Berisso, T. (2008) 'Working with farmers and local institutions to improve soil quality in sub-Saharan Africa', *Current Issues in International Rural Development*, vol. 43, pp. 7–11.

Kebede, Y., Gunjal, K. and Coffin, G. (1990) 'Adoption of new technologies in Ethiopian agriculture: the case of Tegulet-Bulga district Shoa province', *Agricultural Economics,* vol. 4, no. 1, pp. 27–43.

Kumar, K. (1987) *Conducting Group Interviews in Developing Countries*, USAID, Washington, DC.

Latham, M. C. (1997) *Human Nutrition in the Developing World*, FAO, Rome.

Lorgen, C. C. (2000) 'Villagization in Ethiopia, Mozambique, and Tanzania', *Social Dynamics: A Journal of the Centre for African Studies University of Cape Town*, vol. 26, no. 2, pp. 171–198.

Mkumbira, J., Chiwona-Karltun, L., Lagercrantz, U., *et al.* (2003) 'Classification of cassava into "bitter" and "cool" in Malawi: from farmers' perception to characterisation by molecular markers', *Euphytica*, vol. 132, no. 1, pp. 7–22.

Nichols, P. (1991) *Social Survey Methods: A Field Guideline for Development Work*, Oxfam, Oxford.

Survival International (1988) *For Their Own Good: Ethiopia's Villagisation Programme,* Survival International, London

Taylor-Powell, E. and Renner, M. (2003) *Analyzing Quantitative Data*, Cooperation Extension and University of Wisconsin, Madison, WI.

# Part IV

# Communicating and disseminating complex knowledge

# 19 Walking the impact pathway

## CIALCA's efforts to mobilize agricultural knowledge for the African Great Lakes region

*Boudy van Schagen*[1], *Guy Blomme*,[2] *Bernard Vanlauwe*,[3] *Piet van Asten*[4] *and Patrick Mutuo*[5]

[1] *Bioversity International, Burundi*
[2] *Bioversity International, Uganda*
[3] *International Institute for Tropical Agriculture, Kenya*
[4] *International Institute for Tropical Agriculture, Uganda*
[5] *International Center for Tropical Agriculture, Bukavu, DRC*

### CIALCA

The Consortium for Improving Agriculture-based Livelihoods in Central Africa (CIALCA) is a collaborative agricultural research-for-development platform operating in the Great Lakes region of Central Africa, specifically in Burundi, Rwanda and North and South Kivu provinces (also Kisangani and Bas-Congo) in the Democratic Republic of Congo (DRC). It was founded in 2006 by three international agricultural research centres (IARCs): Bioversity International, the Tropical Soil Biology and Fertility Institute of the International Center for Tropical Agriculture (CIAT-TSBF) and the International Institute for Tropical Agriculture (IITA).

Across the three countries, CIALCA operates in ten 'mandate areas', which represent reasonably similar terrains in terms of agro-ecology, poverty profiles and ease of access to markets. Mandate areas each have an estimated population of between 300,000 and 1.2 million people and are subdivided into multiple 'action sites' and 'satellite sites'. In action sites, CIALCA and its partners devise and test – jointly with farmers – promising agricultural technologies.

The translation of the logic of the Consortium's structure from paper into reality is not straightforward. With three IARCs implementing project sub-components in three countries, managing effective communication can be a challenge. Yet the segmented nature of the Consortium also has clear benefits, particularly in terms of being more adaptable and responsive to the unique institutional and partnership landscape in each country (Cox 2011).

## Achieving impact

CIALCA has an explicit focus on valorizing new, farmer-validated agricultural innovations. Specifically, CIALCA intends to deliver science-based agricultural knowledge to farming households in the mandate areas. The approach attempts both to engage farmers in technology development, as well as to act as a coordination mechanism for the diffusion of innovations for realizing development impact. For CIALCA, impact is defined as a direct and measurable change in farmers' livelihoods attributable to the Consortium's research and development interventions, where an assessment of the change in livelihood status is largely provided by *ex-ante* and *ex-post* impact assessment studies. These studies adopt the Sustainable Livelihood Framework (Scoones 1998; DFID 1999, 2000, 2001) as a tool for assessing changes to livelihood parameters.

Achieving impact in the mandate areas through scaling out (and, to a more limited extent, by seeking to influence policy) has been the focus of CIALCA's activities in the second phase. The conceptual framework for achieving impact is based on an 'impact pathways' model.

## Impact pathways

In the simplest sense, impact pathways can be thought of as a series of causal linkages that need to be put in place for research to result in the intended benefits (Briones *et al.* 2004). It is desirable that these linkages are made explicit *ex-ante*, because researchers and stakeholders are required to develop hypotheses about the 'route' between research and impact, to define the changes that need to be realized at each step and to describe how the linking process works (Springer-Heinze *et al.* 2003).

The impact pathway model of Briones *et al.* (2004) fits closely with what CIALCA aspires to achieve in the action sites (Figure 19.1). Basic ideas and frameworks for new agricultural technologies are conceived by scientists in IARCs aligned with the Consortium, but these are introduced at an early stage to development partners (National Agricultural Research Systems (NARS), farmers' organizations) for testing and feedback. Suitably refined innovations are validated by farmers and the final product is distributed by outreach partners for farmers to adopt (and subsequently to adapt).

## Impact in the satellite sites

Satellite sites are geographic areas that have been selected by CIALCA for the dissemination of technology options developed in the action sites. Only suitably refined innovations that have been validated by farmers are delivered to extension partners for dissemination. Activities in satellite sites are led by development partners, usually non-governmental organizations (NGOs) or community-based organizations, and public extension providers such as the Rwanda Agricultural Board (RAB). The balance between each type of outreach

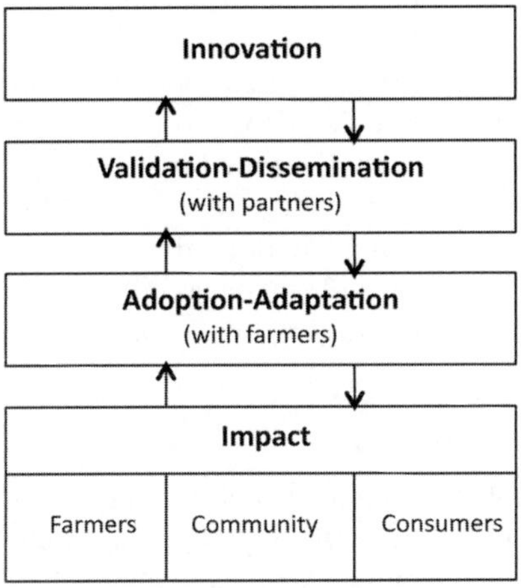


*Figure 19.1* CIALCA's impact pathway schematic (source: adapted from Briones *et al.* 2004).

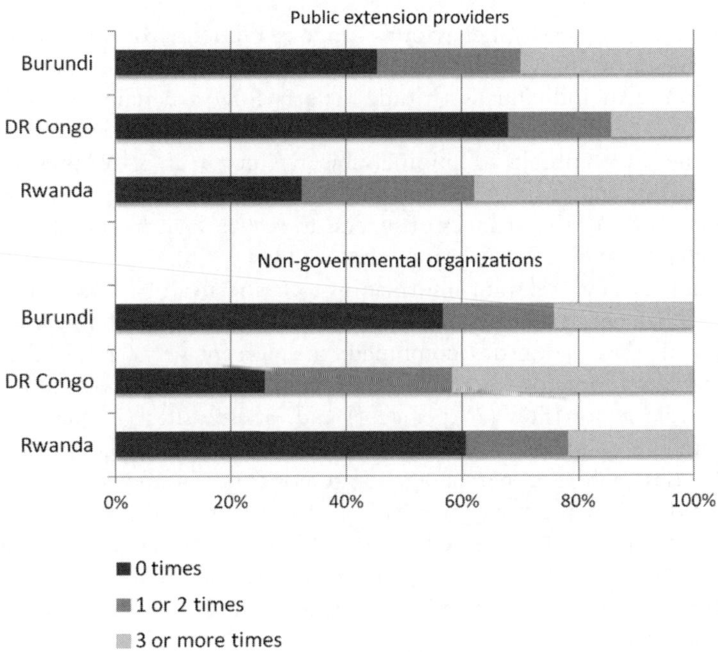


*Figure 19.2* Frequency of contact with agricultural information providers in the last 12 months.

partner varies with the country, depending on the institutional and policy setting.

In 2011 CIALCA implemented a comprehensive impact assessment among 913 households in action, satellite and control sites in the focus countries (Macharia *et al.* 2013). The study included questions related to the provision of agricultural information. The data from the CIALCA impact assessment indicate that public extension providers are relatively active providers of information in Burundi and Rwanda, whereas in DRC (North and South Kivu) NGOs play a much stronger role (Figure 19.2).

## Impact and behaviour

Achieving project impact means realizing a change in farmers' behaviour. A novel technology may be scientifically rigorous, field tested and partner-validated, but if it fails to convince farmers to adopt it – and thus change their behaviour – the technology is useless. The terms 'information' and 'knowledge' are frequently used in a manner that suggests that once farmers are empowered by them, the farmers' behaviour will change. This presumption ignores the basic principles of behaviour change. Actual behaviour is tangible and measurable, but the reasons for this demonstrated behaviour are not. According to van Woerkum *et al.* (1999), behaviour change is a function of how knowledge is influenced by individuals' attitudes, personal effectiveness and their subjective norms. An individual's attitude reflects how a certain action or behaviour is perceived, which can be positive (stimulates adoption of the behaviour) or negative (inhibits adoption). Subjective norms are social pressures encouraging or constraining certain behaviours. Personal effectiveness is the subjective perception of the chances of success in realizing an intended result by way of a change in behaviour (Ajzen and Madden 1986).

The provision of new, relevant information to farmers can be a key driver of learning and behaviour change. Data from the CIALCA impact assessment show that for all three countries combined, the percentage of respondents that had received information about new crop varieties in the past 12 months was higher in the action sites (70 per cent) and satellite sites (71 per cent) compared with the control sites (63 per cent). For new knowledge on crop pests and diseases, a higher percentage of respondents in the action sites indicated that they had received information in the past 12 months (57 per cent) when compared with the satellite sites (47 per cent) and control sites (48 per cent). These results suggest that the areas in which CIALCA operates either directly or through development partners have a higher level of information penetration.

## The CIALCA Knowledge Resource Centre

At the meeting to launch the second phase of CIALCA in 2008, it was agreed that there was a need for both research and extension partners to increase the level of integration with CIALCA's activities and staff (CIALCA 2008). One of the agreed actions was to implement a regional Knowledge Resource Centre (KRC), with a communications specialist to facilitate the knowledge flow between CIALCA and the partners. The KRC became operational in October 2010, based in Bujumbura, Burundi. The KRC aims to provide information and communication support for the needs of partner organizations for technical information. It does this by translating and repackaging technical information and knowledge in various client-specific forms, supporting the scaling-out of research results and monitoring the impact of improved communications. To achieve this, a range of appropriate tools and approaches is used to share knowledge.

## Communication channels and content

CIALCA's outreach strategy is largely based on cascade training (or training-of-trainers; ToT) with interested partner organizations operating in the mandate areas. But there is also a strong desire to reach a much larger audience, both within and beyond the mandate areas. This is done by leveraging mass media tools to help create awareness and stimulate demand for CIALCA's agricultural innovations. Table 19.1 illustrates the range of communication channels typically available to disseminate agricultural messages, along with their relative strengths and weaknesses in terms of reach (how easily audiences over large areas can be informed), depth (the level of complexity that can be addressed using the communication medium), knowledge durability (how well the message 'sticks' and the ease with which the message can be seen or heard again), facilitation and learning (how much external input or interaction the channel requires) and the potential depth of learning as opposed to the level of interactivity or feedback, the relative cost in financial terms and the level of accessibility of the message to farmers.

The results from CIALCA's impact assessment may be compared with Table 19.1. In this impact study, respondents were asked about their main information channels for new knowledge on a number of agricultural topics. Figure 19.3 presents the main sources of information for respondents who indicated they had received information on crop pests and diseases and/or new crop varieties in the last 12 months. Due to the high level of accessibility of radio and farmer-to-farmer interaction, these channels have a high penetration rate and are clearly important in this setting. However, investing in information dissemination via the television and newspapers is not justified, with very low levels of information penetration.

### *Cascade training*

CIALCA has developed many partnerships with development organizations (both local and international NGOs), public extension providers and farmers' organizations. The agricultural technologies developed by the Consortium

*Table 19.1* Communication channels and their relative strengths and weaknesses

| | *ToT/ cascade* | *Autonomous diffusion (farmer to farmer)* | *Factsheet/ poster* | *Video/TV* | *Radio* | *Internet/ online* | *SMS* |
|---|---|---|---|---|---|---|---|
| Reach | Low to medium | High | Low | Low to high | High | High | High |
| Depth (level of complexity able to be conveyed) | High | ? | Low to medium | Medium | Low | Medium to high | Low |
| Knowledge durability (per intervention) | Low | Medium to high | High | Low | Low | High | Low |
| Facilitation/learning | High | Low to medium | Low | Low to medium | Low | Low to high | Low |
| Interaction/feedback | High | Low? | None | None | None (unless interactive radio) | Medium to high | High |
| Cost | High | None | Medium | Medium | Low | Low to high | Low |
| Accessibility for farmers | Low | High | Low | Low to medium | High | Low | Medium to high |

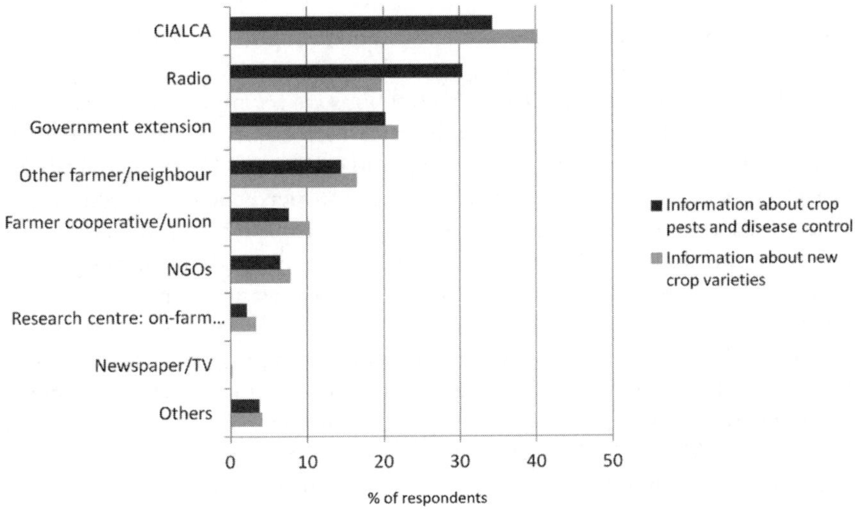


*Figure 19.3* Main sources of information for respondents that received new agricultural information in the last 12 months.

resonate with these partners' visions and objectives and so they receive training from CIALCA on how best to implement these technologies with their beneficiaries. The training is most often done by way of interactive cascade training. A small number of agronomists or other suitable members of staff from the organization receive intensive training from CIALCA's staff, usually agronomists but also socio-economists for entrepreneurial and business training. These 'master trainers' are then expected to train others, thus yielding a potentially significant knowledge reach. Follow-up surveys tentatively indicate that the number of beneficiaries reached through secondary training (and beyond) lies between 48 (for a small NGO) and more than 1,000 for some of the larger NGO partners (Macharia *et al.* 2013). Knowledge materials for the initial training facilitated by CIALCA are often used to support the subsequent training and for future reference to improve the durability of knowledge. These are usually packaged so as to be context-specific and in an appropriate local language. Print materials are always made available to partners in digital form for further reproduction, if desired.

Although resource-intensive, cascade training facilitated by CIALCA's staff is likely to provide the best chances of the desired change in behaviour and adoption of technology. The communication lines are short, providing an ample opportunity for follow-up, feedback and the rectification of teething problems. This outreach approach is superior for the communication of technically complex messages, such as the progressively complex knowledge required for advanced integrated soil fertility management (ISFM).

### *Farmer-to-farmer diffusion*

Horizontal spread via autonomous diffusion through the social networks of farmers is perhaps the best indicator for the appropriateness and acceptability of CIALCA's innovations. CIALCA's trainings are an important motivator for further information spread via farmer-to-farmer contact, farmers' organizations and cooperatives, on-farm trials and NGOs. The results from the impact assessment indicate that, on average, 16 per cent of the respondents receive information from other farmers about crop pests and disease control (17 per cent) and about new varieties (15 per cent), and another 9 per cent receive the same information via farmers' organizations and cooperatives (10 per cent for crop pests and disease control and 9 per cent for new crop varieties). Although these figures are moderately high, there is a significant further potential for farmer-to-farmer dissemination of information (particularly given its low cost yet large reach). As well as continuing to invest in the training of trainers, additional efforts should be invested in helping farmers and their organizations to build and strengthen their local networks for an improved exchange of information and the broader impact of innovations.

### *Video and television*

In the past, video has not been a favoured platform for extension messages, with more attention being devoted to radio. This is despite the fact that video has a huge potential to promote learning among farmers, because the visual element is very powerful, particularly when combined with audio in a local language (Van Mele 2010). The limited availability of television sets and programming in rural locations in the focus countries suggests that the best way to introduce farmers to video is through DVDs to support in-field training. This is supported by data from impact assessment, as none of the respondents indicated that they rely on television as one of their main sources of agricultural information.

CIALCA has developed 11 short, technical extension videos on topics related to banana production, with six more in development on various subjects. Some of these videos have been translated and repackaged by partners, notably by RAB in Rwanda and FAO in Burundi. CIALCA anticipates several forthcoming projects that foresee a much greater reliance on the available videos to support the adoption by farmers of CIALCA's innovations. It is recognized that group discussions and additional facilitated learning will be needed to complement viewing sessions for extension messages to be adequately understood.

### *Rural radio*

Radio has been used successfully in many parts of the world as an agricultural extension platform. CIALCA has used rural radio programming in Burundi and eastern DRC, mostly for the purposes of creating awareness about encroaching banana diseases. This is reflected in survey data, in which radio is rated the second most important channel for information on crop pests and disease control; 48 per cent of Burundians and 24 per cent of Congolese (North and South Kivu) cite radio as one of their main sources of information on crop pests and disease control.

Radio has a relatively wide reach, with many poor farmers owning a radio set or a mobile telephone with an FM receiver. Despite this, relying on radio for encouraging technology adoption may be challenging. Van Mele (2010) contends that this is partly because many radio broadcasters do not have a background in agriculture and the fact that agricultural technologies are difficult to explain orally. Despite limited experience so far with rural radio, the KRC intends to leverage its key benefits (wide reach and low cost) to further explore opportunities using this medium of communication, particularly for awareness-raising purposes, or for training farmers in relatively basic agronomic or pest management techniques.

### *Internet and web-based tools*

In CIALCA's first project phase, a website was developed to keep stakeholders updated on developments and interesting news items, and to provide an online resource portal for access to relevant information resources (technical reports, fact sheets, posters and brochures). The website remains, for now, a simple non-'social' platform, and content is obtained by information pull (demand) rather than push. Online information-seeking behaviour appears limited in Central Africa. This is revealed both by website metrics and anecdotal discussions with partners and stakeholders. In 2011 www.cialca.org received a monthly average of 691 visitors, who each viewed an average of 3.6 pages on the website. The geographic location of visitors is highly skewed towards Europe, North America and Kenya. At least a part of this is explained by the 'digital divide' – the inaccessibility or high cost of internet and information and communication technologies. For the time being, it is apparent that web-based tools have only limited value for the dissemination of CIALCA's agricultural technologies in the focus countries.

## Conclusion: communicating complex knowledge

Since its inception in 2006, CIALCA has made a positive impact on the livelihoods of poor farmers in Burundi, Rwanda and DRC by improving household nutrition and increasing income (Macharia *et al.* 2013). The pathway for impact appears to be well established and key CIALCA technologies are of major interest to the target population. CIALCA has become a widely known information source on agriculture throughout the mandate areas, and three in four individuals are aware of CIALCA and one or more of its technical products. But adoption of new technologies by farmers lags behind.

As CIALCA refocuses some of its future efforts on agro-ecological intensification and ISFM, the Consortium must reflect on why crop productivity and the adoption of some technologies have been limited in CIALCA's interventions to date. Outreach approaches and the supporting knowledge-sharing tools and channels need to be evaluated to address weaknesses and identify new options for reaching (and convincing) farmers. One of the hurdles will be how to clearly convey increasingly complex knowledge to farmers that wish to progressively adopt the complete portfolio of ISFM innovations (improved germplasm + fertilizer + organic resource management + local adaptation). This will be very likely to require a concerted outreach effort with partners, supported by mixed-media learning content. Campaigns may be needed to create awareness and demand (rural radio will be likely to provide a suitable platform), followed by the intensive training of farmers supported by print materials and video. There are face-to-face training approaches that may be better suited than cascade training for promoting the learning and understanding of complex agro-ecology, such as farmer field schools (Godtland *et al.* 2003; David and Asamoah 2011).

Finally, there are novel and innovative communication approaches that could be explored to drive a new demand for innovations, such as the 'social marketing' of CIALCA's technology. This approach, successfully championed by organizations in the public health sector,[1] combines effective communication to achieve a change in behaviour with the low-cost pricing of important items such as mosquito nets and iodized salt. It could be of great interest to learn whether a suitably adapted approach could also work for seeds and fertilizer.

## Note

1 Social marketing for public health is an approach that combines effective communication with the provision of important products to motivate the adoption of desirable practices and behaviour. Products are subsidized but not distributed free of cost to the end user, leveraging local supply chains to market and sell branded, affordable products. This approach has been used with success by the NGO Population Services International (among others) to reduce malaria and HIV/AIDS transmission, improve mother and child health and to promote other life-saving products and clinical services (PSI 2009).

## References

Ajzen, I. and Madden, J. T. (1986) 'Prediction of goal directed behaviour: attitudes, intentions and perceived behavioural control', *Journal of Experimental Psychology*, vol. 22, pp. 453–474.

Briones, R., Dey, M. M., Ahmed, M., Stobutzki, I., Prein, M. and Acosta, B. O. (2004) 'Impact pathway analysis for research planning: the case of aquatic resources research in the WorldFish Center', *NAGA, WorldFish Center Quarterly*, vol. 27, no. 3, pp. 51–55.

CIALCA (2008) *Launching Meeting of CIALCA-II* (CIALCA Technical Report number 4), CIALCA Consortium.

Cox, T. P. (2011) *Describing the CIALCA Organizational Model* (CIALCA Technical Report number 16), CIALCA Consortium.

David, S. and Asamoah, C. (2011) 'The impact of farmer field schools on human and social capital: a case study from Ghana', *The Journal of Agricultural Education and Extension*, vol. 17, no. 3, pp. 239–252.

DFID (1999, 2000, 2001) *Sustainable Livelihoods Guidance Sheets, Numbers 1–8*, UK Department for International Development, London.

Godtland, E., Sadoulet, E., de Janvry, A., Murgai, R. and Ortiz, O. (2003) 'The impact of FFS on knowledge and productivity: a study of potato farmers in the Peruvian Andes', *CUDARE Working Paper 963*, Department of Agricultural and Resource Economics, University of California, Berkeley, CA.

Macharia, I., Garming, H., Ouma, E., Birachi, E., Ekesa, B., de Lange, M., Blomme, G., Van Asten, P., Vanlauwe, B., Kanyaruguru, J., Lodi-Lama, J., Manvu, J., Bisimwa, C., Katembo, J. and Zawadi, S. (2013) *Assessing the Impact of CIALCA Technologies on Crop Productivity and Poverty in the Great Lakes Region of Burundi, the Democratic Republic of Congo and Rwanda*, CIALCA Consortium.

PSI (Population Services International) (2009) *PSI: Social Marketing for Healthy Lives*. http://mim.globalhealthstrategies.com/blog/wp-content/uploads/2009/10/Social-Marketing-For-Healthy-Lives.pdf (accessed 17 June 2012).

Scoones, I. (1998) 'Sustainable rural livelihoods: a framework for Analysis', *Working Paper 72*, Institute of Development Studies, Brighton.

Springer-Heinze, A., Hartwich, F., Henderson, J., Horton, D. and Minde, I. (2003) 'Impact pathway analysis: an approach to strengthening the impact orientation of agricultural research', *Agricultural Systems*, vol. 78, no. 1, pp. 267–285.

Van Mele, P. (2010) *Video-Mediated Farmer-to-Farmer Learning for Sustainable Agriculture: A Scoping Study for SDC*. www.agroinsight.com/downloads/articles-divers/Farmer-to-farmer-video-FINALREPORT-Van-Mele-2011.pdf (accessed 15 June 2012).

van Woerkum, C. M. J., van Kuiper, D. and Bos, E. (1999) *Communicatie en Innovatie: Een Inleiding*, Samson, Alphen aan den Rijn.

# 20 Scalability and farmer heterogeneity

## Implications for research on sustainable intensification

*John Lynam*

*World Agroforestry Centre*

Global development agendas have put farming systems research back at the centre of agricultural R&D, particularly with respect to smallholder, rainfed production systems. Adaptation to climate change, payment for ecosystem services, and agro-ecological intensification under growing resource constraints[1] all have farming systems at the core of their R&D strategy. Farming systems can be closely integrated with livelihood strategies and as such can be easily extended into such areas as vulnerability/resilience, rural poverty, nutrition and gender equity, that are outcomes for which a singular commodity or natural resource management (NRM) focus are too limited in achieving. The latter maps directly into performance monitoring and results frameworks that are increasingly being applied to investments in agricultural R&D and allows research programmes to directly track intermediate impact on such development objectives.

To many this trend can be viewed as a return to the farming systems research (FSR) methods of the 1970s and 1980s (Collinson 2000). To the contrary, this chapter will argue that this is a very different casting and implementation of FSR, probably more accurately called agro-ecological intensification (AEI) or eco-efficient agriculture. Nevertheless, AEI must accommodate two central programme design issues that the earlier FSR was not able to achieve, namely how to organize agricultural research with a focus on production systems and, second, how to take the outputs of that research process to a relevant scale within a smallholder context, particularly so that the investments in the research remain cost-effective. The chapter assesses these two questions by briefly reviewing the history of agricultural research focused on smallholder development in rainfed systems, primarily within the CGIAR. This review is then contrasted with the evolution of extension approaches in taking technologies to scale, drawing lessons on how they either reinforced or were incompatible with dominant research approaches. Lastly, scalability of AEI research is then assessed within the current thinking on revitalizing extension and the role of market-led technology dissemination.

## Farming systems and the organization of agricultural research

International agricultural research has over the last four decades been organized around conceptual models, primarily related to the understanding at a given time of the dominant drivers of sustained productivity growth (see Table 20.1). In the colonial period, agricultural research was structured around disciplines, for example plant biology, plant pathology and chemistry, very much patterned on the organization of research at Cambridge University, where colonial officers did their training. The organizational history of the CGIAR, on the other hand, has had much more of a problem focus. The creation of the first international centres, IRRI and CIMMYT, by the Rockefeller and Ford Foundations, were around principal staple food commodities. However, the creation of the next two centres, CIAT and IITA, focused on agricultural development in the lowland tropics and the initial design was to focus on farming systems (Roberts and Hardin 1966). Nevertheless, the creation of the CGIAR has its roots in the Green Revolution in Asia and the growth multipliers derived from widespread adoption of improved staple food technologies. CIAT, and to a more limited extent IITA, evolved into a set of multidisciplinary commodity research programmes, with plant breeding at the core of the programme, namely an organizational model derived from the Green Revolution experience. At CIAT's tenth anniversary, Hardin (1984) reflected on the shift in organizational design:

> Because the areas served are so diverse, it was impossible to develop improved whole-farm systems that were widely relevant. So CIAT decided to concentrate on generating better commodity components which could then be integrated into whole farm systems via local institutions. On-farm research on commodity systems was continued, but the focus was shifted to component technology which rarely includes the whole farm.

The development and evolution of FSR in the late 1970s and 1980s was very much a recognition of the difficulty in moving the Green Revolution into smallholder, rainfed agriculture. These systems dominated the agricultural sectors of sub-Saharan Africa (SSA) and the concerns about population and food production were shifting from Asia to Africa. The origins of FSR were in the cropping systems work at IRRI, which demonstrated the critical importance of crop management on farm productivity of rice, and the systems work of David Norman and Mike Collinson in SSA (Collinson 2000). Yet the challenges inherent in Hardin's assessment would condition the development of FSR during this period. Researchers could see the value of a systems approach to increasing smallholder productivity, but they were overwhelmed by the heterogeneity problem. Thus, the small farmer research programme at CIAT and similar FSR programmes at IITA and ICRISAT were eventually disbanded and testing of technology components on-farm became the focus

*Table 20.1* Stages in the evolution of international agricultural research

| *Period* | *Stage* |
|---|---|
| 1960s to 1970s | Green Revolution: interdisciplinary commodity research programmes |
| 1970s to 1980s | Farming systems research: adaptive research and farmer participatory research |
| 1980s to 2000 | Sustainability: commodity versus natural resource management research |
| Beyond 2000 | Agro-ecological intensification: research organized around production systems |

of FSR, which was consolidated in what was eventually termed adaptive research. Adaptive research has become a critical step in the development of technology for smallholder, rainfed agriculture, and this has been an important legacy of that period of FSR. This was a fertile period in the development of methods, as attention turned to the relative involvement of farmers in trial design and management and the whole field of farmer participatory research (Chambers *et al.* 1989). These participatory methods have become a conventional part of adaptive research in agricultural R&D.

The publishing of the Brundtland Report (United Nations 1987) and the subsequent expansion of the CGIAR, in which 'natural resource management' centres were added, marked the next phase of organizational change in the CGIAR and what is commonly termed the sustainability agenda. This was a period of two very different paradigms in organizing agricultural R&D. The rapid advances in molecular biology significantly expanded the research possibilities and scientific rigour inherent in plant breeding. Commodity research programmes shifted to an almost exclusive focus on plant breeding, at the same time expanding scientific capacity in biotechnology. NRM research programmes followed a very different track. NRM had its roots in the science of ecology, and heterogeneity and scale were core elements of the methods. The intent in this period was to link NRM research directly to impact on smallholder productivity, including soils, water, trees on farms and aquaculture, and the methods were framed in what became integrated NRM (Science Council 2003).

As the term implies, NRM technologies are defined in terms of improved management and these in turn depend on context, particularly such factors as initial status of the natural resource base, market integration, farmer investment in technology and the time lag on effective return, and labour constraints. Improved management was thus conditional on a range of characteristics of the farming system. NRM research focused on determining the functional role of core contributors to productivity, e.g. in the case of soil organic matter defining thresholds in impact on productivity and interactions with other sources of improved productivity. The task, however, was to convert such new knowledge into improved management of soil organic matter under heterogeneous farmer conditions. Thus, such research was often extended through simulation models, decision-support systems, diagnostics and on-farm methodologies. The efficacy of these methods in improving farmer management was not often tested, nor understood in terms of farmers' learning and improved decision-making (McCown 2002).[2]

The current initial movement towards organizing research around AEI is best exemplified by the three CGIAR Research Programs (CRPs) on production systems (humid tropics; drylands; and aquatic-agricultural systems). Within the CGIAR reform process these CRPs provide a framework to integrate commodity and NRM research within the original vision of farming systems research for the lowland tropics (and other agro-ecologies), although importantly (and this is the innovation) research organization is hierarchical and distributed rather than attempting to integrate both commodity and NRM research programmes into a single research institute. At the same time AEI has a new and evolving research agenda, including enhanced resource efficiency in this period of rising nitrogen and phosphorus prices and increasing demand on water resources, the associated potential of payment for ecosystem services, and adaptation to climate change and to enhanced rainfall variability. The heterogeneity problem still remains a dominant issue in how to organize the research, but appropriate scaling, characterization, diagnosis and targeting have all evolved since FSR in the 1980s, facilitated by the advances in geographic information systems (GIS), remote sensing and database management. That said, these techniques are still not rigorously applied in site selection, multi-locational evaluation, farm survey design and testing of interventions. Increased rigour in the application of these tools will be essential in effective management of the heterogeneity problem and in turn in understanding farmer adoption, designing effective dissemination methods and evaluating impact on outcomes such as reducing rural poverty and sustainably managing the natural resource base. Critically, farming systems research is integrally linked to an understanding of farmer decision-making, and while there has been significant research in that area, it remains inadequate for the purposes of sustainable intensification of smallholder production systems.

## The farmer in sustainable intensification

There is a large and varied literature on farmer adoption of improved technology, but the subject has had increased attention recently in the economics literature in terms of understanding the behavioural responses of farmers that underlie adoption and in the use of randomized control trials as the methodology most appropriate for studying farmer behaviour (Foster and Rosenzweig 2010). Much of the focus has been on understanding farmer learning as it is associated with such factors as farmer education, participation in social networks and participation in credit and insurance markets. At the same time, as Foster and Rosenzweig note, these studies have been based on component technologies, usually either fertilizer or improved seed, and in general have not accounted for the heterogeneity of technology response and profitability across farmer samples. In an African context, and particularly in the Great Lakes region, the larger question is why there has not been more widespread adoption of improved technologies and, where there has been adoption, why the productivity increases are not greater. Farming system intensification offers a reframing of these questions around those factors that motivate system change. While this adds to the complexity of understanding farmer decision-making, it also expands the potential entry points to motivate system intensification.

*Path dependency, the food staple and market integration.* The structure of African farming systems is centred around the dominant food staple, which is determined primarily by agro-ecology and culture (although recognizing the historical shift over several centuries to food staples from the Americas, and to a lesser extent Asia). Farmers balance food security and market participation objectives depending on the relative costs of provisioning the household through either purchases in the market or on-farm, subsistence production. There are significant costs to farmers in participating in markets involving transport, contacts with traders, bags, price information, storage and other organizational and coordination costs involving the time of the farmer. These transaction costs vary geographically depending on road and market infrastructure, market organization and urban demand. Moreover, with the market liberalization process of the late 1990s, farmers face a significant degree of price volatility both seasonally and in terms of year-to-year fluctuations. The costs of market participation result in a significant spread between the farm-gate price that a farmer can sell his food staple for compared to the purchase price, including transaction costs. This price spread also varies geographically, and often results in a rational choice by farmers to focus on subsistence production (Omamo 1998a, 1998b). However, this leads in many regions like the Great Lakes to what Barrett (2008) refers to as a semi-subsistence poverty trap.

Sustainable intensification in the farming systems of the East African highlands that moves households out of this semi-subsistence poverty trap involves resolving a basic dilemma, namely reducing the transactions costs inherent in improving farmer participation in markets while at the same time improving the productivity of staple food systems that either generates the surpluses that allow increased market participation or releases the land and labour resources that permits planting of a cash crop. Increasing farm sales involves the export of nutrients from the farming system at the same time that increasing nutrient supply is essential for improving crop and livestock productivity. In high population density areas, such as the East African highlands, transport of nutrients onto the farm as manure, fuelwood or vegetative biomass is limited as off-farm access to these resources has declined with increasing population density and loss of common property resources. Access to the import of inorganic fertilizer eventually becomes necessary for sustainable intensification and that in turn involves increased market participation in both input and output markets.

On the other hand, the costs of fertilizer to farmers are high, potentially resulting in lack of profitability (Guo *et al.* 2011) and reinforcing the poverty trap. There is thus a set of interactions and feedback loops between the evolving structure of the farming system and the development of input and output markets, which is both spatially defined and dependent on the state of development of the agricultural economy.

This path dependence of farming system change is thus dependent on the development of alternatives in how the household meets its subsistence objectives, which is in turn dependent on relative efficiency in staple food markets. This path dependence is presented schematically in Table 20.2, which posits a series of stages in farming system evolution and market development. The farmer's first response to improved market access is to diversify. With land constraints the limits on diversification are provided by the productivity of staple food production and the cost of purchasing food. Thus, poor households tend to have simpler, low productivity systems, where they are forced to buy expensive food staples from primarily selling their labour (Tittonell *et al.* 2010). Diversification, however, can reach limits, as in the Kenyan central highlands, where early investment in market infrastructure and nearness to the Nairobi market have led to adoption of a succession of higher value crops running from coffee to dairy to horticulture. Better resourced farmers are investing in additional land and moving towards greater specialization in the most profitable farm activities, although this is only beginning (Kimenju and Tschirley 2008). Defining research strategies for system intensification in regions such as the East African highlands thus requires an understanding of market context, farmer objectives and associated intensification pathways.

*System dynamics and farmer investment strategies.* As the term implies, system intensification is a dynamic concept and yet in SSA there is little data collected that will track system change, much less understand the factors driving that change and how farmers respond. Farmer adoption (and disadoption) of improved technologies is not well understood, at least from the perspective of how best to design, target and disseminate such technologies, namely the key decisions within agricultural R&D programmes. However, ad hoc evidence from adaptive research programmes suggest that the increased income that initial adopters gain from new staple food technologies are most often reinvested in other production activities, typically livestock, or used to finance consumption, including school fees. Yet understanding farmer investment patterns is critical to understanding system intensification.

Investment capital is a significant constraint on smallholder farming systems. This constraint affects farmers' ability to purchase inputs, especially fertilizer, to adopt technologies such as conservation agriculture where the returns have a time lag, and especially influence the ability to integrate livestock in the system. Rural credit, including microcredit programmes, have expanded significantly in East Africa over the last decade, but risk, market access and collateral remain significant hindrances to effective deployment by finance agencies and broad-scale uptake by smallholders. Tying credit to weather-indexed crop insurance

*Table 20.2* Markets as a driver for farming systems evolution

| *Stage in farming system evolution* | *Farmer objective* | *Principal driver* |
|---|---|---|
| Static equilibrium | Subsistence dominates | • Rural population growth |
| Diversification | Both income and subsistence | • Shifting farm-gate terms of trade<br>• Staple food productivity |
| Specialization | Income dominates | • Price signals in efficient markets<br>• Regional competition and comparative advantage |

has been one approach to the problem, but these are only in the pilot stage and require broad spatial application to diversify the risk to the insurer. Smallholders, especially in the highlands of East Africa, will continue to primarily rely on cash surpluses generated from marketable surpluses and sales as the source of farm investment. Introducing improved technologies through adaptive research programmes and improving market access are then a principal vehicle in increasing the working capital cycling in the system.

Understanding the interaction between farmer adoption of improved technology and farmer reinvestment strategies would help to inform the design of adaptive research and dissemination strategies. It is not clear whether the initial entry point should be the staple food crop or a cash crop, in terms of generating improvements in productivity of the overall system and how different market conditions may influence farmers' investment choices. Also, in scaling up programmes that focus on system intensification, the design revolves around phasing of technology interventions that reinforce farmers' expected investment strategies and at the same time are complementary in improving system productivity. Gender becomes an important element in such design, as it influences not only technology choice, for example between cash and food crops and what traits are preferred in improved varieties, but in management of cash surpluses and eventual investment choices.

## Scalability and production systems

A convergence of interests has put scalability on the agricultural R&D agenda in SSA. Prominent among these are the comparability in scaling out of technologies in the health sector with such interventions as bed nets and vaccines, the increased importance given to performance monitoring and outcome evaluation which requires achieving a level of impact at scale, and the importance now given to scalability in project selection for development funders, especially the Bill & Melinda Gates Foundation. The Bill & Melinda Gates Foundation and others such as USAID's Feed the Future programme have the resources to fund at scale. Implicit in this is that adoption at a relevant scale, particularly if linked to delivery through markets, will lead to a process of autonomous diffusion of the technology. In turn, adoption at scale and the associated productivity effects will generate the growth multipliers that have so far been missing from agricultural change processes on the continent. However, to date these programmes, as in the health sector, have been based on very simple technologies, particularly fertilizer, animal vaccines and improved seed. The dilemma is how to design dissemination strategies for research programmes organized around production systems.

Scalability has traditionally been conceptualized in terms of the design of agricultural extension programmes. The organization and methods of these extension systems have to a fair degree mirrored the organization of the research system, in the sense that both were searching for compatibility with the other. The Green Revolution and multidisciplinary commodity research programmes were very closely tied to what was termed training and visit extension. As Swanson and Rajalahti (2010: 14) note,

> Beginning in the mid-1970s and continuing until the mid-1990s, the World Bank introduced the Training and Visit [T&V] extension system into about 70 countries. The stimulus for these investments was to speed up the dissemination of Green Revolution technologies to farmers, mainly in Asian and African countries.

T&V was based on a technology transfer model, was commodity focused and was supply driven. Methods were designed around reaching individual farmers through contact or lead farmers, on support given by disciplinary specialists and on demonstration plots and extension bulletins. T&V was a very hierarchical, top-down, command-and-control approach with a particular focus on organizational efficiency. As a result the linkages between research, as was done in the National Agricultural Research Institute (NARI), and extension organized within the Ministry of Agriculture were often weak and contentious, dependent on memoranda of understanding between directors at the highest level and with few organizational linkages at the field level.

Farming systems research generated a period of significant creativity in exploring dissemination methods. On-farm adaptive research became a bridge between research and extension, especially as these methods evolved into farmer participatory research (FPR). FPR put a particular focus on farmer learning as part of problem diagnosis, trial design and trial evaluation, and where understanding key principles was utilized in farmers' interpretation of differences in productivity. Learning by doing was thus a core focus of participatory research. Farmer groups became a key methodological approach where group formation was a vehicle for integrating both gender and poverty objectives. These methods were distilled into a number of overall methodological approaches, possibly the best known being farmer field schools (FFS). These approaches were rarely institutionalized into formal extension systems, as T&V dominated in the ministries of agriculture during this period. Rather, they were explored through NGOs, more downstream projects of CGIAR centres and FFS programmes such as that in Indonesia. However, the methods were to endure, as they touched on some essential elements in farmer adoption, such as learning by doing.

The late 1990s saw the collapse of T&V extension, liberalization of agricultural markets and the consolidation of climate change and ecosystem services as global agendas. These changes reinforced the bifurcation that characterized research into a duality in the design of approaches to dissemination. Commodity research became much more product focused, moving away from a focus on crop management. With liberalization of input markets there were programmes oriented towards development of agrodealer networks for delivering improved seed and fertilizer. Project development was increasingly cast within a value chain framework, which further reinforced organizing research and dissemination around a commodity orientation. On the other hand, NRM R&D moved away from a focus on integrating commodity and resource management to achieve improved productivity (Science Council 2003) to more of a focus on ecosystem services and land use management at a landscape scale. Dissemination approaches were designed around knowledge into action programmes (Clark *et al.* 2011), where improved management was promoted through pilot sites, the community was the locus of the dissemination process, and decision-support tools were instrumental in adaptive management of the resource base. NRM approaches built on the participatory methods developed earlier, but extended these outside the farming system to common property resources and management across a landscape. However, scalability of these approaches was not yet on the NRM agenda.

The collapse of T&V extension highlighted the capacities that had developed in the community-based organizations and NGO community and in highly specific areas of the private sector, e.g. export horticulture or smallholder dairy, and gave legitimacy to the idea of pluralistic extension systems. Coordination and coverage were obvious issues in such systems and dissemination programmes were largely project driven. This was also a period of application of information and communication technologies in agriculture, including mobile phones, video and some experience with village-based web access. Access to information became a dominant feature in dissemination programmes, often to the exclusion of the experience with farmer learning-by-doing in FFS and adaptive research programmes. ICTs reinforced the role of access to information that was central to innovation-diffusion theory (Abbott and Yarbrough 1999). However, information alone only sensitized the farmer to the problem or the technology and, as with decision-support systems, had little relation with farmers' cognitive behaviour, much less farmers' evaluation of improved management options (McCown 2005).

The end result of reliance on projects as the means for technology dissemination is that the interaction between the design of dissemination programmes, farmer adoption at scale and technology diffusion within the context of heterogeneous, smallholder agriculture in Africa is still not understood

*Table 20.3* Extension and the adoption process

| *Stages in adoption* | *Extension methods* | *Relative cost* |
|---|---|---|
| Sensitization | Mass media, radio | + |
| Farmer search | Farmer networks, participatory video, mobile phones | + + |
| Evaluation/adaptation | Farmer field schools, input delivery systems | + |

in any relevant detail. Dissemination projects are not of long enough duration to evaluate farmer adoption, much less diffusion. This has left little space for the development of evaluation methods for understanding the phasing and implementation of dissemination strategies and the eventual impact on diffusion. A schematic of the interaction between stages in the farmer adoption process and the application of dissemination methods is shown in Table 20.3. Coordinating the phasing of the different methods in a scaling programme and evaluating the results has rarely been done, as it would involve incorporating a research capacity and this is usually not considered in what are ostensibly development projects.

Most adoption studies focus on relatively simple technologies, such as improved varieties, or purchased inputs. Management technologies such as integrated soil management or integrated pest management are more difficult to assess, although crop management in rainfed contexts makes the major contribution to increased crop productivity. Other, more system-based technologies, such as agroforestry, crop–livestock integration or conservation agriculture usually involve an adaptive research phase in dissemination and even then adoption is not assured. Knowler and Bradshaw (2007: 25) review the existing adoption studies on conservation agriculture and find that virtually no variables provide universal explanation for the adoption of conservation agriculture. They conclude that 'efforts to promote conservation agriculture will have to be tailored to reflect the particular conditions of individual locales'. Douthwaite *et al.* (2001), in an evaluation of a range of technologies, find that 'as technology and system complexity increase so does the need for interaction between the originating R&D team and the key stakeholders (those who will directly gain and lose from the innovation) when the latter first replicate and use the new technology'. Location specificity, market and institutional context and complexity of the technology will in turn apply to research outputs organized around sustainable intensification of production systems, which leads to the question of how to think about dissemination and scalability of products from such research.

Research on sustainable intensification of production systems merges with the problem of dissemination, especially when research is evaluated against a set of intermediate development outcomes. Organizing the research on production will be integrated with the design of dissemination systems. In essence, there will be no segmentation of the R&D continuum into research and extension functions, but rather these will interact across the continuum but importantly within large and well-defined benchmark sites (Douthwaite *et al.* 2005). Such place-based research will incorporate heterogeneity in the design and implementation of research and dissemination activities. Integrated data and analysis will be essential to ensure efficiency in the allocation of limited R&D resources. Spatial analytical techniques will be used to develop efficient designs for locating research sites, technology testing, farm household surveys and targeting dissemination platforms. Systematic characterization, diagnosis, testing and farmer evaluation generate data that in turn feed into development of a network of adaptive research sites and monitoring of farmer adoption and testing of pilot dissemination methods. An efficient network of adaptive research sites will be central to managing heterogeneity, whether in terms of variation in farming systems or agro-ecologies or differences in market access. Sampling, experimental design and information flows are essential to each stage of an integrated research and dissemination programme. There is thus a clearer integration between research on farmer decision-making and productivity response, often supplemented by simulation modelling at various system levels. In turn, monitoring systems will track and analyse adoption and diffusion patterns. Scale may be achieved through such evolving approaches as innovation systems and innovation platforms (World Bank 2012), which provide a framework for integrating the set of critical agents essential to effective innovation and diffusion, but where these platforms have access to a range of information products. However, both the research and the dissemination methods and models for sustainable intensification are currently in a very formative period and the institutional arrangements under which they will develop are also evolving.

## Conclusions

In this second decade of the new century, sustainable intensification of smallholder, rainfed farming systems remains at the top of the development agenda in SSA. At the same time it is being melded with other global agendas that have farming systems as their operational core. Agricultural research, particularly within the CGIAR, has over the last several decades had to integrate an expanding array of development objectives from rural poverty to sustainable management of the natural resource base to improved nutrition to climate change resilience. However, aligning research strategies and technology design with complementary research on how to achieve such development outcomes is still only in its infancy through still emergent frameworks such as agricultural research for development. Organizing agricultural R&D within a systems framework expands the pathways by which to reach this expanding array of development challenges. At the same time, an FSR focus better integrates farmer decision-making and deepens an understanding of farmer adoption or lack of adoption of improved technologies in smallholder, rainfed systems. As argued in this chapter, organizing research at a farming systems scale was premature in past efforts. Nevertheless, the progressive development of system methodologies and now the institutional reform of the CGIAR offer the potential to fully explore agricultural research organized around systems. However, the global agendas driving the increasing focus on sustainable intensification of agriculture in SSA are in many respects not matched by the agendas driving approaches to dissemination.

In the past there has been a significant complementarity in dominant research approaches and associated methods for dissemination. The current focus on scalability, particularly through market-driven approaches, lends itself to simpler, input-based technologies. At the same time, the R&D community does not have a deep understanding of farmer adoption and diffusion of improved technology, particularly the lack of diffusion of 'adopted' technologies beyond the operational area of development projects. Production systems research should have scalability at the centre of its research agenda and in that regard needs to consider how best to integrate a research continuum from technology design to diffusion of improved practices. Such an integration not only meets a current research gap in agricultural R&D, but at the same time will provide a framework for better evaluation of the performance and impact of investments in agricultural research. Possibly the most difficult hurdle in achieving this vision will be in the reconfiguration of institutional arrangements that will be required to effectively integrate across this continuum, but that is a topic for another time.

## Notes

1 Agro-ecological intensification in this chapter is used interchangeably with eco-efficient agriculture as defined by Keating *et al.* (2010), namely 'a multi-dimensional concept relating the efficiency with which a bundle of desired outputs [food, fibre, environmental services] is produced from a bundle of inputs [land, labour, nutrients, energy], with minimal generation of undesired outputs [soil loss, nutrient loss, greenhouse gas emissions]'.

2 McCown (2005) analyses farmer interaction with decision-support systems and argues that such systems, rather than providing normative results for improved farmer management, play a facilitative role 'in which farmers construct personal, subjective knowledge that is relevant to practical action'. The latter requires a more contextualized approach to deployment of DSS. McCown concludes by arguing that 'mass delivery of a DSS is highly feasible, but generally of low value. The mediated process is generally of high value, but does not lend itself to mass delivery.' This reflects the classic problem in scaling NRM research.

## References

Abbott, E. A. and Yarbrough, J. P. (1999) 'Re-thinking the role of information in diffusion theory: an historical analysis with an empirical test', Paper submitted to Communication Theory and Methodology Division, Association for Education in Journalism and Mass Communication, for its annual convention, New Orleans, LA.

Barrett, C. (2008) 'Smallholder market participation: concepts and evidence from Eastern and Southern Africa', *Food Policy*, vol. 33, pp. 299–317.

Chambers, R., Pacey, A. and Thrupp, L. A. (eds) (1989) *Farmer First: Farmer Innovation and Agricultural Research*, IT Publications, London.

Clark, W. C., Tomich, T. P., van Noordwijk, M., Guston, D., Catacutan, D., Dickson, N. M. and McNie, E. (2011) 'Boundary work for sustainable development: natural resource management at the Consultative Group on International Agricultural Research (CGIAR)', *Proceedings of the National Academy of Sciences of the United States of America*. doi:10.1073/pnas.0900231108.

Collinson, M. (ed.) (2000) *A History of Farming Systems Research*, CABI Publishing, London, and FAO, Rome.

Douthwaite, B., Keatinge, J. D. H. and Park, J. R. (2001) 'Why promising technologies fail: the neglected role of user innovation during adoption', *Research Policy*, vol. 30, pp. 819–836.

Douthwaite B., Baker, D., Weise, S., Gockowski, J., Manyong, V. M. and Keatinge, J. D. H. (2005) 'Ecoregional research in Africa: learning lessons from IITA's benchmark area approach', *Experimental Agriculture*, vol. 41, pp. 271–298.

Foster, A. D. and Rosenzweig, M. R. (2010) 'Microeconomics of technology adoption', *Economic Growth Center Discussion Paper no. 984*, New Haven, CT, Yale University.

Guo, Zhe, Jawoo Koo and Wood, S. (2011) *Spatial Patterns of Profitability of Fertilizer Use in Maize Production Systems in Northern and Central Corridors of East Africa*. http://harvestchoice.org/sites/default/files/GuoKooWood_AddisProductivity.pdf (accessed 27 August 2012).

Hardin, L. (1984) 'CIAT as originally conceived and CIAT today: mandate, objectives, and achievements', in *Centro Internacional de Agricultura Tropical: Proceedings of the 10th Anniversary*, Cali, Colombia.

Keating, B. A., Carberry P. S., Bindraban, P. S., Asseng, S., Meinke, H. and Dixon, J. (2010) 'Eco-efficient agriculture: concepts, challenges and opportunities', *Crop Science*, vol. 50, pp. 109–119.

Kimenju, S. and Tschirley, D. (2008) *Agriculture and Livelihood Diversification in Kenyan Rural Households*, Tegemeo Institute of Agricultural Policy and Development, Nairobi.

Knowler, D. and Bradshaw, B. (2007) 'Farmers' adoption of conservation agriculture: a review and synthesis of recent research', *Food Policy*, vol. 32, no. 1, pp. 25–48.

McCown, R. L. (2002) 'Changing systems for supporting farmers' decisions: problems, paradigms, and prospects', *Agricultural Systems*, vol. 74, pp. 179–220.

McCown, R. L. (2005) 'New thinking about farmer decision makers', in J. L. Hatfield (ed.), *The Farmer's Decision: Balancing Economic Successful Agriculture Production with Environmental Quality*, Soil and Water Conservation Society, Ankeny, IO, pp.11–44.

Omamo, S. W. (1998a) 'Transport costs and smallholder cropping choices: an application to Siaya District, Kenya', *American Journal of Agricultural Economics*, vol. 80, no. 1, pp. 116–123.

Omamo, S. W. (1998b) 'Farm-to-market transaction costs and specialization in small-scale agriculture: explorations with a non-separable household model', *Journal of Development Studies*, vol. 35, no. 2, pp. 152–163.

Roberts, Lewis M. and Hardin, Lowell S. (1966) *A Proposal for Creating an International Institute for Agricultural Research and Training to Serve the Lowland Tropical Regions of the Americas*, Rockefeller Foundation and Ford Foundation, New York.

Science Council (2003) *Towards Integrated Natural Resource Management: Evolution of NRM Research within the CGIAR*, CGIAR Science Council Secretariat, Rome

Swanson, B. and Rajalahti, R. (2010) *Strengthening Agricultural Extension and Advisory Systems: Procedures for Assessing, Transforming, and Evaluating Extension Systems*, The World Bank, Washington, DC.

Tittonell, P., Muriuki, A., Shepherd, K. D., Mugendi, D., Kaizzi, K. C., Okeyo, J. and Verchot, L. (2010) 'The diversity of rural livelihoods and their influence on soil fertility in agricultural systems of East Africa: a typology of smallholder farms', *Agricultural Systems*, vol. 103, no. 2, pp. 83–97.

United Nations (1987) *Report of the World Commission on Environment and Development: Our Common Future*, United Nations, New York.

World Bank (2012) *Agricultural Innovation Systems: An Investment Sourcebook*, The World Bank: Washington, DC.

# 21 Integrated Agricultural Research for Development (IAR4D)

## An approach to enhance the livelihoods of smallholder farmers in the Lake Kivu region

*R. A. Buruchara,*[1] *M. Tenywa,*[2] *J. D. G. Majaliwa,*[2] *W. Chiuri,*[3] *J. Mugabo,*[4] *S. O. Nyamwaro*[1] *and A. Adewale*[5]

[1] *International Centre for Tropical Agriculture, Uganda*
[2] *Makerere University, Faculty of Agriculture, Uganda*
[3] *International Centre for Tropical Agriculture, Rwanda*
[4] *Rwanda Agricultural Board, Rwanda*
[5] *Forum for Agricultural Research in Africa, Ghana*

## Introduction

Sub-Saharan Africa (SSA) faces several challenges (FAO 2005) despite considerable progress in research and outreach, and the apparent socio-economic transformation of some environments. Declining productivity, food insecurity, poverty and land degradation are still on the increase (Batjes 2001) due to a combination of technical, socio-economic, policy, environmental and market challenges (Barbier 2000). These include the inadequacy and/or lack of technical skills, unavailability of credits, weak extension services, poor infrastructure, lack of supportive policies and poor linkages to input and output markets (Buruchara *et al.* 2011a). Nevertheless, agriculture is still the best pro-poor option for moving people out of poverty and an important engine for economic growth and development (Thirtle *et al.* 2003), contributing significantly to rural livelihoods, increasing food security and ensuring sustainable natural resource management (NRM) (Pengali 2006). Linking agricultural production to industrialization in SSA can boost non-farm and urban employment and increase agricultural production.

In the past 50 years, there have been several innovative attempts to address the complex SSA agricultural systems. These include the following: farming systems research (FSR), farmer participatory research (FPR); rapid appraisal of agricultural knowledge systems (RAAKS), sustainable livelihoods approach (SLA); and integrated natural resources management (INRM) (Chambers and Conway 1992; Norman *et al.* 1994). Although these 'innovations' or approaches made useful contributions, they have generated limited successes in addressing multiple scales and interactions within and between physical and social subsystems (Campbell *et al.* 2001). The result has been generally low returns to investment in agricultural R&D in SSA and continued under-performance in using agriculture for development. This is argued to be due to the way conventional research and the use of research outputs have been organized with insufficient participation from the stakeholders. In addition, agricultural innovation is not just about adopting new technologies; but requires finding the adequate flexible and responsive capacity to resolve interlinked issues around market, productivity, NRM and policies affecting farmers' production. An approach that combines these elements and puts them to use in a holistic manner is the Integrated Agricultural Research for Development (IAR4D), whose emphasis is on utilizing synergies from linkages among stakeholders (FARA 2005). This approach, which operates on the basis of geographically defined 'innovation platforms' (IPs), integrates key multiple actors, technological, policy and institutional components responding to changing market and policy conditions and provides commercial, social and institutional solutions that achieve broad and multiple objectives. Its strength lies in its ability to capture policy and market aspects in addition to fostering systemic linkages and communication among the actors. The participatory identification of key constraints and opportunities to be addressed by research, a multidisciplinary approach, capacity building and learning to continuously improve the use of agricultural research for development are the keys in IAR4D. However, despite its attractiveness, there is no evidence about the effectiveness of the approach, which was needed before it could be widely used. Therefore the objective of the study was to demonstrate (proof of concept) that using the IAR4D approach was advantageous and delivers more benefits than other R&D approaches.

## Methodology

Assessing the benefits of the IAR4D approach was carried out in a region bordering three countries, the Democratic Republic of Congo (DRC), Rwanda and Uganda, constituting the Lake Kivu Pilot Learning Site (LKPLS), one of three under the Sub-Saharan Africa Challenge Programme (SSA-CP). The 'proof of concept' that IAR4D is better than conventional approaches and non-intervention areas was based on the premise that IAR4D has the following six characteristics: (1) a functional linkage point between farmers, the private sector and service organizations, (2) the integration of productivity, NRM, markets and policy, (3) an efficient modality for organizing farmers, (4) an effective mechanism for knowledge transfer to farmers, (5) action that is research oriented towards problem-solving and impact, and (6) a bottom-up organizational development (FARA 2008).

### *Site selection*

The process of site selection in the LKPLS consisted of seven steps described in detail by Farrow *et al.* (2009). To attribute results to IAR4D interventions, the process used was based on guidelines defined in the research methodology for the 'proof of concept' of IAR4D, and factored in a number of criteria that included representativeness, the current state of IAR4D, market access, and accessibility using GIS-based spatial information (FARA 2008). Market access (good or poor) was a key variable considered in interventions and was used in the stratification of sites and selection. Sites were of three categories: action sites consisted of areas where less research and fewer development activities had been implemented and where IAR4D interventions were to be made; counterfactual sites consisted of places where previous R&D work had been implemented within the last 2–5 years (also known as conventional sites). Clean sites were where very little or no development work had been implemented. Counterfactual sites were the basis for the comparison of IAR4D interventions and were chosen for each action site where IPs had been established. These were meant to be as similar as possible to the action sites with respect to the agro-ecology, farming system, market linkages, culture and demography. Given the limited number of districts (third-level administrative units) within the LKPLS and the difficulty of finding suitable counterfactual sites, the most appropriate size for a site was the fourth-level administrative unit. This is a sub-county in Uganda, a *secteur* in Rwanda and a *groupement* in DRC. A randomized control trial (RCT) design was used to assess IAR4D and examine three questions, namely: whether IAR4D is effective; whether it is better than the traditional linear approaches; and whether the results it generates could be scaled out.

Implementation of IAR4D was done on the basis of geographically defined IPs, which is a forum for conducting action research and facilitating different stakeholders around a common entry point or theme. The IP also served as a forum through which stakeholders: identify and prioritize issues or constraints to be addressed or opportunities to be exploited; develop joint action plans; share roles, responsibilities and resources; exchange information; track progress in the implementation of action plans; and monitor the process and outcomes of their interactions. Steps in establishing and facilitating the functioning and assessment of the IPs included electing sites; building partnership and stakeholder alliances; making baseline surveys; conducting action research focusing on productivity-enhancing and value-adding technology–NRM–market–policy interfaces to overcome existing and emergent challenges; facilitating the IPs to improve the performance of innovation systems; and making the end-line surveys (Tenywa *et al.* 2011).

Before IAR4D was implemented, a survey of 2,365 households was carried out to establish the baseline status and determine the initial socio-economic conditions of the selected study sites. Parameters measured included human, natural, physical, social and fiscal capitals; IP activities consisted of research and facilitation. Research was carried out on-farm to address identified interface issues such as assessing new production technologies, e.g. alternative options for enhancing soil fertility, crop performances and market preferences, particularly of potatoes, beans, bananas and cassava.

The establishment and functioning of IPs were assessed by using over 17 monitoring and evaluation (M&E) tools, which included the IP registers, activity reports, types of innovations, and information flow. One of the key assessments of the performance of IPs was the institutionalization of the systemic process of generation of innovations to solve the identified and emergent challenges. These could be categorized into linkages and partnerships, technological and market-oriented policy and institutional arrangements.

### *Assessing benefits due to IAR4D*

After about two years of implementation, a survey of 2,245 households in the LKPLS was conducted to determine the household-level benefits of IAR4D. Households were selected using a stratified and random sampling design to enable the comparison of households and communities between IAR4D, conventional and clean villages. The number of villages and households selected was based on the methodology of Miguel and Kremer (2004), which randomizes treatments across districts and village communities and not by individual households, because this method captures spillovers, externalities and benefits.

Assessment of the benefits of IAR4D was done by considering pathways through which the approach had a positive impact on poverty (on the basis of a mean poverty line at $1.25/person/day) and food security. The latter included assessing what the impact had been of IAR4D on adoption and awareness, productivity and market participation. Poverty was measured using a headcount, the poverty gap and severity of poverty. Food security was determined by food consumption score – dietary diversity and coping strategy index – components of the coping mechanism. Productivity was assessed by the productivity of land, labour and capital. However, there was no information on land area for major crops, and market participation was estimated from the value of crop produce sold (intensity of participation) and level of participation (proportion of farmers marketing crop produce). Analysis was carried out in two parts, community level and household level. The analytical methods used included descriptive statistics and econometric methods for the attribution of impacts.

## Results and discussion

Results reported here were derived from two years of work. This apparently was inadequate for conclusive answers to some of the key questions on the benefits of IAR4D and so these should be considered as preliminary. On the other hand, very useful lessons were learned on the IAR4D effect and process.

### Site selection

In each of the 12 sites selected (four per country), one IP was established and a corresponding counterfactual site was selected (Figure 21.1).

A mixture of methods, tools and data was used in site selection and to ensure consistency across the three countries. Consultation with local participants and policy-makers on the choice of candidate sites enabled the articulation of local needs and the expression of critical issues. As a result there was a more nuanced set of information on which to base the choice of action sites and ensure, as much as possible, similarity with counterfactual sites (Farrow *et al.* 2009).

### Facilitation of IAR4D sites

After site selection, IPs were established through the mobilization of social capital of multi-stakeholders around commodity value chains and by linking the IPs

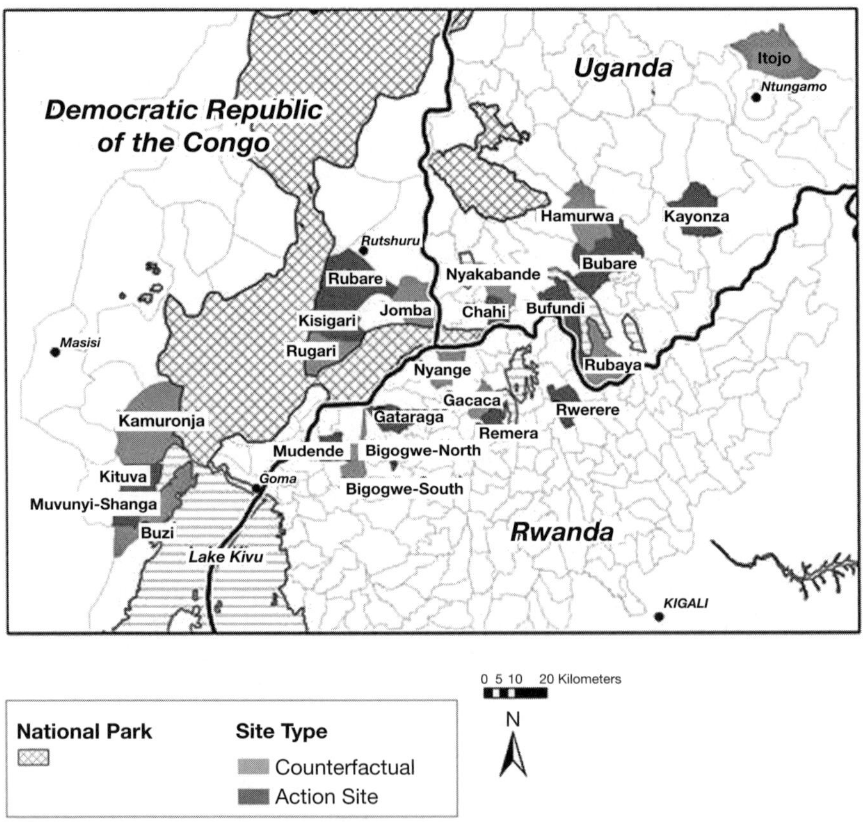


*Figure 21.1* Distribution of action and counterfactual sites in the Lake Kivu Pilot Learning Sites.

to markets. Stakeholders included representatives from local government, non-governmental organizations (NGOs), processors, micro-finance institutions, farmers' groups/associations, public sector agricultural research and extension organizations (NARESs), agricultural training colleges and universities, civil society, CGIAR centres and advanced research institutes. Operationalization of IPs involved the identification of farmers' challenges and prioritization and IP facilitation, which consisted of strengthening governance, forging partnerships, increasing the information flow to support informed decision-making, capacity building and establishing participatory learning, as well as mechanisms for M&E. Each IP had a focus on a specific crop or livestock enterprise as an entry point and the most pressing challenges as identified by farmers. Governance of IPs was strengthened through their institutionalization, the organization of regular stakeholders, periodic planning and monthly IP meetings and participatory M&E activities. The IPs were facilitated to assist in obtaining access to credit for inputs and linked to local and regional markets. Farmers were trained on post-harvest handling and especially in grading, hygiene and sanitation and the packaging of products to meet the market's standards.

### *Functioning of innovation platforms*

Facilitation of the IPs resulted in interactions between stakeholders, actions, activities, processes, knowledge, information and technology diffusion within the IPs. The functioning of IPs was assessed by a well-defined M&E system (Pali 2010) that used a choice of tools to track perceptions interactions and the functioning of IPs so as to draw lessons from them and guide their workings. Altogether, the perceptions of the 93 stakeholders sampled with regard to the level of the satisfaction of IP outcomes showed that the greatest change in IPs was in interactions among actors (score 8 out of 10) and in acquired knowledge of the IAR4D approach. This was followed by knowledge on markets (score 7.1), NRM issues (score 6.4) and the achievement of planned activities (score 6.2), in that order. The understanding of policy issues had a relatively low score (5.7) and this was attributed to its complex nature. Most members were generally satisfied with the performance of IPs with an average score of 7.6 out of 10. Their involvement in IP activities was scored (8.2), whereas relatively weaker areas included conflict resolution in IPs in DRC (5.5 out of 10) and information sharing within IPs in Uganda (6.9 out of 10) (Buruchara *et al.* 2011b).

### *Institutional innovations*

A significant institutional arrangement was the establishment of IPs as tools that fostered and enhanced new institutional linkages and partnerships (Tenywa *et al.* 2011).

Some IPs built value addition and agro-processing networks that strengthened commodity supply chains. For example, the Gataraga IP in Rwanda identified

*Figure 21.2* Gataraga IP potatoes in Nakumart supermarket, Kigali, Rwanda.

and prioritized the market–technology–policy interface of low prices, poor harvests and post-harvest handling procedures as the challenge with the Irish potato value chain. IP partners considered value addition (potato washing, sorting, grading and packaging); rapid multiplication and access to planting material of market-preferred variety (*Kinigi*); dehaulming before harvest; linking to credit and niche markets; and exploiting the favourable policy environment. With facilitation by one of the stakeholders (a lady trader) who buys Irish potatoes from IP members and with input from research, farmers adopted the use of clean seeds, inputs and crop management practices to provide good-quality potatoes for the Kigali market. In 2010 the IP delivered 700 kg/month of potatoes, but in 2012 this increased to 9,000 kg/week through a number of outlets (hotels and supermarkets) (Figure 21.2). This has created 15 jobs in the small production area, and facilitated credit access to traders and farmers (to allow them to dehaulm and wait for the potatoes to cure and hence increase shelf life). Farmers receive RWF 20–30 per kilogram above the going market prices (Nyamwaro *et al.* 2011). Outcomes include improved potato quality, increased yields, improved shelf life and increased prices and incomes.

Similarly, Bubare Sorghum IP in Uganda partnered with Huntex Ltd, a private agro-processing company, to add value through modern processing, packaging and branding of a locally popular traditional non-alcoholic sorghum drink. As a way of marketing it, this has now been branded and registered for the *Mamera* trademark. The drink now sells in new up-market outlets, attracts better prices and appeals to middle-class and urban consumers. This value addition is becoming the driver for the sorghum value chain in the region (Buruchara *et al.* 2011a; Nyamwaro *et al.* 2011).

### *The benefits of IAR4D: preliminary evidence*

The results from the initial baseline survey (Nkonya *et al.* 2009) showed that most of the human and physical capital endowment in the selected villages was not significantly different across treatment sites (IPs and counterfactuals), suggesting that there was no significant bias in site and household selections. For example, using the Gini coefficient, comparison of income inequality showed no significant difference across the treatment in each country. However, there were differences between countries with greater income inequality in DRC and Rwanda than in Uganda. Countries also exhibited differing levels of technology use.

Surveys conducted two years after the start of IAR4D interventions were aimed at demonstrating their impacts relative to clean and conventional sites on some outcomes of interest, such as market participation, poverty, productivity and food security. Regression results for a pooled data set, regressing three key impact proxies (headcount ratio, a food consumption score (FCS) and a coping strategy index) for poverty and food security on the IAR4D are summarized in Table 21.1. These results show mixed evidence of some impact, but the evidence is not as robust across the sites (FARA 2010). At a

*Table 21.1* IAR4D impact on selected outcomes

| *Outcome* | *OLS as compared with* | | *OLS interacted as compared with* | | *2SLS as compared with* | |
|---|---|---|---|---|---|---|
| | *Clean* | *Conventional* | *Clean* | *Conventional* | *Clean* | *Conventional* |
| *Poverty* | | | | | | |
| Incidence of poverty (headcount ratio) | –9.61 (2.48)*** | –7.55 (2.56)*** | 9.65 (2.52)*** | –8.16 (2.57)*** | 7.46 (2.74)*** | –6.44 (2.48)** |
| *Food security* | | | | | | |
| Food consumption score as indicator | 5.18 (1.24)*** | 1.80 (1.37) | 5.26 (1.27)*** | 1.60 (1.41) | 5.45 (1.28)*** | 1.71 (1.38) |
| Coping strategy index as indicator | 0.08 (1.29) | 0.022 (1.45) | 0.23 (1.370) | 0.61 (1.48) | 0.086 (1.33) | 0.020 (1.46) |

Source: FARA 2010.

*Notes*: ** Significant at 5 per cent level. *** Significant at 1 per cent level
These are statistical regression methods, with different complexities, used to analyse significant differences between variables or outcomes. OLS: Ordinary Least Squares Regression. 2SLS: Two-Stage Least Squares.

country level, the IAR4D approach did seem to reduce the incidence of poverty to a greater extent than conventional extension efforts in Rwanda and Uganda, but there was no significant effect for DRC (Table 21.2), where poverty fell equally in clean, conventional and IAR4D sites (FARA 2010). Similarly, the impact of IAR4D on awareness, adoption and factor productivity showed mixed results on awareness, and a very limited result on the adoption of NRM. There was no significant impact of IAR4D on market participation. This was expected, given the short time the IPs have had to implement IAR4D. The latter is a complex and relatively new concept to the majority of the stakeholders and its impact is contingent on their ability to synthesize the concept, adapt it to the environment and then implement it. While institutional linkages are expected to reduce the time lag between technology development and use,

*Table 21.2* Summary of changes in poverty status per country

| *Poverty indicator* | *Clean* | *Conventional* | *IAR4D* |
|---|---|---|---|
| *DRC* | | | |
| Baseline headcount ratio | 98 | 100 | 98 |
| End line headcount ratio | 93 | 93 | 91 |
| Change in poverty incidence | –5 | –7 | –7 |
| Population at the baseline period | 19,025 | 28,906 | 12,317 |
| *Number of individuals having escaped poverty* | *951* | *2,023* | *862* |
| *Rwanda* | | | |
| Baseline headcount ratio | 83 | 92 | 95 |
| End line headcount ratio | 86 | 90 | 89 |
| Change in poverty incidence | 3 | –2 | –6 |
| Population at the baseline period | 8,524 | 13,584 | 19,704 |
| *Number of individuals having escaped poverty* | *–256* | *272* | *1,182* |
| *Uganda* | | | |
| Baseline headcount ratio | 95 | 93 | 95 |
| End line headcount ratio | 90 | 89 | 84 |
| Change in poverty incidence | –5 | –4 | –11 |
| Population at the baseline period | 7,201 | 7,848 | 14,489 |
| *Number of individuals having escaped poverty* | *360* | *314* | *1,594* |
| *Lake Kivu* | | | |
| Baseline headcount ratio | 95 | 97 | 97 |
| End line headcount ratio | 90 | 94 | 88 |
| Change in poverty incidence | –4.46 | –3.82 | –8.62 |
| Population at the baseline period | 34,753 | 50,338 | 46,510 |
| *Number of individuals having escaped poverty* | *1,550* | *1,923* | *4,009* |
| Difference IAR4D/conventional | | (–4.8)*** | |
| Difference IAR4D/clean | (–4.16)** | | |

Source: FARA 2010.

*Note*: ***, ** Denote significant differences at 1 per cent and 5 per cent levels.

the establishment of these institutions as well as mechanisms to ensure the institutions operate in accordance with the principles of IAR4D require a considerable period of time. It is thus apparent that more time was necessary to achieve developmental impacts on the farm level and an impact on incomes. It is on this basis that a review panel commissioned by the Science Council observed that IPs required a few more years to assess the impact after this institutional innovation and therefore recommended an extension of the research phase of the SSA-CP for at least two years, but within the context of some key revisions to the research plan (Lynam *et al.* 2010).

One of the key observations made during the implementation of IAR4D is that, because of the potential effects of productivity, market, NRM and policy innovations on one another, there is a need to consider them simultaneously rather than in isolation. One main lesson from this is that to meaningfully respond to constraints or opportunities associated with the four interlinked components – NRM, productivity, markets and policy – a self-regulating mechanism/system involving IPs and facilitators needs to be in place and operational (Buruchara *et al.* 2011b). In industrial machinery or systems, a central processing unit (CPU) monitors, regulates and harmonizes the operations of several different components (or subsystems) to deliver the desired outputs and products. Similarly in a system based on the IAR4D principles, there is need for a 'mechanism' or 'process' to regulate the four different components

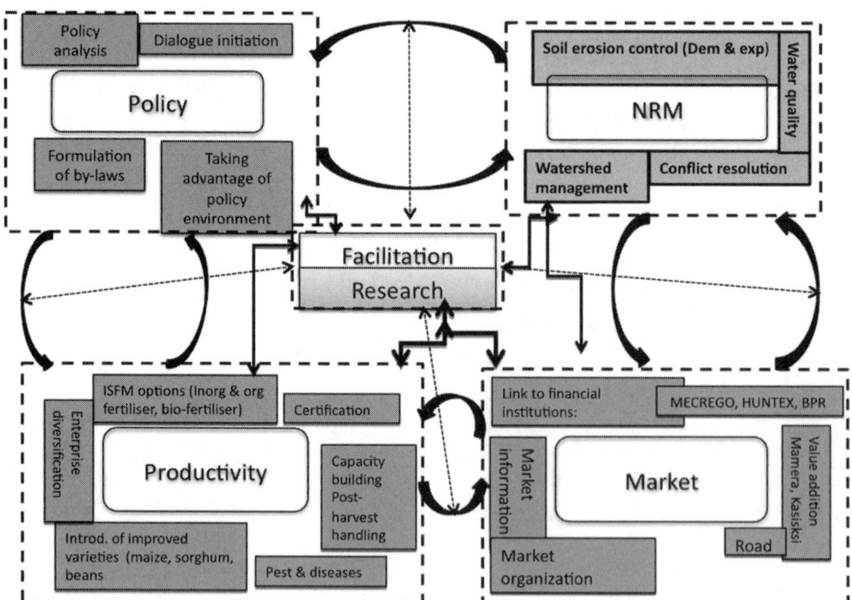


*Figure 21.3* Self-organizing and regulating IP model in IAR4D (source: Buruchara *et al.* 2011b).

to be able to deliver the desired outputs and impact to the end-users (Figure 21.3). At the centre of such a system are the individuals, actors or systems that play key roles in addressing issues (in advance or as they evolve) through both research and facilitation.

## References

Barbier, E. B. (2000) 'The economic linkages between rural poverty and land degradation: some evidence from Africa', *Agriculture, Ecosystems and Environment*, vol. 82, pp. 355–370.

Batjes, N. H. (2001) 'Options for increasing carbon sequestration in West African soils: an explanatory study with special focus on Senegal', *Land Degradation & Development*, vol. 12, pp. 131–142.

Buruchara, R., Tenywa, M. M., Mugabo, J., Chiuri, W., Fatunbi, A. O., Adewale, A., Nyamwaro, S. O. and Majaliwa, J. G. M. (2011a) 'Establishment and implementation of integrated agricultural research for development in Eastern and Central Africa: some operations and lessons learnt from the Lake Kivu Pilot Learning Site', in IAR4D: *From Concept to Practice: Experiences from the Lake Kivu Learning Site*, in press.

Buruchara, R., Tenywa, M. M., Tukahirwa, J., Kashaija, I., Farrow, A., Rao, K. P. C., Wanjiku, C., Adewale, A., Fatoumbi, W., Nyamwaro, S. O., Majaliwa, J. G. M. *et al.* (2011b) 'Design, principles and processes of Integrated Agricultural Research for Development in Africa: the case for LKPLS. IAR4D of the SSA-CP', *African Journal of Agriculture and Resource Economics*, in press.

Campbell, B. M., Sayer, J. A., Frost, P., Vermeulen, S., Ruiz Pérez, M., Cunningham, A. and Prabhu, R. (2001) 'Assessing the performance of natural resource systems', *Conservation Ecology*, vol. 15, no. 2. www.consecol.org/vol5/iss2/art20 (accessed 2 January 2009).

Chambers, R. and Conway, G. R. (1992) *Sustainable Rural Livelihoods: Practical Concepts for the 21st Century*, Institute of Development Studies, Cambridge.

FAO (2005) *Contribution of Farm Power to Small-holder Livelihoods in Sub-Saharan Africa*, FAO, Rome.

FARA (2005) 'LKPLS validation report 2003', in *Forum for Agricultural Research in Africa Sub-Saharan Africa Challenge Program. Findings of the Lake Kivu Pilot Learning Site Validation Team. A Mission Undertaken to Identify Key Entry Points for Agricultural Research and Rural Enterprise Development in East and Central Africa 5 to 30 October 2005*, FARA, Accra.

FARA (2008) *Sub-Saharan Africa Challenge Program (SSA CP): Securing the Future for Africa's Children – Medium-Term Plan 2009–2010*, Forum for Agricultural Research in Africa, Ghana.

FARA (2010) *Interim Proof of Concept of Integrated Agricultural Research for Development (IAR4D): East and Central Africa*, FARA, Accra.

Farrow, A., Opondo, C., Tenywa, M., Rao, K. P. C., Nkonya, E., Njeru, R. and Lunze, L. (2009) *Selecting Sites to Prove the Concept of Integrated Research for Agricultural Development: LKPL Site Selection Annual Report (2008/9)*, CIAT Africa, Kampala.

Lynam, J. K., Harmsen, K. and Sachdeva, P. (2010) *Report of the Second External Review of the Sub-Saharan Africa Challenge Program (SSA-CP)*, Independent Science and Partnerships Council of the CGIAR, FARA, Accra.

Miguel, E. and Kremer, M. (2004) 'Worms: identifying impacts on education and health in the presence of treatment externalities', *Econometrica*, vol. 72, no. 1, pp. 159–217.

Nkonya, E., Pali, P., Oduol, J., Andam, K. S. and Kato, E. (2009) *Lake Kivu Pilot Learning Site Baseline Study: Socio-economic Baseline Study Report*, CIAT, Kampala.

Norman, D., Frankenberger, T. and Hildebrand, P. (1994) 'Agricultural research in developed countries: past, present and future of farming systems research and extension', *Journal of Production Agriculture*, vol. 7, no. 1, pp. 124–131.

Nyamwaro, S. O., Buruchara, R., Tenywa, M. K., Kalibwani, R., Mugabo, J., Wanjiku, C., Tukahirwa, J. M. B. *et al.* (2011) 'Success stories', in IAR4D: *From Concept to Practice: Experiences from the Lake Kivu Learning Site*, in press.

Pali, P. (2010) *Monitoring and Evaluation Indicator Report for LKPLS 2008–09*, FARA, Accra.

Pengali, P. (2006) 'Agricultural growth and economic development: a view through the globalization lens', Paper presented at the Presidential address to the 26th International Conference of Agricultural Economists, Gold Coast, Australia, 12–18 August.

Tenywa, M. M., Rao, K. P. C., Tukahirwa, J. B., Buruchara, R., Adekunle, A. A., Mugabe, J., Wanjiku, C., Mutabazi, S., Fungo, B., Kashaija, N. I., Pali, P., Mapatano, S., Ngaboyisonga, C., Farrow, A., Njuki, J. and Abenakyo, A. (2011) 'Agricultural innovation platform as a tool for development oriented research: lessons and challenges in the formation and operationalization', *Learning Publics Journal of Agriculture and Environmental Studies*, vol. 2, no. 1, pp. 117–146.

Thirtle, Colin, Lin, Lin and Piesse, Jennifer (2003) 'The impact of research-led agricultural productivity growth on poverty reduction in Africa, Asia and Latin America', *World Development*, vol. 31, no. 12, pp. 1959–1976.

# 22 Communication channels used in dissemination of soil fertility management practices in the central highlands of Kenya

*S. W. Kimaru-Muchai,*[1] *M. W. Mucheru-Muna,*[2] *J. N. Mugwe,*[3] *D. N. Mugendi*[2] *and F. S. Mairura*[4]

[1] *Department of Environmental Studies (Community Development), Kenyatta University, Kenya;* [2] *Department of Environmental Science, Kenyatta University, Kenya;* [3] *Department of Agricultural Resource Management, Kenya;* [4] *Tropical Soil Biology and Fertility, Kenya*

## Introduction

Declining soil fertility is a major cause of low per capita food production in the smallholder farms of sub-Saharan Africa (SSA). To address this challenge, studies in the central highlands of Kenya and other areas in SSA have identified multiple management interventions that can reverse this decline and mitigate problems of food scarcity (Bationo *et al.* 2003). In central Kenya, large increases in maize yield with the application of *Tithonia, Calliandra* and *Leucaena* biomass have been reported (Mugendi *et al.* 1999; Mugwe *et al.* 2008). However, past research shows that the adoption of new agricultural technologies among the smallholder farmers, including soil management practices, has generally lagged behind scientific and technological advances, and hence their impact on agricultural production has been low (Okuro *et al.* 2002). Some of the challenges to the accessibility and utilization of integrated soil fertility management (ISFM) include inadequacies in communication, dissemination and approaches to scale-up the technologies.

Evidence shows that farmers on many occasions use diverse sources of information to meet their soil technology needs. Ekoja (2003) explained that the most commonly used sources of information in Nigeria include extension agents, neighbours, other farmers, leaders of opinion, and organized groups. Rogers (1995) explained that the most commonly used channels of communication include the mass media (radio and television), print media (pamphlets, brochures, labels and magazines) and inter-personal media (seminars, demonstrations, field days, exchange visits and agricultural shows). Different sources of information are important as they create the awareness of technological alternatives. The change agents, researchers, extension workers and policy-makers need to identify the sources of information that farmers prefer as this helps in appraising effective communication pathways in the dissemination of soil fertility management practices.

Effective communication is a necessity for extension services to achieve the broad-set goals of farmers acquiring knowledge, skills and attitudes (Okunade 2007). Extension workers use a variety of methods to disseminate information regarding soil fertility management practices. According to Farouque and Takeya (2009), the more methods are used in presentation of a topic, the more quickly will people tend to grasp the subject matter. However, limited financial resources may force extension agents to choose among teaching methods and approaches. In such situations understanding the target audience, including the methods by which farmers prefer to receive information, allows agents to select the most effective methods and events and to transfer information efficiently (Richardson and Mustian 1994). This study therefore endeavoured to assess the availability and reliability of information sources and find out the methodologies for communication and the approaches that are effective in the promotion and scaling up of knowledge on soil fertility management technologies. It also endeavoured to assess the factors that affect the reliability of the farmers' sources of information.

## Materials and methods

### *Study area*

The research was carried out in Maara district (Mbeere South district in the central highlands of Kenya). The study area was chosen because several ISFM research projects on soil fertility management practices had been ongoing in the region.

### *Sampling and sample size*

Ganga location in Maara district and Mbita location in Mbeere South district were purposively selected for the study. The researcher obtained all the household names from the sub-chiefs of the two villages which constituted the sampling frame. Systematic random sampling technique was used to select 120 farmers from each location, a total of 240 respondents. The selected farmers were interviewed in May 2010 using structured and unstructured questionnaires. Pre-testing and training of enumerators on the questionnaires were carried out to ensure that valid and reliable data were collected.

### *Data analysis procedure*

Data were first summarized and a database template containing the collected information was made using Statistical Package for Social Sciences (SPSS).

Descriptive statistics such as frequencies, percentages and means were used to summarize the data. Kendal's tau correlation (*r*) analysis was used to determine the relationship between the dependent variable (approach preference) and independent variables. Pearson correlation was used to assess the relationship between farmers' characteristics and extension approaches.

## Results

About 39.6 per cent of the farmers were aged between 31 and 45 years and only 13.8 per cent were above 60 years; most households in the study were male headed – 173 (72 per cent). Of the respondents, 143 (59.6 per cent) had primary education; 62 (25.8 per cent) had secondary education; and 22 (9.2 per cent) had attained post-secondary education. Approximately 50 per cent of the farmers had more than 20 years of experience in farming.

Results showed that approximately 22.7 per cent of the farmers received information on animal manure from government extension officers, while 45 per cent utilized their own farming experience (Table 22.1). Based on the findings, 37.7 per cent who used green manure obtained information from government extension officers. About 43.2 per cent who used inorganic fertilizers obtained information from other farmers, while 28.4 per cent received information from government extension officers. Almost half (49.1 per cent) who practised erosion control obtained information from other farmers. The results revealed that 41.1 per cent who used compost utilized their own farming experience, with 38.4 per cent obtaining information from other farmers (Table 22.1).

Farmers were asked to score the availability of various sources of communication. The rank orders on availability were identified using the scores 1 = never available, 2 = least available, 3 = available, 4 = most available. The source of information that got the highest score was regarded as the most available. Other farmers (mean = 3.8), radio/TV (3.6) and government extension officers (2.2) were scored as the three most available sources of information; the print media (1.5) was scored as the least available (Figure 22.1). Other farmers, (mean = 3.04), government extension officers (2.61) and researchers (1.94), were scored as the first three reliable sources of information; agro-input dealers were scored as the least reliable with a mean of (1.47) (Figure 22.1).

Extension methods may be classified according to their functional nature, which includes individual contact approaches, group contact approaches and mass contact approaches.

The majority of the farmers (67.1 per cent) strongly preferred methods of individual interaction (Table 22.2). About 55 per cent did not prefer the mass media; 48.3 per cent mildly preferred group approaches. Individual farmer interaction was the best scored method (mean = 2.58), followed closely by group approaches (2.37); mass media (1.58) scored the lowest (Table 22.2). There were significant differences among the approaches preferred by farmers ($P = 0.001$) (Table 22.2).

*Table 22.1* Sources of information utilized by farmers to get access to information on ISFM in Mbeere South and Maara Districts in central Kenya

| *Sources of information* | *Animal manure* | *Green manure* | *Inorganic fertilizers* | *Combined organic and inorganic fertilizers* | *Erosion control measures* | *Compost* |
|---|---|---|---|---|---|---|
| Government extension officers | 50 (22.7) | 20 (37.7) | 65 (28.4) | 78 (36.4) | 43 (19.9) | 9 (12.3) |
| NGO extension officers | 2 (0.9) | 0 (0) | 3 (1.3) | 17 (7.9) | 10 (4.6) | 3 (4.1) |
| Researchers | 4 (1.8) | 13 (24.5) | 6 (2.6) | 10 (4.7) | 2 (0.9) | 2 (2.7) |
| Agro-input dealers | 1 (0.5) | 1 (1.9) | 23 (10) | 5 (2.3) | 2 (0.9) | 1 (1.4) |
| Radio/TV | 2 (0.9) | 3 (5.7) | 4 (1.7) | 3 (1.4) | 3 (1.4) | 0 (0) |
| Exhibitions | 1 (0.5) | 0 (0) | 3 (1.3) | 4 (1.9) | 0 (0) | 0 (0) |
| Other farmers | 61 (27.7) | 4 (7.5) | 99 (43.2) | 68 (31.8) | 106 (49.1) | 28 (38.4) |
| Own experience | 99 (45) | 12 (22.6) | 26 (11.4) | 29 (13.6) | 50 (23.1) | 30 (41.1) |
| *Total* | *220 (100)* | *53 (100)* | *229 (100)* | *214 (100)* | *216 (100)* | *73 (100)* |

Source: author's survey
*Note*: Numbers in parentheses are percentages of respondents (percentages calculated column-wise). Values outside parentheses are numbers of respondents.

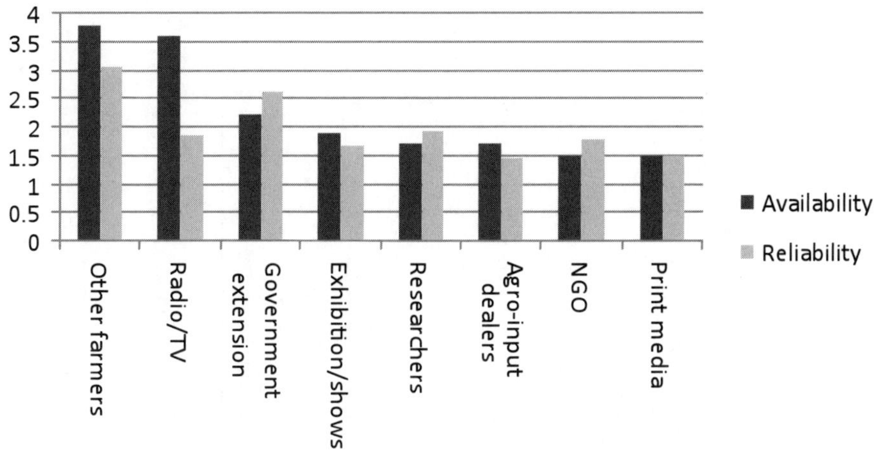


*Figure 22.1* Comparison of availability and reliability of information sources on ISFM practices as scored by farmers in Mbeere and Maara South districts in central Kenya (source: author's survey).

*Table 22.2* Extension approaches preferred by farmers in Mbeere South and Maara districts

| *Approach* | *Percentage of respondents/frequencies* | | | |
|---|---|---|---|---|
| | *Do not prefer* | *Mildly prefer* | *Strongly prefer* | *Mean scores* |
| Individual interaction | 21 (8.8) | 58 (24.2) | 161 (67.1) | 2.58 |
| Group approach | 18 (7.5) | 116 (48.3) | 106 (44.2) | 2.37 |
| Mass media approach | 132 (55) | 76 (31.7) | 32 (13.3) | 1.58 |

| *Paired Samples Test* | *t* | *P Sig. (two-tailed)* |
|---|---|---|
| Individual farmer interaction – group approach | 3.226 | 0.001 |
| Individual farmer interaction – mass media | 14.740 | 0.001 |
| Group approach – mass media | 13.242 | 0.001 |

Source: author's survey.

*Note*: Values in parentheses are row percentages.

*Table 22.3* Communication methods preferred for practices by farmers in Mbeere South and Maara districts, central Kenya

| *Communication method* | *Green manure* | *Fertilizer* | *Compost* | *Animal manure* | *Combined fertilizer and manure* | *Soil erosion* | *Mean* |
|---|---|---|---|---|---|---|---|
| Demonstrations | (2) 2.91 | (1) 3.21 | (2) 3.00 | (1) 3.13 | (1) 3.43 | (2) 2.90 | 3.1 |
| Farmer-to-farmer extension | (1) 3.00 | (2) 2.90 | (3) 2.98 | (2) 2.90 | (2) 3.20 | (1) 3.10 | 3.01 |
| Workshops | (3) 2.16 | (4) 2.49 | (6) 1.93 | (3) 2.30 | (3) 2.46 | (3) 2.50 | 2.31 |
| FFS | (7) 1.67 | (5) 2.18 | (1) 3.49 | (7) 1.83 | (7) 1.93 | (7) 2.13 | 2.21 |
| Field days | (4) 1.99 | (3) 2.50 | (5) 2.13 | (4) 2.10 | (4) 2.15 | (4) 2.30 | 2.2 |
| Teaching aids | (4) 1.99 | (5) 2.18 | (4) 2.16 | (5) 2.03 | (5) 2.12 | (6) 2.18 | 2.11 |
| Exchange visits | (6) 1.86 | (7) 2.06 | (7) 1.87 | (6) 2.01 | (6) 2.02 | (5) 2.21 | 2.01 |

Source: author's survey

*Note*: Numbers in parentheses are ranks of extension methods, based on magnitudes of mean scores. Values outside parentheses are means.

A demonstration was the most preferred method in the training on animal manure, fertilizers and combined fertilizer + manure as it scored highest with means of 3.1, 3.2 and 3.4, respectively. Farmer-to-farmer extension was the approach scored highest in training on green manure (3.0) and soil erosion control measures (3.1). The farmer field school (FFS) was the most preferred method in training on compost use (mean = 3.5). The demonstration was the most preferred method, followed by the farmer-to-farmer extension and workshops in the dissemination of ISFM practices. The use of teaching aids and exchange visits were scored lowest in dissemination methods (Table 22.3).

The age of the household head positively influenced the preference for individual contact and the mass media but was negatively linked with group approaches. Although the effect was statistically not significant, it implied that older farmers preferred individual contact or mass media approaches. There was positive and significant correlation ($P \leq 0.01$) between education and individual farmer interaction. The positive correlation implies that the higher the education level the greater the preference for this approach. Farm size was positively and significantly ($P \leq 0.05$) correlated with the preference for the mass media. Conversely, wealth status was negatively and significantly ($P \leq 0.01$) correlated with the preference for the mass media. This implied that the wealthier the farmers, the less they preferred the mass media approach in teaching the use of ISFM practices. Additionally there was a negative and non-significant relationship between group membership and the preference for the methods of mass media and individual contact. This implied that farmers who belonged to groups were more likely to prefer group approaches for dissemination. On the other hand, there was positive and significant ($P < 0.05$) correlation between gender and the preference for the group approach (Table 22.4). This implies that the group approach was preferred more by female farmers than by male farmers.

*Table 22.4* Correlation between socio-economic factors and preference for extension approaches

| *Socio-economic factors* | *Communication approach* | | |
|---|---|---|---|
| | *Individual farmer interaction* | *Group approach* | *Mass media* |
| Gender | –0.041 | 0.123* | –0.078 |
| Age | 0.105 | –0.042 | 0.102 |
| Educational level | 0.154** | –0.007 | 0.006 |
| Wealth status | 0.108 | –0.117 | –0.129* |
| Membership of groups | –0.002 | 0.096 | –0.048 |
| Farm size | –0.058 | 0.124* | 0.153** |

*Note*: *significant at $P \leq 0.05$, **significant at $P \leq 0.01$.

## Discussion

Based on the research findings, farmers obtained information from government extension agents, researchers, NGOs, other farmers, the radio, agro-input dealers, friends and exhibitions, and also utilized their own knowledge. They preferred farmer-to-farmer approaches as the most available source of information (mean score = 3.8). This agrees with the findings of Oladoja *et al.* (2008) that farmer-to-farmer modes play a vital role in information exchange. The radio was considered the second most available source of information (mean = 3.6). The implication was that since most respondents possess a radio, it was then a readily available source of information. In addition, the radio has many advantages, key among them being that it reaches a large audience, conveys messages or news very quickly, is particularly effective in rural areas and non-literate cultures, is portable, stimulates the imagination and carries authority (Norrish *et al.* 2001). Rogers (2003) observed that mass media channels such as the radio could be very important at the knowledge stage in the decision-innovation process of diffusion as it is the point that awareness of an innovation and understanding of how the innovation functions are first gained. However, the radio can be less interactive, and does not have capacity to transfer practical skills.

Other farmers were also perceived as the most reliable source of information as this was ranked first. This agrees with the findings of Maddox *et al.* (2003), who found other farmers to be a major source of information. Radio and TV were ranked second on availability but fourth on reliability. This suggests that more reliable information related to soil fertility should be broadcast through the radio to make use of its availability. The farmers also indicated that researchers were not very accessible as they were ranked fifth but were more reliable (ranked third). Thus, researchers and extension agents should improve their interaction with farmers for the better delivery of soil fertility knowledge. Rezvanfar *et al.* (2009) reported that access to information sources and communication channels and an adequate number of extension education courses with relevant content may increase awareness.

Different extension methods had been used to teach farmers ISFM practices. Demonstration was perceived as the most preferred method as it was ranked first, followed by farmer-to-farmer extension, workshops, field days, FFS, the use of teaching aids and exchange visits. Other studies looking specifically at farmers' information sources on environmental issues found on-farm demonstrations as the most preferred communication channel (Bruening 1991). The individual farmer interaction approach was strongly preferred by the majority of the farmers. According to Farouque and Takeya (2009), individual teaching methods are superior for instilling conviction and motivating action. The individual (inter-personal methods) approaches engender interaction which may enhance understanding and technology uptake (Okunade 2007). Therefore, the farmers' preference for the individual teaching method for the adoption of ISFM practices was rational. However, it has been reported that these approaches have been slow and have not resulted in better farm management (Thomas *et al.* 1997).

In this study education was found to be significantly correlated with the preference for individual farmer interaction approaches. This implied that the more educated farmers had more preference for this method. This agrees with the findings of Bukenya *et al.* (2008) that the more educated farmers are often more reluctant to learn with other farmers or in groups in Uganda. Gender was found to be significantly but negatively correlated with a preference for group approaches. Davis and Negash (2007) found that groups can be useful vehicles for linking farmers, especially women, to extension and other sources of information in Kenya. Conversely, in another study by Kirumba (2009), membership in farmers' groups proved to be a highly effective strategy for collective action in the dissemination of information to farmers, regardless of their gender. This study also established that farm size had a significant positive correlation with the farmers' preference for the mass media as an approach in teaching ISFM practices. A large farm size is also likely to be a proxy for wealth, and as a result, richer farmers are more likely to have access to and utilize the mass media than poor farmers. Farmers with large parcels of land are also more likely to seek information on how to improve soil fertility than those with smaller pieces of land. Farouque and Takeya (2009) found that farmers in Bangladesh with large portions of land had a high preference for the mass media.

## Conclusion

Farmers obtained information on soil fertility management from different sources. Other farmers were scored as one of the most available and reliable sources of ISFM information. The demonstration was considered the most preferred extension teaching method, followed by the farmer-to-farmer extension approach in the dissemination of soil fertility management practices. However, FFS was scored the highest in teaching about compost and the farmer-to-farmer extension method was the most preferred in teaching about soil control measures, green manure and crop rotation. In the selection of extension approaches in training on ISFM, researchers and extension agents should consider the farmers' socio-economic characteristics. Further research is essential to assess the cost-effectiveness of the dissemination methods used by researchers and extension agents. This will provide more guidance to agricultural stakeholders in the selection of communication channels, and hence will improve the delivery of soil fertility management research findings to the farmers who are the end-users. Additional studies should also be conducted to further review the competencies of extension agents and researchers in all the extension delivery methods.

## Acknowledgements

For this study, the authors acknowledge the financial support of the Association for Strengthening Agricultural Research in East and Central Africa (ASARECA) in the project on Accelerated Uptake and Utilization of Soil

Fertility Management Best-Bet Practices in Eastern and Central African sub-region. Special thanks are also due to the staff of Kamurugu Agricultural Development Initiative and the Ministry of Agriculture in Maara, Meru South, Mbeere South and Embu districts for their cooperation during data collection. The interviewed farmers are also appreciated for participating in the study and contributing their time.

## References

Bationo, A., Mokwunye, U., Vlek, P. L., Koala, S. and Shapiro I. (2003) 'Soil fertility management for sustainable land use in the West African Sudano-Sahelian zone', in M. P. Gichuru, A. Bationo, M. A. Bekunda, P. C. Goma, P. L. Mafongoya, D. N. Mugendi, H. M. Murwira, S. M. Nandwa, P. Nyathi and M. J. Swift (eds), *Soil Fertility Management in Africa: A Regional Perspective,* TSBF-CIAT, Nairobi, Kenya.

Bruening, T. H. (1991) 'Communicating with farmers about environmental issues', *Journal of Applied Communications*, vol. 75, no. 1, pp. 34–41.

Bukenya, M., Bbale, W. and Buyinza, M. (2008) 'Assessment of the effectiveness of individual and group extension methods: a case study of Vi-Agroforestry Project in Uganda', *Research Journals of Applied Sciences,* vol. 3, no. 3, pp. 250–256.

Davis, K. and Negash, M. (2007) 'Gender, wealth and participation in community groups in Meru Central District, Kenya', *CAPRi Working Paper Series 65*, International Food Policy Research Institute, Washington, DC.

Ekoja, I. (2003) 'Farmers' access to agricultural information in Nigeria', in *Bulletin of the American Society for Information Science and Technology, Aug./Sep. Evaluation*, World Bank, Washington, DC, pp. 21–23.

Farouque, M. G. and Takeya, H. (2009) 'Adoption of integrated soil fertility and nutrient management approach: farmers' preferences for extension teaching methods in Bangladesh', *International Journal of Agricultural Research*, vol. 4 no. 1, pp. 29–37.

Kirumba, E. G. (2009) 'Gender differentials in adoption of soil nutrient replenishment technologies in Meru South District, Kenya', MSc thesis, Department of Environmental Sciences, Kenyatta University, Kenya.

Maddox, S. J., Mustian, D. and Jenkins, D. M. (2003) 'Agricultural information preferences of North Carolina farmers', Paper presented to the Southern Association of Agricultural Scientists, Agricultural Communications Section, Mobile, AL.

Mugendi, D. N., Nair, P. K., Mugwe, J. N., O'Neill, M. K. and Woomer, P. (1999) 'Alley cropping of maize with *Calliandra* and *Leucaena* in the subhumid highlands of Kenya. Part 1: Soil fertility changes and maize yield', *Agroforestry Systems*, vol. 46, pp. 39–50.

Mugwe, J., Mugendi, D., Mucheru, M., Merkx, R., Chianu, J. and Vanlauwe, B. (2008) 'Determinants of the decision to adopt integrated soil fertility management practices by small holder farmers in the central highlands of Kenya', *Experimental Agriculture*, vol. 45, pp. 61–75.

Norrish, P., Morgan, L. K. and Myers, M. (2001) 'Improved communication strategies for renewable natural resource research outputs', in *Socio-economic Methodologies for Natural Resources Research: Best Practice Guidelines*, Natural Resources Institute, Chatham.

Okunade, E. O. (2007) 'Effectiveness of extension teaching methods in acquiring knowledge, skills and attitudes by women farmers in Osun state', *Journal of Applied Sciences Research*, vol. 3, pp. 282–286.

Okuro, J. O., Muriithi, F. M., Mwangi, W., Verjuikl, H., Gethi, M. and Groote, H. (2002) *Adoption of Maize Seed and Fertilizer Technologies in Embu District, Kenya*, CIMMYT Mexico DF and Kenya Agricultural Research Institute (KARI), Kenya.

Oladoja, M. A., Adeokun, O. A. and Fapojuwo, O. E. (2008) 'Determining the socio-economic factors affecting farmer's use of communication methods for information sourcing in Oluyole local government area of Oyo State, Nigeria', *Pakistan Journal of Social Sciences*, vol. 5, no. 1, pp. 51–56.

Rezvanfar, A., Samiee, A. and Faham, E. (2009) 'Analysis of factors affecting adoption of sustainable soil conservation practices among wheat growers', *World Applied Sciences Journal*, vol. 6, no. 5, pp. 644–651.

Richardson, J. G. and Mustian, R. D. (1994) 'Delivery methods preferred by the target clientele for receiving specific information', *Journal of Applied Communications*, vol. 8, pp. 22–31.

Rogers, E. M. (1995) *Diffusion of Innovations. Fourth Edition.* Free Press, New York.

Rogers, E. M. (2003) *Diffusion of Innovations. Fifth Edition.* Free Press, New York.

Thomas, D. B., Eriksson, A., Grunder, M. and Mburu J. K. (eds) (1997) *Soil and Water Conservation Manual for Kenya*, Soil and Water Conservation Branch, Ministry of Agriculture, Livestock Development and Marketing, Nairobi.

# 23 Targeting farmers' priorities for effective agricultural intensification in the humid highlands of Eastern Africa

*Jeremias Mowo,*[1] *Charles Lyamchai,*[2] *Joseph Tanui,*[2] *Zenebe Adimassu,*[3] *Kenneth Masuki*[4] *and Athanase Mukuralinda*[5]

[1] *World Agroforestry Centre, Kenya*
[2] *Directorate of Research and Development, Tanzania*
[3] *Ethiopia Agricultural Research Organization, Ethiopia*
[4] *World Agroforestry Centre, Uganda*
[5] *World Agroforestry Centre, Rwanda*

## Introduction

The humid highlands of Eastern Africa (altitude ≥1,500 masl) are among the most populated areas in Africa, with densities ranging from 100 to 600 people/km$^2$ (German *et al.* 2012). Available land/person ranges from 0.07 to 0.23 ha and although rainfall is high (>1,000 mm/year), agricultural production is low and poverty levels are high (Stroud and Peden 2005). In response to the dwindling land resources, smallholder farmers in the region are practising some form of agricultural intensification. However, most have failed to observe the basic principles of agricultural intensification, including the use of improved germplasm, fertilizers and the appropriate component integration that would lead to optimal system productivity. Intensification is seen only as increased frequency of cultivation and the growing of multiple crops on the same piece of land. As a result, yields of most crops are very low compared to what potentially can be produced (Mowo *et al.* 2007). The humid highlands of Eastern Africa are also characterized by steep slopes, hence soil conservation is vital; however, most farmers hardly ever practise soil conservation. Meanwhile, indiscriminate cutting of trees is common. Poor land preparation practices leave the soils exposed to agents of erosion, leading to excessive runoff, poor infiltration and limited recharge of underground water. Siltation of valley bottoms is common, rendering them unproductive as fertile soil is covered by infertile subsoils from the upper slopes. Springs which were once full of water have dried out due to poor land management and the removal of vegetation. Despite concerted efforts by both research and development partners to reverse the trend (Mansuri and Rao 2004), success has been limited to a few areas.

The experiences of the African Highlands Initiative (AHI)[1] show that successful agricultural intensification is possible by adopting a stepwise technology development where increasing levels of intensification correspond to the stepwise integration of various technologies over time to achieve integrated natural resource management (INRM) (German *et al.* 2012). Experiences in AHI also showed that effective INRM can be achieved faster when the target community is fully committed to the introduced technological options. R&D teams realized early that introduced natural resource management (NRM) options were not necessarily addressing the priority concerns of the communities. In most cases, the ranking of watershed constraints by communities gave higher scores to domestic water supply (both in quantity and quality) while issues such as soil erosion and poor soil fertility were mainly ranked low, contrary to the expectations of the teams (German *et al.* 2012). It was clear that without addressing the highest ranked issues it would be difficult to attract farmers' commitment to the introduced technologies. Domestic water supply does not fall under the mandate of agricultural R&D agencies. Therefore, the R&D teams needed to use imagination in coming up with strategies that considered the priorities of the farmers, without these being construed as out of their mandate, and still ensuring the support of funding agencies and partner organizations.

In this study, addressing domestic water supply was considered an important entry point for attracting farmers' buy-in and commitment to introduced NRM technologies. The hypothesis was that addressing a high-priority constraint as part of an integrated watershed management would lead to multiple system benefits, such as soil conservation, improved valley bottom productivity and water recharge.

## Methodology

### *Location of the study sites*

Figure 23.1 shows the location of the study sites in central Ethiopia (Ginchi–Galessa Watershed) and northeastern Tanzania (Lushoto–Baga Watershed).

### *Identification of watershed issues*

The study was conducted between 2004 and 2007. Identification of watershed issues was done by selected farmers working closely with a multidisciplinary R&D team to generate a list of issues from all social groups (women, men and children) taking into account wealth classes and location in the landscape. Individuals representing the different social groups ranked the watershed issues to produce a socially disaggregated understanding of watershed priorities. In this way, the interests of the different groups were taken into consideration. The process for the identification of watershed issues took about two months and analysis of the information from farmers, cross-checking for accuracy with

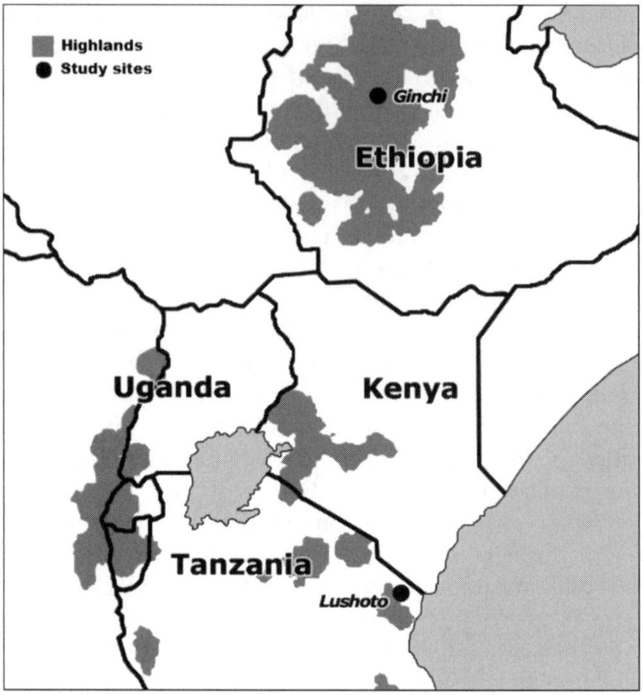


*Figure 23.1* Map showing the humid highlands of Eastern Africa and the study sites in Ethiopia and Tanzania.

focus groups and the community at large to ensure a consensus took another two months. Further investigation on the domestic water issue, which was ranked highest by most groups, was done using focus group discussions (FGDs) to obtain information on the status of the water sources, the reasons for their deterioration and the opportunities for reversing this trend. Key informants included the elders with long experience on the situation in the watersheds, retired government officials and government and religious leaders.

### *Addressing highly ranked watershed issues*

The R&D teams developed strategic action plans that linked highly ranked issues with those that are within the mandate of R&D organizations but ranked low, such as soil conservation, soil fertility improvement and tree planting. The action plan was discussed with farmers for their inputs before it was accepted by both parties. The following steps were taken in the implementation of the action plan:

- Identification of all the springs in the watershed (those still active and those that had dried out). The task, which took one month, involved documenting the status of the water sources and reasons why some had dried and others were still active.

- Capacity building of farmers in the conservation of land near water sources and the conservation of the larger catchment area.
- Assessment of materials required for the construction of structures to protect water sources.
- Mobilization of farmers for in-kind contribution towards the cost of construction (mainly labour).
- Establishment and capacity building of water users' groups and spring management committees to ensure water sources are well managed and protected.
- Frequent follow-up by R&D teams and monitoring and evaluation.

The project provided materials that could not be obtained locally, such as cement and sand. Stones were obtained locally. The construction of structures to protect water sources was done in parallel with capacity building of the farmers in soil conservation around springs and the formation and capacity building of water user groups, all of which took 7–8 months. Impact studies were conducted towards the end of 2007 to find out the extent to which the livelihoods of farmers participating in the project had been affected by the AHI interventions, including water source rehabilitation. In this chapter the emphasis is on the results related to water source rehabilitation.

Apart from building structures to protect the springs, the rehabilitation process also included the planting of water-friendly trees such as *Ficus-Vallis chaudae*, *Albizia schiniperiana*, *Albizia gummifera*, *Rauvolfia caffra* and *Hallea rubrastepulata* around the springs, and enacting by-laws such as those barring cultivation close to the water sources to ensure they were not abused. Design work and cost estimation for the structures was done by water engineers from the district offices. AHI covered 50 per cent of the costs, while farmers covered the rest through their labour in collecting stones (available locally) and doing masonry work.

## Findings

### *Priority watershed issues*

The four topmost ranked issues are shown in Table 23.1. Water availability in terms of amount and quality was ranked high in both watersheds. In the Baga Watershed (Tanzania) the quality of water for domestic use was a major concern for upstream farmers, high-income farmers and the youth; water quantity for domestic use was a major concern for all social groups. The fourth ranked issue in Baga was the negative effects of boundary trees on crops; *Eucalyptus camaldulensis* was mostly blamed for this with the effect on soil moisture and maize plant height being observed up to 28 m from the trees.

*Table 23.1* Ranking of watershed constraints: average for the watersheds

| *Watershed issues* | *Group* | *Rank* |
|---|---|---|
| *Baga Watershed, Lushoto (Tanzania) (n = 72)* | | |
| Poor water quality (domestic use) | Upslope farmers, high income, youth | 1 |
| Declining water quantity (domestic use) | All | 2 |
| Declining amounts of irrigation water | High income, men, youth | 3 |
| Negative effects of boundary trees | High income | 4 |
| *Galessa Watershed, Ginchi (Ethiopia) (n = 36)* | | |
| Loss of indigenous tree species | All | 1 |
| Poor water quality and quantity (domestic use) | All | 2 |
| Land shortage | Youth, women | 3 |
| Declining soil fertility | All | 4 |

All the social groups in Galessa Watershed (Ethiopia) ranked water quality and quantity for domestic use the second most important issue, with the latter being more critical during dry seasons. Loss of indigenous tree species (declining abundance) was ranked first in Galessa, where the abundance of two major tree species, *Hagenia abyssinica* and *Olea africana*, was mentioned as being reduced. This was mainly attributed to extensive cutting for fuelwood and other tree-based services. Land shortage came third and declining soil fertility in Galessa came fourth. According to the Central Statistical Agency of Ethiopia (CSA 2006), land per capita has declined from 0.4 ha to 0.33 ha. Evidence of declining soil fertility was also reported by Haileselassie *et al.* (2005), who observed a negative annual balance of nitrogen, phosphorus and potassium (NPK).

In both sites, poor water quality was attributed to the abuse of water sources, such as washing and animals drinking directly from the water sources, cultivation around them and the disposal of solid waste nearby.

### Rehabilitation of water sources

In Baga, 26 out of the 30 springs scheduled for rehabilitation by 2007 were completed. The communities planted a total of 400 water-friendly trees around the water sources. In Galessa, three springs and a cattle trough were rehabilitated. The rehabilitated water sources led to improved water availability. The time taken to collect water ranged from 30 minutes to two hours before rehabilitation and was reduced to 12–16 minutes after rehabilitation in Baga. Corresponding figures for Galessa were 55 minutes before rehabilitation and 24 minutes afterwards (Figure 23.2).

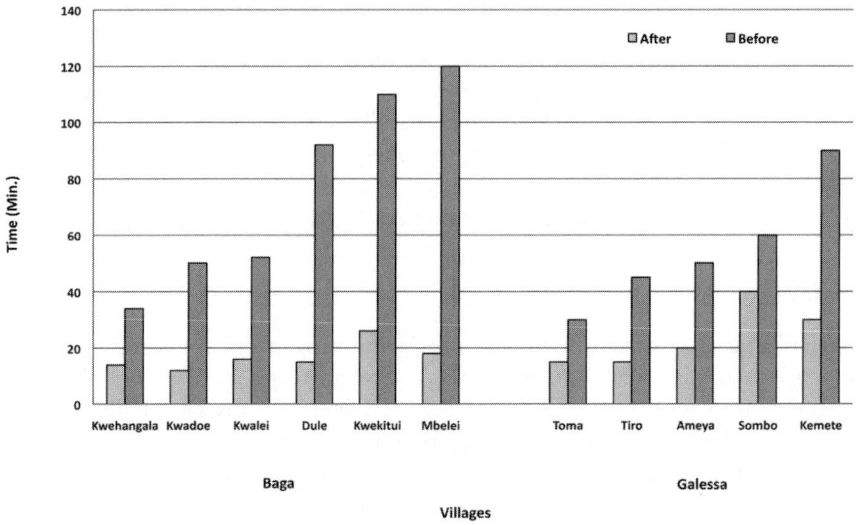


*Figure 23.2* Time (minutes) spent in collecting water before and after interventions in Baga and Galessa.

In Galessa, the rehabilitated springs provide water throughout the year. In the 2007 survey, 84 per cent of the respondents ($n$ = 45) confirmed that time for fetching water decreased after the rehabilitation of the water sources. Improvement in water quality was deduced from the decline in the number of patients treated for waterborne diseases. For example, in Mbelei Village of the Baga Watershed, cases declined from 77 in the 2006 long rains to 22 in 2007. In Galessa, 83 per cent ($n$ = 45) of respondents indicated that the prevalence of waterborne diseases decreased after management of the springs.

## *Implications of the interventions for adoption of NRM*

The impact survey conducted in 2007 showed that water source rehabilitation greatly benefited participating and non-participating farmers. In Baga it was voted by 30 per cent of the respondents as the third most popular intervention after improved crop varieties (banana, tomato, sweet pepper) and soil conservation practices, and 97 per cent of respondents ($n$ = 44) reported improved agricultural productivity. In Galessa, the rehabilitation of water sources was voted by 39 per cent of respondents as the second most popular intervention after the Irish potato; 87 per cent ($n$ = 45) reported an improvement in agricultural productivity due to enhanced access to quality water, while 80 per cent reported a positive impact on health. All the 170 households in the watershed were benefiting from clean water; there was a drop of 83 per cent in incidences of waterborne diseases and 30 per cent of the households were participating in soil conservation activities.

Labour constraints in the form of time and energy wasted in collecting water had reduced the time farmers could devote to NRM. Also, the quality of labour before the rehabilitation of water sources (frequent illness and labour withdrawn to cater for the sick) had contributed to reducing the time available for NRM, including soil conservation. With the rehabilitation of water sources more time was devoted to NRM. Quantitative evidence on the impact of water source rehabilitation and other interventions on agricultural productivity is a subject of a meta-study currently being conducted by AHI.

## Discussions

Agricultural intensification aims at producing more from the same land area, an option that is more important where land shortage is a major problem, as in the humid highlands of Eastern Africa. Given the myriad challenges that farmers in these highlands are facing, they rarely practise sustainable intensification. This has led to negative impacts on the natural resource base and hence on agricultural productivity and environmental health. However, agricultural intensification should strive to minimize the negative environmental impacts and contribute to increasing the natural capital and flow of beneficial environmental services. According to Piras (2011) it is possible to intensify agriculture while preserving the environment. Although poor land management was the major cause of water source degradation, it is difficult to promote soil conservation and soil fertility improvement unless farmers' most pressing needs are met. Similar observations were made by Amede (2003), who noted that the sustainable implementation of watershed agendas is possible only if supported by interventions that give immediate benefits to farmers. In this study, farmers' commitment to NRM was high when introduced technologies were linked to the availability of domestic water. Farmers observed an improvement in agricultural productivity and livelihoods attributed to the enhanced quality and quantity of the domestic water supply (Mekuria *et al.* 2008).

Quality water supply contributed significantly to the reduction of waterborne diseases (Adimassu *et al.* 2008). A healthy community will provide quality labour for the timely adoption of introduced technologies and hence achieve an improvement in agricultural production and livelihoods.

## Conclusion

The study shows that the effective management of natural resources in the humid highlands of Eastern Africa, which is critical for achieving sustainable agricultural intensification, requires innovative approaches that do not overlook the priority concerns of the target communities. The rehabilitation of water sources significantly reduced the time farmers spend in collecting water and managing waterborne diseases. This resulted in more time and quality labour for NRM. A major gap identified in this study is the lack of information as to why farmers fail to invest their resources (financial, physical and human) in issues that have a high benefit to them without external intervention. More research is required to identify mechanisms for motivating communities to invest their labour in managing priority watershed issues as a stimulant for achieving sustainable agricultural intensification.

## Acknowledgement

The authors would like to thank the national partners and farmers from Baga (Tanzania) and Galessa (Ethiopia) Watersheds for their active participation in the project.

## Note

1 Until 2008 AHI was an eco-regional programme of the World Agroforestry Centre (ICRAF) and the Association for Strengthening Agricultural Research in East and Central Africa (ASARECA). AHI is now a programme under ICRAF. Its focus is the development of methods and approaches for NRM in the humid highlands of Eastern Africa.

## References

Adimassu, Z., Ayele, S., Mekonnen, K., Alemu, G., Amare, G., Gojjam, Y., Tsegaye, M., Berhane, K., German, L. and Amede, T. (2008) 'Spring management for integrated watershed management in Galessa', in Z. Adimassu, K. Mekonnen and Y. Gojjam (eds), *Working with Communities on Integrated Natural Resources Management*, EIAR.

Amede, T. (2003) *Differential Entry Points to Address Complex Natural Resource Constraints in the Highlands of Eastern Africa,* AHI.

CSA (Central Statistical Authority of Ethiopia) (2006) *Agricultural Sample Survey*, CSA, Addis Ababa.

German, L., Mowo, J., Amede, T. and Masuki, T. (eds) (2012) *Integrated Natural Resource Management in the Highlands of Eastern Africa: From Concept to Practice*, Earthscan, London.

Haileselassie, A., Priess, J., Veldkamp, E., Teketay, D. and Lesschen, J. P. (2005) 'Assessment of soil nutrient depletion and its spatial variability on smallholders' mixed farming systems in Ethiopia using partial versus full nutrients balances', *Agriculture, Ecosystems & Environment*, vol. 108, pp. 1–16.

Mansuri, G. and Rao, V. (2004) 'Community-based and -driven development: a critical review', *The World Bank Research Observer*, vol. 19, no. 1, p. 19.

Mekuria, M., La Rovere, R. and Szonyi, J. (2008) *External Review and Impact Assessment of the African Highlands Initiative (AHI)*, Program Evaluation Report, Impact Targeting and Assessment Unit (ITAU), International Maize and Wheat Improvement Center.

Mowo, J. G., Nabahungu, L. N., Dusengemungu, L. and Sylveri, S. (2007) 'Opportunities for overcoming soil fertility constraints to agricultural production in Gasharu watershed, Southern province, Rwanda', in Njeru *et al.* (eds), *Sustainable Agricultural Productivity for Improved Food Security and Livelihoods, Proceedings of the National Conference on Agricultural Research Outputs*, Kigali, Rwanda, 26–27 March 2007, pp. 506–514.

Piras, N. (2011) 'Sustainable agricultural intensification: technologies that respect people and nature'. www.agriculturesnetwork.org/news/technology-people-nature (accessed 19 October 2011).

Stroud, A. and Peden, D. (2005) 'Situation analysis for the intensively cultivated highlands of East and Central Africa Part A: 2005'; An input into the AHI strategy for ASARECA 2005–2010. Project report to IDRC. African Highlands Initiative (AHI), International Centre for Research in Agroforestry (World Agroforestry Centre), Nairobi, Kenya.

# Index

Locators in **bold** refer to figures/diagrams